Advances in Ion Exchange for Industry and Research

# Advances in Ion Exchange for Industry and Research

Edited by

**P. A. Williams**
*North East Wales Institute, Wrexham, UK*

**A. Dyer**
*University of Salford, UK*

ROYAL SOCIETY OF CHEMISTRY

The Proceedings of the Fifth International Conference on Ion Exchange Processes held on 5–9 July 1998 at Plas Coch College, The North East Wales Institute, Wrexham

The front cover illustration is taken from the contribution by A. Clearfield, p. 147.

Special Publication No. 239

ISBN 0-85404-744-1

A catalogue record for this book is available from the British Library

Published by The Royal Society of Chemistry,
Thomas Graham House, Science Park, Milton Road,
Cambridge CB4 0WF, UK

For further information see our web site at www.rsc.org

Printed and bound by MPG Books Ltd, Bodmin, Cornwall, UK

# Preface

This book is a record of the proceedings of ION-EX 98, the fifth in a series of conferences held at The North East Wales Institute, Wrexham, UK. It covers a very broad aspect of the study of ion exchange. Sections address the use of ion exchange in the clean-up of aqueous effluents from the nuclear industries, developments in the theory of ion exchange, employment of the technique to environmental problems in drinking water treatment and industrial waters. The analytical aspects of ion exchange are covered in depth including the latest developments in ion exchange chromatography, gradient and ion exclusion chromatography, and capillary electrophoresis protein separations. Finally, novel organic and inorganic exchangers are described together with their potential industrial uses.

Each major section is headed by a review from an international expert which describes recent progress in their area of expertise.

Professor M. Streat covers the prospectives for the 20th century employment of ion exchange technologies, Dr W. Höll describes the surface complexation approach to multispecies ion exchange equilibria, Professor A. Clearfield reviews progress in tunnel inorganic materials, Professor P. Haddad discusses ion exclusion and capillary electrophoresis and Dr K. Jones considers the new technology of combinatorial libraries.

Professor Alan Dyer

# Members of the Organising Committee

# Contents

## Analytical Methods

## Novel Inorganic Ion Exchangers

## Metal Ion Removal and Separation

**Ion Exchange for the Environment**

# ION EXCHANGE – A TECHNOLOGIST'S PERSPECTIVE OF THE 20th CENTURY

Michael Streat
Department of Chemical Engineering
Loughborough University
Loughborough
Leics
LE11 3TU
UK

## 1 INTRODUCTION

Ion exchange is now so well established that it needs little fundamental introduction to an audience of academics, practitioners and specialists in the field. Ion exchange as we know it today, is defined as the counter-exchange of dissociated ions in solution with fixed functional groups of opposite charge within a porous matrix. The origins of ion exchange have been traced back to the holy scriptures (1) and a great leap forward in understanding came in the last century due the work of Thomson (2) and Way (3) who studied the exchange of ions in cultivated soils. The application of ion exchange as a viable industrial process depends on the availability of robust, durable materials capable of working at extremes of ionic concentration and temperature, and under arduous conditions in the presence of radioactivity and/or strong oxidants/reductants. Calmon has listed the development of both natural and synthetic materials in the 20th century and this is given in Table 1. During the first 30 years of this century, ion exchange was restricted to use of naturally occurring inorganic materials such as aluminosilicates or natural organic materials such as coal products, bone char and activated carbon. Much pioneering work, largely physico-chemical, was performed in the early years and this laid the foundations for present-day ion exchange process technology.

Adams and Holmes, chemists working in England, are attributed with the discovery of the first family of synthetic organic ion exchange resins which revolutionised the approach to materials development (4,5). Their discovery that crushed gramophone records (their work is based on phenol-formaldehyde resins) exhibited ion exchange properties was the stimulus for the spontaneous creation of a thriving ion exchange resin industry. This single discovery in 1935 lead to the production of sulphonated and aminated condensation polymers capable of treating water to remove hardness, alkalinity and trace impurities. Synthetic organic ion exchange resins opened the door to a host of potential separation projects, e.g. the separation, purification and recovery of lanthanide and transuranic elements during the Manhattan project in the period 1940-45 and the discovery of element 61, promethium, which was separated from fission products at the same time (6).  The second industrial revolution, which commenced after the end of the second world war, benefitted from advances in ion exchange.  The surge in energy requirements prompted a dramatic increase in the demand for demineralised feed water for high pressure boilers.  The recovery of uranium to provide the fuel for thermal nuclear power stations also depended on the successful application of new synthetic organic resins and ion exchange technology.  Commercial polymeric ion exchange resins were widely available for industrial use from the late 1940's/early 1950's.  Much history has already been written about the origins of the ion exchange resin industry from its earliest beginnings in condensate polymers to modern-day macroporous polymers.  The reader is referred to a recent review by Abrams and Millar (7) which gives a comprehensive account of ion exchange resin developments in the second half of this century.

| 1906 | fused aluminosilicates |
| 1918 | processed natural inorganic products |
| 1927 | porous synthetic aluminosilicates |
| 1935 | processed natural organic (coal) products |
| 1937 | sulphonated and aminated condensation polymers |
| 1946 | sulphonated addition polymers |
| 1949 | aminated (quaternary) addition polymers |
| 1950 | carboxylic addition polymers |
| 1951 | molecular sieves |
| 1952 | ion selective membranes |
| 1956 | chelate resins |
| 1957 | ion retardation resins |
| 1959 | macroporous resins |
| 1960 | isoporous resins |
| 1973 | thermally regenerated resins |
| 1973 | polymeric adsorbents |
| 1973 | enzyme bound polymers |
| 1974 | specific ion exchangers |
| 1975 | extractant containing polymers |

*Cal Calmon, Closing Address to the 1976 Society of Chemical Industry International Conference, "Theory and Practice of Ion Exchange" held at the University of Cambridge, England, 1976*

**Table 1. Development of Ion Exchanger Materials**

It is particularly fascinating to re-read the writings of Dr Robert Kunin, a pioneer of ion exchange technology, who presented a personal and detailed account of the advances in ion exchange technology in "Amber-hi-lites", a Rohm and Haas Company publication.  These regular leaflets, which started in April 1949, contain an encyclopaedia of educational and practical guidance for the benefit of their customers and the wider community.  Some landmark issues are given in Table 2 to illustrate the scope and diversity of industrial and intellectual activity that started in 1949 and continued until the late 1980's when Dr Kunin retired.  The foundations of modern day ion exchange were laid down about 50 years ago and it is striking to note the topics as they unfolded in the second half of the century. Some of the exciting developments of today have their origins in the work performed in industry and by researchers in 1949-55, e.g. biochemical and medicinal applications, water treatment, chromatography, catalysis and hydrometallurgy.

| Issue No | Date | Topic |
|----------|------|-------|
| 1 | Apr 1949 | Bacteria binding, protein purification |
| 2 | Jun 1949 | Production of blood proteins |
| 3 | Aug 1949 | Penicillin purification |
| 5 | Feb 1950 | Demineralisation, water softening |
| 12 | Aug 1951 | Ion exchange chromatography |
| 13 | Oct 1951 | Catalysis with ion exchange resins |
| 18 | Sept 1952 | Ion exchange in the food industry |
| 23 | Jul 1953 | Metal concentration and recovery |
| 31 | Sept 1955 | Sugar refining |
| 32 | Dec 1955 | Uranium processing |
| 35 | Jul 1956 | Resins in wine making |
| 36 | Oct 1956 | Rare earths |
| 52 | Jul 1959 | Water, ion exchange and electronics |
| 78 | Nov 1963 | Macroreticular ion exchange resins |
| 86-89 | Mar-Sept 1965 | Deionization of water |
| 92-100 | Mar 1966-Jul 1967 | Applications-Water treatment |
| 102-105 | Nov 1967-Mar 1968 | Industrial applications of hydrometallurgy |
| 106-112 | Jul 1968-Jul 1969 | Sugar processing |
| 117-124 | May 1970-Sept 1971 | Pollution abatement and control |
| 142-145 | Sept 1974-Mar 1975 | Ion exchange in medicine and pharmaceutical industry |
| 164-166 | Winter 1980-Winter 1981 | Role of silica in the water industry |
| 167-170 | Summer 1981-Summer 1982 | Role of organic matter in water |
| 173-178 | Summer 1983-Winter 1985 | Acrylic-based ion exchange resins and adsorbents |
| 182 | Spring 1988 | Biotechnology with ion exchange resins and polymeric adsorbents |

*Extracted from the back copies of AMBER-HI-LITES edited by Dr Robert Kunin (Rohm and Haas Company, Philadelphia, USA)*

**Table 2. The Advance of Ion Exchange Technology using Polymeric Resins**

There is great emphasis on environmental pollution control as we reach the millennium. However, this was anticipated by Kunin in 1970, when he wrote a number of thought-provoking articles in Amber-hi-lites and introduced the idea of applying ion exchange resins for a variety of environmental problems. Key issues addressed at that time were pollution arising in agriculture, municipal and industrial water treatment and the wider field of noxious gaseous dispersions. Ion exchange was offered as one possible technique to remediate the effects of pollution in water and effluents. Many potential techniques were proposed but little industrial exploitation was evident at that time. Today, legislation is in place to enforce environmental pollution control and this has focused attention once again on the application of cost-effective technologies for the mitigation of problems arising in liquid effluents and gaseous discharges. It is timely, therefore to review the potential for adsorption and ion exchange technology for the removal of toxic metal and harmful organic micropollutants in water, waste-water and effluents arising in industrial process operations.

## 2   ENVIRONMENTAL POLLUTION CONTROL

Adsorption and ion exchange are widely used for the removal of trace micropollutants from water, waste-water and industrial effluents. Adsorption involves the binding of molecules to the surface of porous materials by physisorption whereas ion exchange involves chemisorption of charged ionic species onto functional ionogenic groups linked to the matrix of an adsorbent. Activated carbon has been widely used for the removal of taste, odour and organic micropollutants from potable water. Porous polymeric sorbents with similar attributes to activated carbon are also commercially available although these materials are expensive and therefore less widely applied in water and effluent treatment. Ion exchange is the preferred technology for the removal of toxic metals such as Pb, Cu, Ni, Zn, Cd, Cr, Hg and arsenic (as arsenate) from water. There are several classes of ion exchange material, naturally occurring or synthetic, inorganic or polymeric. The range of materials is vast and it is now possible to tailor suitable materials for a particular process operation.

Adsorption and ion exchange do not in themselves solve a liquid phase environmental pollution problem. Whereas these technologies are capable of removing toxic micropollutants from water and concentrating the contaminant onto a solid surface, there is an associated problem of treatment and disposal of the secondary waste. Immobilisation of harmful micropollutants onto the surface of porous solids and ultimate disposal by landfill is a costly and environmentally unacceptable option. Adsorbents must be regenerated, recycled and reused and toxic micropollutants must be disposed in a benign way. This poses some challenging technological problems.

The regeneration and reactivation of activated carbon involves high temperature processing. It is extremely effective, but costly, due to energy consumption and the aggressive attrition of the adsorbent materials. More cost-effective regeneration and reactivation techniques are under investigation involving low temperature solvent stripping or supercritical fluid extraction of micropollutants. Ion exchange resins are regenerated with chemical eluents, usually mineral acids and/or bases. This represents a significant processing cost and adds to the electrolyte burden of recovered eluents and may restrict subsequent recovery of concentrated toxic metals. It is desirable to use ion exchange materials containing weakly ionising functional groups rather than ion-specific, complexing or chelating functional groups which are more difficult to regenerate.

At Loughborough University, we have been developing a range of adsorbents and ion exchange materials tailored for the removal of toxic metals and organic micropollutants from water at neutral and near-neutral pH values. Under these conditions, it is possible to use weak acid/base ion exchange materials to remove dissociated ions from aqueous solution. This is an important advantage since it permits efficient elution and regeneration of ion exchangers using relatively cheap reagents with great chemical efficiency, for example, using carbon dioxide or ammonia. Two current projects will be described to illustrate our approach.

## 3  THE DEVELOPMENT OF FUNCTIONALISED ACTIVATED CARBON FOR THE REMOVAL OF TOXIC METALS

The maximum admissible concentrations of undesirable and toxic metals in water are given in Table 3.

| Substance | Maximum Admissible Concentration ($\mu$g/L) | Comments |
|---|---|---|
| Arsenic | 10 | |
| Antimony | 10 | |
| Cadmium | 5 | |
| Chromium | 50 | |
| Copper | (3,000) | guide level after standing 12 hours at point of consumption |
| Cyanides | 50 | |
| Iron | 50 | |
| Lead | 50 | in running water |
| Mercury | 1 | |
| Nickel | 50 | |
| Selenium | 10 | |
| Zinc | (5,000) | guide level after standing 12 hours at point of consumption |

**Table 3. Maximum Admissible Concentrations of Undesirable and Toxic Metals in Water (taken from the EU Directive)**

The application of activated carbon for the removal of trace toxic metals from water has been studied by numerous investigators (8-12). The selectivity and capacity of conventional activated carbons towards heavy metals is not particularly high. In order to enhance selectivity, weakly acidic functional groups have been introduced by surface oxidation. Oxidised activated carbons possess some unique properties, due mainly to the large proportion of oxygen containing groups located at the surface. These carbons are readily distinguished from conventional carbons activated with water, steam or carbon dioxide by their increased capacity, selectivity, lyophilic surface and acidic character. Activated carbon supplied by Chemviron (designated F-400) has been oxidised in hot nitric acid with constant agitation. The surface characteristics of as-received and oxidised activated carbons are given in Table 4.

Water containing traces of Pb, Cu, Ni and Cd (pH=4.6) has been treated in a mini-column containing 0.6 cm$^3$ Chemviron F-400 at a flow rate of 1.5-2 cm$^3$/min. The breakthrough curve for as-received unoxidised carbon is shown in Fig 1. Negligible uptake capacity or selectivity for the trace metals is shown. The breakthrough curve for oxidised activated carbon is given in Fig 2 and this shows a significant uptake capacity for Pb and Cu and selective uptake of the trace metals in the sequence Pb>Cu>Ni>Cd. The presence of excess concentrations of alkaline earth metals (Na, K, Ca and Mg) did not affect the sorption of lead. Divalent toxic metals are sorbed by weakly acidic surface functional groups, e.g. carboxylates and phenolates. Sorption of metals is fully reversible and can be achieved with two bed volumes of mineral acid yielding a highly concentrated eluent.

| Carbons | F-400 (as-received) | F-400 (oxidised) |
|---|---|---|
| BET surface area (m$^2$/g) | 944.9 | 800 |
| Micropore area (m$^2$/g) | 717.2 | 513.46 |
| Micropore volume (cm$^3$/g) | 0.340 | 0.269 |
| BJH adsorption average pore diameter (Å) | 63.17 | 54.19 |
| BJH desorption average pore diameter (Å) | 58.73 | 53.33 |

**Table 4. Surface Characteristics of Chemviron F-400 Activated Carbon**

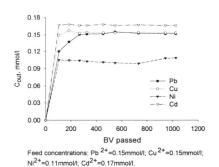
Feed concentrations: Pb$^{2+}$=0.15mmol/l; Cu$^{2+}$=0.15mmol/l; Ni$^{2+}$=0.11mmol/l; Cd$^{2+}$=0.17mmol/l.

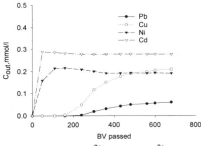
Feed concentrations: Pb$^{2+}$=0.072mmol/l; Cu$^{2+}$=0.23mmol/l; Ni$^{2+}$=0.19mmol/l; Cd$^{2+}$=0.28mmol/l

**Figure 1** Breakthrough curve for Pb$^{2+}$ on F400(unoxidised) in the presence of Cu$^{2+}$, Ni$^{2+}$ and Cd$^{2+}$

**Figure 2** Breakthrough curve for Pb$^{2+}$ on F400(oxidised) in the presence of Cu$^{2+}$, Ni$^{2+}$ and Cd$^{2+}$

Selectivity and capacity of trace toxic metal sorption onto activated carbon can be improved by locating heteroatoms within the surface of the adsorbent. For example, phosphorus-containing functional groups are known to bind divalent metals selectively and therefore we have prepared novel spherical activated carbons containing (hydrated) phosphorus atoms at the surface. These materials are produced from porous spherical phenol-formaldehyde resin, phosphorylated by refluxing in phosphorus oxychloride and carbonised at about 700°C. A typical breakthrough curve for the phosphorus containing activated carbon (PGP-P) is shown in Fig 3. A high uptake capacity and strong preference for Pb over Cu, Ni and Cd is indicated. Regeneration is carried out with mineral acid and over 90% lead recovery is achieved in one sorption/desorption cycle.

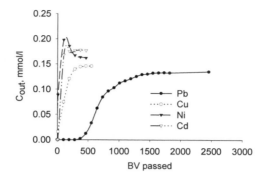

**Figure 3** Breakthrough curve for $Pb^{2+}$ on PGP-P in the presence of $Cu^{2+}$, $Ni^{2+}$ and $Cd^{2+}$

## 4  DEVELOPMENT OF HYPERCROSSLINKED POLYMER PHASES FOR THE REMOVAL OF PESTICIDES

The EU and national standard for any individual pesticide in drinking water at the point of supply is 0.1 µg/L (0.1 ppb) with a maximum of 0.5 µg/L for all detected pesticides. The widespread use and complex chemistry of some pesticides causes drinking water levels to frequently exceed the UK legal limit. The six most frequently detected pesticides in 1991 and 1992 were atrazine, simazine, diuron, isoproturon, chlorotoluron and mecoprop. The average concentrations of pesticides in drinking water supplies in 1993 were about 0.03 µg/L or less, although some sites exceeded 0.1 µg/L for long periods whereas others were well below the limits of detection. The maximum concentrations of pesticides found in drinking water supplies in 1992/93 are given in Table 5.

Styrene/divinylbenzene copolymers are well established for the recovery of phenols from effluent streams due to cheap, effective regeneration (13). However, many polymers offer relatively low surafce area at a high capital cost. Hypercrosslinked polymers, e.g. Hypersol-Macronet$^{TM}$, produced by Purolite International Ltd, possess a high surface area (800-1200 $m^2$/g) and a well defined pore structure comparable to activated carbon.

Experiments have been performed with Hypersol-Macronet MN-200, a commercial hypercrosslinked polymeric adsorbent, for the treatment of water contaminated with a trace mixture of heterocyclic pesticides. The breakthrough curve obtained in a mini-column run using a feed water containing 20 µg/L each of atrazine, simazine, diuron, isoproturon and chlorotoluron is shown in Fig 4. The column contained approximately 0.65 $cm^3$ of polymer and was operated at a flow-rate of 7.5 BV/min at a pH of about 7.

| Pesticide | 1992 | 1993 |
|-----------|------|------|
| Atrazine | 1.0 | 5.5 |
| Simazine | 56.3 | 0.73 |
| Diuron | 0.64 | 1.42 |
| Isoproturon | 3.4 | 2.3 |
| Chlorotoluron | 0.78 | 0.34 |

**Table 5. Maximum Concentration (ppb) of Pesticides in Drinking Water Supplies in 1992 and 1993**

Breakthrough of atrazine to the EU legal limit occurred after about 130,000 BV passed and there is evidence of significant selectivity in the sequence isoproturon>diuron>atrazine>chlorotoluron>simazine. These observations have been comprehensively reported by Streat and Sweetland elsewhere (14-17). The great advantage of polymeric adsorbents over activated carbon is the ease of regeneration. Fig 5 shows the low temperature solvent stripping of the pesticides using ethanol at a flow-rate of 0.5 BV/min. Complete removal of pesticides is achieved in about 5 BV. The solvent can be recovered by evaporation, leaving a crystalline deposit of concentrated pesticides, and recycled. The potential to develop large scale, cost-effective water treatment processes is most encouraging using hypercrosslinked polymers as an alternative to conventional activated carbon.

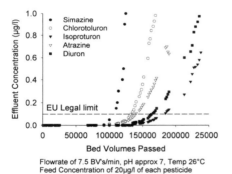
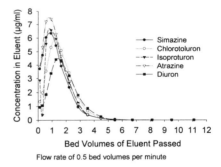

**Figure 4.** MN-200 Breakthrough Curve

**Figure 5.** Regeneration of MN-200 using HPLC Ethanol

## 5   CONCLUSIONS

The second half of the 20th century has seen major advances in ion exchange resin development as well as improved process technology, such as the introduction of counterflow regeneration, improved mixed bed operation and the advent of continuous countercurrent ion exchange contacting. This has established ion exchange as the foremost technique for the production of high-quality boiler feed water for the power industries, pure water for the chemical, food and beverage industries, ultra-pure water for the nuclear, semiconductor and pharmaceutical industries. Adsorption and ion exchange have an important rôle to play in environmental pollution control, especially for the removal of toxic metals and harmful organics from potable water, waste-water and effluents.

The ability to tailor polymeric exchangers with ion specific ligands enables new separation processes to be developed, e.g. in the hydrometallurgical and biochemical industries. Structured hypercrosslinked polymers are now available for adsorptive applications and this offers new opportunities to develop novel processes technology in aqueous and non-aqueous media. Much research and development remains to be done to meet the challenges of the next century.

## 6  REFERENCES

1.  R. Kunin and R.J. Myers, "Ion Exchange Resins", John Wiley and Son, New York, 1950.
2.  H.S. Thomson, *J. R. Agric. Soc. Engl.,* 1850, **11**, 68.
3.  J.T. Way, *J. R. Agric Soc Engl.,* 1850, **11**, 313 and *ibid*, 1855, **15**, 491.
4.  B.A. Adams and E.L. Holmes, *J. Soc. Chem. Ind.,* 1935, **54**, 1.
5.  B.A. Adams and E.L. Holmes, *Chem Age,* 1935, **38**, 117.
6.  J.A. Marinsky, L.E. Glendenin and C.D. Coryell, *J. Am. Chem. Soc.,* 1947, **69**, 2781.
7.  I.M. Abrams and J.R. Millar, *Reac. Func Polymers,* 1997, **35**, 7.
8.  N. Petrov, T. Budinova and I. Khavesov, *Carbon,* 1992, **30**, 135.
9.  T.K. Budinova, K.M. Gergova, N. Petrov and V.N. Minkova, *J. Chem. Technol, Biotechnol.,* 1994, **60**, 177.
10. R.M. Taylor and R.K. Kuennen, *Environ. Progress,* 1994, **13**, 65.
11. B.E. Reed and S. Arunachalam, *Environ. Progress,* 1994, **13**, 60.
12. M.O. Corapcioglu and C.P. Huang, *Water Res.,* 1987, **21**, 1031.
13. T.F. Speth and R.J. Miltner, *J. Am. Water Works Assoc.,* 1980, **82**, 72.
14. M. Streat and L.A. Sweetland, *Trans. IChemE.,* 1998, (B2), **76**, 115.
15. M. Streat and L.A. Sweetland, *Trans. IChemE.,* 1998, (B2), **76**, 127.
16. M. Streat, L.A. Sweetland and D.J. Horner, *Trans. IChemE.,* 1998, (B2), **76**, 135.
17. M. Streat, L.A. Sweetland and D.J. Horner, *Trans. IChemE.,* 1998, (B2), **76**, 142.

# AN ION EXCHANGE-BASED PROCESS FOR COAGULANTS RECOVERY FROM WATER CLARIFIER SLUDGE

D.Petruzzelli[*], A.Volpe[*], G.Tiravanti[*], and R.Passino[+]

[*] Istituto di Ricerca sulle Acque, National Research Council,
5, Via De Blasio, 70123 Bari, Italy.
[+] Istituto di Ricerca sulle Acque, National Research Council,
1, Via Reno, 00198 Rome, Italy.

## 1   INTRODUCTION

Progressive impairment of the quality of natural waters and the increased demand for human consumption calls for stricter water potabilization operations[1,2].   Treatment of natural waters for potable uses has long been accomplished by using inorganic coagulants to remove turbidity (suspended solids, colloids), trace contaminants (metals) and colour-causing organics (humics).   Coagulation-flocculation, sedimentation and sand filtration operations are typically used in the de-stabilisation, agglomeration and removal of the mentioned contaminants that pose threats to public health and/or affect the quality of distributed tap waters[2].

As a result of this operation several million tons per year of clarifier sludges are produced in Europe with previews of doubling the figure by the next decade.   End disposal of the reference inorganic sludge is mostly based on controlled landfilling, after conditioning operation, which is carried-out in acidic ($H_2SO_4$) or alkaline media (NaOH, CaO).   The conditioning operation, intended to minimise the volume of solids to be disposed-of, allows also for coagulants recovery (Al, Fe chemicals).   The purity of the coagulants recovered may be not sufficient to justify their reuse to the water potabilization operations[3].   On the other hand, the cost of landfilling operations is continuously raising with site saturation and with the increased concern on landfilling site locations, especially in densely populated areas[4].

To improve the quality of the coagulants recovered the clarifier sludge acidic leachate (pH 3.5) is eluted onto carboxylate weak electrolyte cation resins (i.e., Purolite C106), retaining aspecifically the metal species during the exhaustion step.   Selective separation and recovery of the aluminium and iron species of interest, from other species (Ca, Mg, Mn) and trace metals (Zn, Pb, Cr, Cu) is obtained during the resin regeneration step, carried-out by controlled elution of 0.4M NaOH.

Experimental data referring to laboratory scale pilot plant (15 L/d) investigation are reported in the paper to verify technical feasibility of the process.   Some mechanistic interpretation of the ion exchange phenomena on polyvalent ionic systems is also reported in the paper to better substantiate basic principles of the process.

## 2 METHODS

Figure 1 shows the layout of the automatic ion exchange-based laboratory pilot plant (15 L/d) for recovery and separation of coagulants from water clarifier sludges. C1 is a plexiglas column (i.d.=2 cm; h=120 cm) containing 250 cm$^3$ of resin bed, a commercial acrylic matrix, macroporous carboxylate weak cation exchanger (i.e., Purolite C106 from Purolite Co., Pontyclun, UK) in partially hydrolysed Na$^+$/H$^+$ form (see later). S1 is the clarifier sludge extractor, where controlled acid additions (HCl 36% or H$_2$SO$_4$ 98% w/v) to pH 3.5 allows for the quantitative leaching of soluble metal substrates of interest. S2 is the service vessel allowing for continuous feed of the leachate during the resin exhaustion step; S3 is the precipitator for the trace metals impurities from recovered coagulants; S4 to S6 are regeneration vessels containing 0.4M NaOH, to carry-out the stepwise regeneration of the exhausted column.

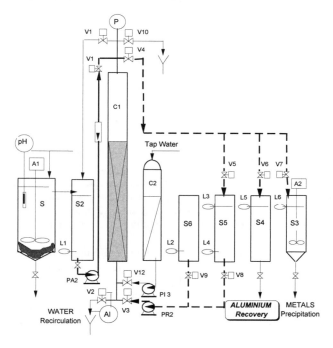

**Figure 1** *Schematic flowsheet of the ion exchange-based process for coagulants recovery from water clarifier sludge.*

The clarifier sludge, periodically withdrawn from the Sinni River Water Potabilization Station of Apulian Water Authority, EAAP, SE Italy, is leached with concentrated sulphuric or hydrochloric acid, and the leachate (pH 3.5) fed downward at a flow-rate, $F_{es}$ 5 BV/h (BV = Bed Volumes). The resulting detoxified solid residue can be applied safely to land since no more metals release is observed. Table 1 shows the average composition of the raw clarifier sludge, its acidic leachate and the solid residue resulting from the leaching operation.

**Table 1** *Average composition of the clarifier sludge, its acidic leachate (pH 3.5) and solid residue after leaching of the Sinni River Water Potabilization Station sludge( °).*

| Parameter | Clarifier sludge (g/kg$_{dry}$) | Acidic leachate (mg/L) | Solid residue (g/kg$_{dry}$) |
|-----------|------------------|------------------|------------------|
| Al | 150-300 | 300 - 500 | 100-200 |
| Ca | 45-60 | 200 - 400 | 10-20 |
| Mg | 3-5 | 25 - 100 | 0-10 |
| Mn | 2-5 | 5 - 10 | 1-3 |
| Fe | 20-50 | 0.5 - 2 | 15-40 |
| Cu | 1-5 | 0.05 - 0.2 | 0 |
| Zn | 2-3 | 0.1 - 0.5 | 1-2 |
| Pb | 1-2 | 0.1 - 1.0 | 0 |
| Si | 200-300 | | 300-500 |
| TOC (*) | | 200-300 | |

Leachant: Sulphuric acid 98% w/v;
(*) Total Organic Carbon
(°) Apulian Water Authority, EAAP, S.E. Italy.

Every 50 BV of treated effluent (exhaustion time: 10h) (Figure 2), the resin is regenerated counter-current (flow-rate, F$_{reg}$ 5BV/h) with 30 BV of 0.4M NaOH (regeneration time: 5h). To minimise metals precipitation in the column, during the exhaustion step, the regenerated (Na-form) carboxylate resin is back-washed by controlled elution of softened water (100 BV, F = 10 BV/h, washing time: 10h), produced by column C2. After the back-washing operation the resin is partially hydrolysed, thus converted in the mixed Na$^+$/H$^+$-form.

The counter-current mode of operation controls pressure build-up in the column and related packing of the resin bed as a consequence of the large volume variation of resin passing from the exhausted to the regenerated form. One complete cycle per day (exhaustion-regeneration-backwashing) is performed in these conditions.

After the exhaustion step, the resin bed is regenerated in a stepwise mode to save on chemicals consumption. Accordingly, based on elution curves (Figure 4), the first 5 BV of the spent regeneration eluate, containing most of the impurities, are collected in the vessel S3 (Figure 1) where, after pH adjustment to 8 metals precipitation is induced. The resulting solids are properly disposed-of in controlled landfilling, the supernatant solution is recycled to the subsequent regeneration step. The second fraction (10 BV) of the spent regeneration eluate, containing >98% of the exchanged aluminium species is collected in vessel S4 (Figure 1) ready for reuse as coagulating agent. A third polishing fraction of regenerant solution (~15BV) is needed for cleaning-up the residual ion species from the resin. This latter fraction, collected in vessel S5, is recycled to the subsequent regeneration step and used as first and second fractions. Overall, only 15 BV/cycle of regenerant solution are used in each step, with related savings in the general running costs of the ion exchange operation.

Analytical determinations of the metal species were carried out by Inductive Coupled Plasma Atomic Emission Spectrometry (ICP-AES) with a Mod.Optima 3000 from Perkin Elmer Corporation, Norwalk, CT, USA, or by Atomic Absorption Spectroscopy (AAS) with a Mod.SpectrAA 250 from Varian, Victoria, Australia.

### 3 RESULTS AND DISCUSSION

#### 3.1 The exhaustion behaviour of the resin

Figure 2 shows the typical exhaustion breakthrough curves after elution of the clarifier sludge acidic leachate at the composition reported in Table 1. 50 BV of leachate can be treated before appreciable aluminium leakage occurs. From the close observation of the breakthrough in Figure 2, a peculiar roll-over the feed composition of the elution curves referring to Ca, Mg, Mn, Zn ions is observed at the breakthrough of the aluminium species. From the mechanistic point of view, it is possible that the more densely charged free aluminium ions "sweeps-out" from the column the weak interacting species which have been retained "kinetically" in the initial stages of the resin exhaustion step. Accordingly, the selectivity scale for the species of reference appears in the order $Al^{3+}>Mn^{2+}>Mg^{2+}>Ca^{2+}$, in full agreement with expectation based on the Eisenman theory[5].

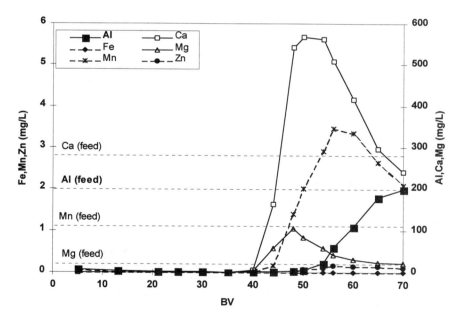

**Figure 2** *Exhaustion breakthrough curves of the carboxylate resin Purolite C106 after elution of the clarifier sludge leachate (pH 3.5; Fes = 5BV/h).*

From Figure 3, where are reported the cyclic performances of the plant in terms of: a) %Al removal; b) influent and effluent Al concentrations; c) exhaustion and regeneration operative capacities toward aluminium; it is observed that plant performances remained fairly constant throughout the investigation with average aluminium removal from the leachate solution exceeding 98% and average output aluminium concentration not exceeding the European limit for discharge into recipient water-bodies (MAC, Maximum Allowable Concentration = 2 mgAl/L)[6].    Although a trend in the aluminium operative exchange capacities is observed, possibly due to the accumulation of manganese species in the resin phase (the mass balance does not close in the corresponding streams), the general agreement between exhaustion and regeneration performances (Figure 3 bottom), ensures that quite satisfying resin performances toward aluminium species are obtained in this conditions.

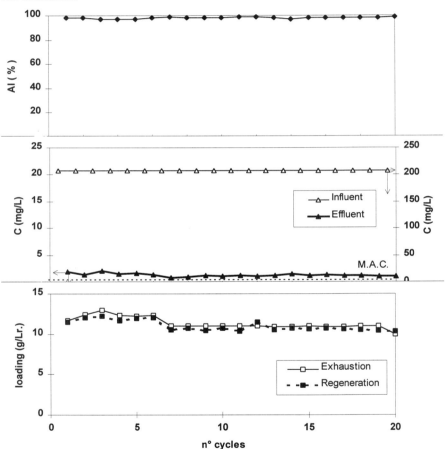

**Figure 3**  *Cyclic performances of the laboratory scale pilot plant: %Al removal (upper); Influent and effluent Al concentration  (middle); Operative Al resin loading during exhaustion and regeneration steps (bottom).*

## 3.2 The regeneration behaviour of the resin

Figure 4 shows the elution curves of metal species from the carboxylate resin by using 0.4M NaOH solution. It is clearly evident that most of the macro and micro components corresponding to the metal impurities in the clarifier sludge leachate (i.e., Ca, Mg, Cu, Fe, Mn, Pb, Zn) are eluted in the first 5 BV dosed to the column, whereas Al and Fe species, the coagulant principle, is eluted almost clean in the subsequent 10 BV of regenerant solution. This particular contingence allows, in practice, for the mentioned "stepwise" regeneration protocol (the three re-circulating fractions of the regenerant solution), for the chromatographic separation of different metal species.

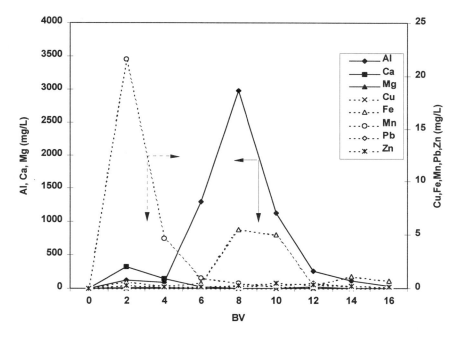

**Figure 4** *Regeneration curves of the carboxylate resin Purolite C106 after exhaustion with the clarifier sludge leachate (pH 3.5; $F_{reg}$=5 BV/h).*

From mechanistic point of view such behaviour could be interpreted in the following terms. Two simultaneous reactions occur during metals elution after the regeneration step: i.e. ion exchange at functional groups, operated by the sodium ion, followed by hydrolysis reaction of metal species in the liquid-phase:

$$R\text{-}(COO)_nMe + nNaOH ---- nR\text{-}COONa + Me[(H_2O)_p]^{m+}$$

$$[Me(H_2O)_p]^{m+} + zOH^- ---- [Me(H_2O)_{p-z}(OH)_z]^{(m-z)-} + zH_2O.$$

Among potentially hydrolysing species, zinc, lead and manganese are more easily released by the functional groups, thus promptly hydrolysed to the anionic species and, consequently, rejected (eluted) by the resin phase by Donnan effect[7].

Aluminium and iron, being held more strongly by the resin functional groups, may not be hydrolysed, thus, as cationic species, are retarded in their column migration. Calcium and magnesium ions, not forming complexes, are eluted after the ion exchange reaction.

### 3.3 The quality of the coagulants recovered and fate of the solid residues after the leaching operation

Table 2 shows a comparison of the average impurities (expressed as mg/kgAl) present in: a) the clarifier sludge leachate at pH 3.5;  b) in the coagulant solutions recovered as spent regeneration eluates from the ion exchange process of reference, and,  c) in a typical commercial product such as aluminium polychloride.

The quality of the coagulants recovered, after the ion exchange operation, results to be better than the average quality of commercial products.

**Table 2**  *Impurities present in the water clarifier sludge leachate (pH 3.5), in the coagulants recovered after the ion exchange operation and in a typical commercial product (mg/kg Al).*

| Species | Sludge leachate | Coagulants recovered by ion exchange ($^+$) | Commercial coagulants (*) (aluminum polychloride) |
|---------|-----------------|---------------------------------------------|---------------------------------------------------|
| Ca | 83,000 | 20 | - |
| Mg | 79 | 2 | - |
| Mn | 22 | 1 | 50 |
| Fe | 5 | 2 | 300 |
| Cu | 0.1 | 0.08 | - |
| Zn | 0.3 | 0.1 | 50 |
| Pb | 0.2 | 0.1 | 10 |

(+) Al: 99.99%; (*)%Al:99.9%

Two residual effluents are formed as a result of the leaching and the ion exchange operation respectively.

An inert easy sedimentable solid fraction (~20% of the initial volume of clarifier sludge, see Tab.1 third column) which is no more metals leachable, thus completely detoxified and safely disposed-of  by land application according to current European legislation.

A minimised solid fraction (1% of the initial clarifier sludge volume) resulting from the precipitation operation of the first fraction of the resin spent regeneration eluates, including most of metals impurities such as Pb, Cu, Zn, Mn.

## 4 CONCLUSIONS

Technical feasibility of an ion exchange-based process for separation and recovery of the coagulant principally from water clarifier sludge has been demonstrated at laboratory scale pilot plant level, by using real effluents from water treatment installations.    The innovation is based on the use of a commercial standard exchanger manifesting no specific selectivity toward the ionic species of interest.    Accordingly, selective separation and purification of the coagulation chemicals from other trace pollutants is obtained by chromatographic elution of limited amounts of cheap chemicals (0.4M NaOH) to the column.

Experiments are in progress on a larger scale demonstration installation    to substantiate economical aspects of the process.

## ACNOWLEDGEMENTS

Mr Nicola Limoni is acknowledged for design and assembling of the pilot plant, Mr Giuseppe Labellarte for the analitical activity, whereas Mr Giuseppe Laera and Mr Wojciech Josef Wojtkow, from the Dept.of Chemistry of the Jagellonian University, Krakow, Poland  are gratefully acknowledged for carrying out the experimental work.

**References**

1.  F. Pontius, 'Water Quality and Treatment'. Am.Wat.Works Ass., McGraw-Hill, New York, NY, 4th edition, 1990.

2.  J. Bratby, 'Coagulation and Flocculation. With an Emphasis on Water and Wastewater Treatment'. Upland Press Ltd. Croydon, UK, 1980.

3.  F.M. Saunders, M.L. Roeder, R.B. Rivers, Y. Magara, 'Coagulant Recovery: A Critical Assessment'. Am.Wat.Works.Ass.Res.Found., Denver, CO, USA, 1991.

4.  S.J. Koorse,  The role of residuals disposal laws in treatment plant design. *Journ.Am.Wat.Works Ass.*, 1993 **57(3)**120-125.

5.  G. Eisenman, The molecular basis of ionic selectivity in macroscopic systems. In L.Liberti, F.G.Helfferich Eds. 'Mass Transfer and kinetics of ion exchange', NATO-ASI Ser.E71, Martinus Nijoff Pub.,1983, p.143-160.

6.  *EU Directive* (1991) n. 91/271/EEC.

7.  F. Helfferich, 'Ion Exchange', McGraw-Hill, New York, NY,1962, p.135.

# ECOLOGICALLY SAFE ION-EXCHANGE TECHNOLOGIES

D.Muraviev[*], R.Kh. Khamizov[**] and N.A.Tikhonov[***]

[*]Autonomous University of Barcelona, Dept of Analytical Chemistry, E-08193 Bellaterra (Barcelona), Spain, [**]Vernadsky Institute of Geochemistry and Analytical Chemistry, Russian Academy of Science, 117975 Moscow, Russia, and [***] Department of Physics, Lomonosov Moscow State University, 117899 Moscow, Russia.

## 1 INTRODUCTION

The industrial application of ion-exchange (IE) processes is growing progressively. In many instances the IE technology can successfully substitute existing large-scale industrial processes which do not satisfy modern ecological standards. The requirements that the IE technology should have to create a competitive process are high efficiency and ecological safety. However, the general scheme of "standard" IE processes remains practically unchanged during the last decades and comprises the following auxiliary operations[1,2]: 1) preparation of stock solution; 2) concentration of solution after the IE treatment (e.g., by evaporation); 3) recovery of the purified product (e.g., by crystallization); 4) regeneration of ion exchangers and auxilliary reagents for reuse; 5) neutralization of aggressive wastes, and some others. Several approaches can be applied to eliminate some of these auxiliary operations to improve the efficiency of the process due to significant saving in chemicals, energy, manpower, minimization of wastes, etc. One of those is based on the application of dual-temperature IE processes[3-6] which allows one to exclude auxiliary operations 4 and 5. Hence, a practically reagentless and waste-free, separation process can be designed[7-12]. Another route to avoid the above auxiliary operations (e.g., 2 and 3) is based on a combination of the IE conversion and concentration processes into one stage. Frontal IE chromatography[13,14] and reverse frontal separation[14,15] are applicable for this purpose. In certain instances, both of those IE separation techniques allow concentrating the target substance up to the level exceeding its solubility at a given temperature. Moreover, this supersaturated solution (SS) remains stable for a long period, and after leaving the column it crystallizes spontaneously. This allows for designing a practically ideal process where a crystalline product is obtained directly after the IE treatment. This phenomenon - known as Ion-Exchange-Isothermal-Supersaturation (IXISS) - has been discovered by Muraviev[16]. This paper reports some of the results obtained by tailored application of dual-temperature IE and IXISS effect to design highly-efficient and ecologically safe IE technologies which can be considered as a successful alternative to existing large-scale industrial processes.

## 1. Manufacture of Chlorine-free Potassium Fertilizers

The production of chlorine-free potassium salts (with minimum $Cl^-$ admixture), such as $K_2SO_4$, is of particular interest since it deals with the problem of the effective cultivation of some chlorophobic plants such as citruses, vegetables, herbs etc., which are adversely affected by high $Cl^-$ concentration. Potassium sulfate is produced in substantial quantities in Europe by the Mannheim process from $K_2CO_3$ and $H_2SO_4$, or by reaction of $H_2SO_4$ with KCl. Both versions of the Mannheim process are complicated by problems of utilizing gaseous wastes ($CO_2$ and HCl)[17].

In the US, and some other countries, $K_2SO_4$ is manufactured by exchange reactions between potassium, sodium and magnesium salts by their dissolution and fractional crystallization. The last process requires utilization of large volumes of liquid wastes[17]. The IE synthesis of $K_2SO_4$ from KCl and $Na_2SO_4 \cdot 10H_2O$ can be based either on cation- or anion-exchange reactions written as follows:

1. *Cation-exchange synthesis*

$$R - SO_3Na + KCl \Rightarrow R - SO_3K + NaCl \qquad (1)$$
$$2R - SO_3K + Na_2SO_4 \Rightarrow 2R - SO_3Na + K_2SO_4 \downarrow \qquad (2)$$

2. *Anion-exchange synthesis*

$$R - N(CH_3)_3Cl + Na_2SO_4 \Rightarrow R - N(CH_3)_3)_2SO_4 + 2NaCl \qquad (3)$$
$$(R - N(CH_3)_3)_2SO_4 + 2KCl \Rightarrow 2R - N(CH_3)_3Cl + K_2SO_4 \downarrow \qquad (4)$$

In the first process, a sulfonate cation exchanger is first converted from the Na- to K-form with dilute KCl solution ($0.1$ mol/dm$^3$), followed by desorption (stripping) of the product ($K_2SO_4$) with concentrated $Na_2SO_4$ solution ($2$ mol/dm$^3$). The second process starts with the conversion of a strong base anion exchanger from the Cl- to $SO_4$-form with dilute $Na_2SO_4$ solution (~$0.25$ mol/dm$^3$). Then, $K_2SO_4$ is produced during the stripping of sulfate ions with concentrated KCl solution (~$3-4$ mol/dm$^3$). The typical concentration-volume history of the last process is shown in Fig.1. As seen, the displacement of $SO_4^{2-}$ ions from the resin with concentrated KCl solution leads to the formation of the pure SS of $K_2SO_4$ with the supersaturation degree $\gamma \approx 2$. Nevertheless, $K_2SO_4$ does not precipitate in the column and remains as a stable SS at least over a period of several hours. At the same time this SS crystallizes spontaneously by following its removal from the column. The maximum efficiency of the process shown in Fig.1 is achieved when carried out at 282 K

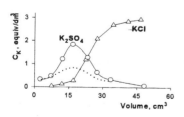

**Figure 1.** *Concentration-volume history of stripping of $K_2SO_4$ from strong base anion exchanger in $SO_4^2$-form with 3M KCl at 282K. Dotted line corresponds to $SO_4^2$ content in supernatant after crystallization of the product.*

to minimize the solubility of $K_2SO_4$ in solution collected. The similar results have been obtained by studying the cation-exchange version of the process.

The necessity to apply dilute KCl and $Na_2SO_4$ solutions at the first stages of both processes follows from the need to shift the ion-exchange equilibrium in both systems to the right. The equilibrium separation factor $\alpha$ for, e.g. $SO_4^{2-}$ - Cl⁻ exchange (see reaction 3) can be written as follows:

$$\alpha_{SO_4}^{Cl} = \frac{q_{Cl}}{q_{SO_4}} \cdot \frac{C_{SO_4}}{C_{Cl}} \tag{5}$$

Here q and C are the concentrations of ions in the resin and solution phases, respectively. If a strong base anion exchanger is equilibrated with dilute solution of $SO_4^{2-}$ - Cl⁻ mixtures, the respective $\alpha$ value < 1, i.e. the resin is more selective for $SO_4^{2-}$. If $K_2SO_4$, displaced by concentrated KCl solution ($C_o$, equiv/dm³), forms a SS, where it exists in an associated (molecular) form at a concentration $C_M$ equiv/dm³ exceeding $\gamma$ times its solubility, $C_S$, at a given temperature, then formation of $K_2SO_4$ molecules can be described by dissociation constant, $K_D$. Hence, equation (5) can be rewritten in the following form[18]:

$$\alpha_{SO_4}^{Cl} = \frac{q_{Cl}}{q_{SO_4}} \frac{K_D \gamma C_S}{C_o(C_o - \gamma C_S)} \tag{6}$$

As follows from equation 6, at constant $C_o$; $C_S$ and $K_D$, $\alpha$ increases with $\gamma$ and may reach sufficiently high values (>> 1), as $\gamma \rightarrow C_o / C_S$. Note, that the same reasoning can be used for interpretation of the selectivity reversal in cation-exchange systems (see, e.g. eq. 1 and 2).

Application of the IXISS effect in the flowsheets of both processes studied improves the product formation stage due to the shift of IE equilibrium in reactions 2 and 4 to the right. Another advantage of the IXISS based synthesis of $K_2SO_4$ deals with the possibility to reuse the supernatants obtained after crystallization of $K_2SO_4$ as displacers in the subsequent stripping cycles. For example, after separation of $K_2SO_4$ crystals, the supernatant obtained in the first process is fortified with $Na_2SO_4$ up to the desired concentration of sulfate ions (2 mol/dm³) and then is directed to the next stripping cycle. Hence the only waste which is produced in both processes is the dilute NaCl solution, which can be readily concentrated by, e.g. reverse osmosis. The desalinated water obtained can also be returned back into the process.

## 2. Dual-temperature Concentration of Magnesium and Bromine from Seawater

At present, around 25% of overall world production of magnesium and 70% of that of bromine is provided from the sea and other hydromineral resources. Traditional methods for producing Mg (lime-process) and Br (air-stripping technique) by processing seawater, despite their profitability do not satisfy growing ecological standards[12]. Consequently, new alternative ecologically clean technologies, based on e.g. IE separation methods, have to be developed. The dual-temperature IE concentration of Mg and Br from seawater is based on the strong temperature dependence of the separation

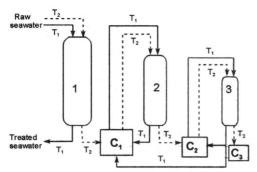

coefficients $\alpha_{Na}^{Mg}$ and $\alpha_{Cl}^{Br}$ (see eq.5) of weak acid (for Mg) and strong base (for Br) IE resins, respectively[8,12]. For example, the $\alpha_{Cl}^{Br}$ value decreases by a factor of ~2, while $\alpha_{Na}^{Mg}$ value increases by a factor of ~1.2 when the temperature of seawater rises from 10 to 90° C. The principle of

**Figure 2.** *Schematic diagram of dual-temperature IE unit for concentration of Mg (or Br) from seawater. Storage tanks for Mg (Br) concentrates ($C_1$-$C_3$) are provided with cooling/heating facilities.*

the dual-temperature concentration of Mg and Br by using a cascade of fixed-bed columns is presented in Fig.2.

In a subsequent treatment of cold and hot seawater in a fixed-bed IE column, the concentration of Br⁻ in the hot stripping solution increases by a factor of 2, while the concentration of Cl⁻ and $SO_4^{2-}$ decreases. The multistage process (using AV-17 resin) enriches Br⁻ concentration in the final concentrate up to acceptable levels for further processing (>5 g/L)[12]. Four dual-temperature sorption-stripping cycles (using Lewatit R 250 K resin) allow for concentrating Mg up to ~0.4 equiv/L (versus 0.11 equiv/L in the initial seawater)[8]. The resins used in both processes do not require any regeneration. Hence, the processes are absolutely waste-free. The reduction in energy expenditure for heating seawater can stimulate the development of the above techniques. This problem can be successfully solved by using conventional or concentrated sunlight (in the areas with high level of solar radiation) as the principal and ecologically clean energy source (sun-boiler systems). An alternative solution can be the use of seawater in the cooling cycles of the steam power stations. Estimation shows that in this case almost 50% of the overall energy costs could be written off[12].

## 3. Recovery of High-purity Magnesium Compounds from Seawater

The traditional magnesium-from-seawater technology includes mixing the raw seawater with $Ca(OH)_2$, filtration of $Mg(OH)_2$ slurry followed by its treatment with HCl, evaporation, drying and electrolysis. The process does not produce sufficiently pure Mg. The possibility of designing an alternative IXISS based process for recovery of high-purity Mg compounds from seawater has appeared after the discovery of IXISS of $MgCO_3$ in the resin bed[12,19]. The IXISS effect is observed during elution of $Mg^{2+}$ from carboxylic resin pre-loaded with decalcinated seawater with solution of $Na_2CO_3$ - $NaHCO_3$ mixture. Magnesium carbonate does not precipitate in the column and remains as a stable SS (with $\gamma \approx 5$) over a period of 72 hours at least. After removal of this solution from the column ~75% of the desorbed magnesium spontaneously crystallizes in the form of well-shaped nesquegonite ($MgCO_3$ $3H_2O$) crystals. The purity of the magnesium compound obtained appears to be >99.9% since, unlike magnesite ($MgCO_3$), nesquegonite crystals are calcium free. The block-scheme of the pilot unit for recovery of

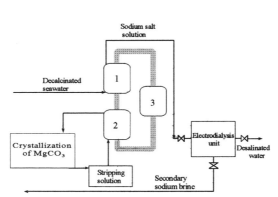

high purity $MgCO_3$ from seawater is shown in Fig.3. The raw seawater is decalcinated by also using the IXISS technique as described elsewhere[12,19]. Ca-free

**Figure 3.** *Block-scheme of pilot plant for recovery of high purity $MgCO_3$ from seawater. Sorption (1); desorption (2), and electrolyte replacement (3) columns.*

seawater is treated with carboxylic resin as the Na-form in sorption Column 1. The resin in the Mg-form is then directed to desorption Column 2 where the stripping of $Mg^{2+}$ with solution of $Na_2CO_3$ - $NaHCO_3$ mixture (containing also a residual $MgCO_3$ from recycled stripping solution) is carried out. The resin is then transferred to Column 3 to replace electrolyte in the porosity volume of the resin bed. The softened seawater (after Column1) is directed to the electrodyalisis unit which produces desalinated water and sodium brine. This brine is used for regeneration of the seawater decalcination column (not shown in Fig.3). The sodium-calcium brine produced during this stage can also be returned back into the process after crystallization and removal of the $CaSO_4$ precipitate. After crystallization of $MgCO_3$ from the supersaturated eluate, the supernatant ($Na_2CO_3$ + $NaHCO_3$ + traces of $MgCO_3$) can be reused as a stripping solution following fortification with desired amounts of $Na_2CO_3$ and $NaHCO_3$. Hence, the process under consideration appears to be essentially waste-free.

## 4. Dual-temperature IE Concentration of Copper from Acidic Mine Waters

The treatment of acidic mine waters (AMW), the natural effluents from pyritic ore deposits, is of great economic and ecological importance. The AMW are characterized by low pH ($\sim2$) and relatively high concentration of metal ions such as $Fe^{3+}$, $Zn^{2+}$, $Cu^{2+}$, $Mn^{2+}$, $Al^{3+}$ and others[9,10]. Results obtained by studying the temperature dependence of IE equilibrium on polyacrylic (Lewatit R 250-K) and iminodiacetic (Lewatit TP-207) resins with the natural AMW of the Rio Tinto area (Huelva, Spain) have shown that both resins can be successfully used for the dual-temperature concentration of copper[9,10]. Acrylic resin has been found to be selective for $Al^{3+}$ and iminodiacetic resin manifests high selectivity towards $Cu^{2+}$. The uptake of $Al^{3+}$ increases while that of $Cu^{2+}$ decreases with temperature for both resins. This effect suggests the possibility for the dual-temperature IE concentration of copper from AMW by using the same set-up and the mode of operation as those shown in Fig.2. The resins are first equilibrated with cold AMW ($20°C$). A selective thermostripping of $Cu^{2+}$ is then carried out by using hot AMW ($80°C$) accompanied by an increase of $Cu^{2+}$ concentration in the eluate by a factor of 1.3 or 1.2 for acrylic or iminodiacetic resin, respectively. The concentration of $Al^{3+}$ in the eluate drops to 50% in the first case and to 70% in the second.

**Acknowledgements**

Part of this work was supported by Research Grants No. ND-200 and ND-2300 from the International Science Foundation and from the Science and Technology Programmes of Russian Federation "Global Ocean" (Grant No. 02.08.1) and "Chemistry and Technology of Water" (Grant No. 195). D.M. thanks the Catalonian Government for the financial support of his visiting professorship at the Universitat Autónoma de Barcelona.

**References**

1.  L.K Arkhangelsky and F.A. Belinskaya, 'Ion Exchange in Chemical Technology', Khimiya, Leningrad, 1982 (Russian).
2.  M. Streat, in: 'Ion Exchangers', K. Dorfner (Ed.), Walter de Gruyter, Berlin, 1991, p.685.
3.  B.M. Andreev, G.K. Boreskov and S.G. Katalnikov, *Khim. Prom-st*, 1961, **6**, 389 (Russian).
4.  V.I. Gorshkov, M.V. Ivanova, A.M. Kurbanov and V.A. Ivanov, *Vestnik Moskov. Univ., Ser. Khim.*, 1977, **5**, 535 (Russian); English translation in *Moscow Univ. Bull.*, 1977, **32**, 23.
5.  M. Bailly and D. Tondeur, *J. Chem. Eng. Symp. Ser.*, 1978, **54**, 111.
6.  M. Bailly and D. Tondeur, *J. Chromatogr.*, 1980, **201**, 343.
7.  D. Muraviev, A. Gonzalo and M. Valiente, *Anal. Chem.*, 1995, **67**, 3028.
8.  D. Muraviev, J. Noguerol and M. Valiente, *React. Polym.*, 1996, **28**, 111.
9.  D. Muraviev, J. Noguerol and M. Valiente, in: 'Progress in Ion Exchange. Advances and Applications', A. Dyer, M.J. Hudson and P.A. Williams (Eds.), Royal Soc. Chem., Cambridge, 1997, p.349.
10. D. Muraviev, J. Noguerol and M. Valiente, *Hydrometalurgy*, 1997, **44**, 331.
11. D. Muraviev, J. Noguerol and M. Valiente, *Environ. Sci. Technol.*, 1997, **31**, 379.
12. R.Kh. Khamizov, D. Muraviev and A. Warshawsky, in: 'Ion Exchange and Solvent Extraction', A. Marinsky and Y. Marcus (Eds.), vol.12, Marcel Dekker, New York, 1995, p.93.
13. O. Bobleter and G. Bonn, in: 'Ion Exchangers', K. Dorfner (Ed.), Walter de Gruyter, Berlin, 1991, p.1187.
14. D. Muraviev, A.V. Chanov, A.M. Denisov, F.M. Omarova and S.R. Tuikina, *React. Polym.*, 1992, **17**, 29.
15. V.I. Gorshkov, D. Muraviev and A. Warshawsky, *Solv. Extr. Ion Exch.*, 1998, **16**, 1.
16. D. Muraviev, *Zh. Fiz. Khim.*, 1979, **53**, 438 (Russian).
17. H.J. Scharf, 'Environmental Aspects of K-fertilizers in Production, Handling and Application. Development of K-fertilizers Recommendations', International Potash Institute, Worblaufen-Bern, 1990, p.395.
18. D. Muraviev, R.Kh. Khamizov and N.A. Tikhonov, *Solv. Extr. Ion Exch.*, 1998, **16**, 151.
19. R.Kh. Khamizov, L.I. Mironova, N.A. Tikhonov, A.V. Bychkov and A.D. Poezd, *Sep. Sci. Technol.*, 1996, **31**(1), 1.

# NITRATE REMOVAL FROM DRINKING WATER BY AN ANION EXCHANGER IN THE HSO₄⁻ FORM WITH REUSE OF ELECTROCHEMICALLY DENITRIFIED REGENERANT SOLUTION

Z. Matějka , D. Nevečeřalová, O. Rejzlova and Š. Gromanova

Institute of Chemical Technology
Power Engineering Department
Prague 16628

## 1 INTRODUCTION

At the present time it is no longer a problem to decrease the concentration of nitrates in drinking water well below the required limit of 50 mg/L and - simultaneously - to achieve a high breakthrough capacity of anion exchanger toward nitrates. Strongly basic anion exchangers with $-N^+Et_3$ or $-N^+Pr_3$ functional groups, exhibiting higher affinity toward nitrate over sulfate, were introduced into industrial practice and thus the adverse effect of sulfates in the raw water on nitrate removal efficiency was eliminated[1,2]

The anion exchanger regeneration step is not efficient enough as yet, because a high excess of a regenerant chemical has to be still applied. This results in the disposal of big quantities of the spent regenerant solutions, containing high concentrations of regenerant chemicals and also nitrates, displaced from anion exchanger, which is the serious environmental problem[3,4]

In order to improve this situation this study tested the concept of reusing the spent regenerant solution after the electrochemical reduction of eluted nitrates.

Applying this concept, two benefits will be gained :

a) the consumption of regenerant chemical will be substantially lowered, because the amount of salt repeatedly added into the spent regenerant is only that which is required to keep its concentration at a constant predetermined level.

b) the presence of $NO_3^-$ in wastewater emanating from the ion exchange system is almost completely eliminated.

### 1.1 Selection of Primary Regenerant Agent

The commonly used NaCl solution is not suitable for this concept because of the harmful effect of $Cl_2$ gas evolved during electrochemical $NO_3^-$ reduction. An additional drawback is the unwanted increase in $Cl^-$ concentration and also the high ratio of $[Cl:HCO_3]$ in the treated water when NaCl is applied for resin regeneration[5].

The main disadvantage of $NaHCO_3$ is the low efficiency of electrochemical $NO_3^-$ reduction in alkaline solution. This electrochemical reduction proceeds fast enough, and without any unwanted side-effects, in the acidic solution containing $HSO_4^-$ anion.

Preliminary experiments have pointed out that an anion exchanger (selective type Ionac SR 7 having $-N^+Prop_3$ moieties) regenerated by $Na_2SO_4$ solution (0.6 M-$Na_2SO_4$ , pH 7.0 , 30 BV) exhibited a poor nitrate uptake efficiency compared to the Cl-form of anion exchanger (Fig.1).

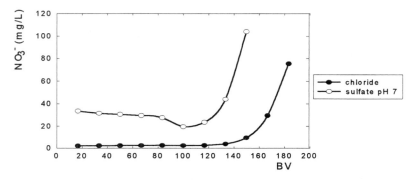

**Figure 1** *Effect of regenerant solution composition on $NO_3^-$ - uptake*

The optimum conditions for application of sulfate solution as a primary regenerant agent were investigated. All laboratory measurements were carried out as dynamic column experiments at a solution flow-rate of 20 $BV.h^{-1}$ using anion exchangers having $-N^+Et_3$ and $-N^+Prop_3$ as functional groups. Loading solution : $NO_3^-$ - 2 meq/L, $SO_4^{2-}$ - 4 meq/L, $Cl^-$ - 1 meq/L, $HCO_3^-$ - 1 meq/L.

## 2 RESULTS AND DISCUSSION

### 2.1 Effect of pH on Nitrate Desorption by Sulfate Solution

It was confirmed (Fig.2), that a $SO_4^{2-}$ solution at pH 7 exhibits a very low desorption efficiency, which explains the high leakage level and low $NO_3^-$ breakthrough capacity achieved under this regeneration condition (Fig.1).
Nitrate desorption improved significantly with a decrease of the solution pH (Fig.2, Fig.3) as a consequence of the increasing amount of $HSO_4^-$ anion in the regenerant solution (Eq.1)

$$SO_4^{2-} + H^+ \Longrightarrow HSO_4^- \qquad (1)$$

In the pH range below 1.4 the prevailing anion in the solution is $HSO_4^-$ . Under these conditions the anion exchanger is regenerated by $HSO_4^-$ solution and will be in the $HSO_4^-$ form [6].
Desorption curves for 0.6M-$Na_2SO_4$ (pH 1.4) and for 0.6M- $KHSO_4$ are almost identical (Fig.4) which proves this hypothesis. At this pH the desorption efficiency of sulfate

solution is higher than that exhibited by NaHCO₃ solution (Fig.5), which is normally considered as the more efficient regenerant agent[3].

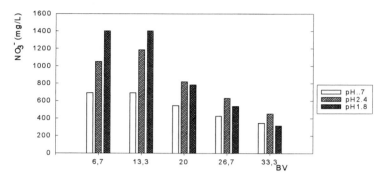

**Figure 2** *Effect of pH of regenerant solutions on NO₃⁻ desorption*

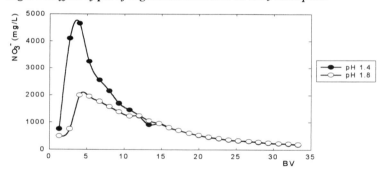

**Figure 3** *Desorption efficiency of sulfate solution at low pH-range*

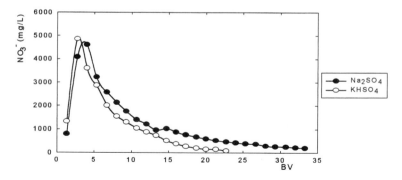

**Figure 4** *Desorption efficiency of Na₂SO₄ (pH- 1.4) and KHSO₄*

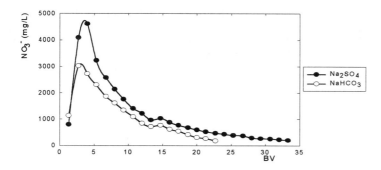

**Figure 5** *Desorption efficiency of NaHCO₃ and Na₂SO₄ (pH 1.4) solutions*

## 2.2 Effect of pH of Regenerant Sulfate Solution on Nitrate Uptake

The nitrate uptake efficiency corresponds well with desorption experiments. The high breakthrough capacity and low leakage are obtained with acidic (pH -1.4) regenerant solution (Fig.6, Fig.7).

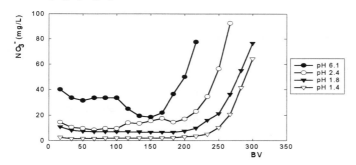

**Figure 6** *Effect of pH of regenerant sulfate solution on nitrate uptake*

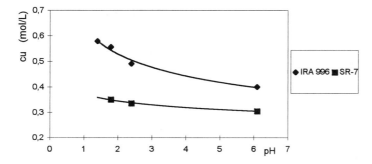

**Figure 7** *Effect of pH of regenerant sulfate solution on nitrate breakthrough capacity*

## 2.3 Effect of volume and concentration of regenerant solution on nitrate uptake

In order to keep the level of nitrate leakage low enough (let us say below 20 mg/L) it is necessary to apply as much as 30 bed volumes (BV) of regenerant solution at a pH value lower than 1.8 . This large volume is derived from the nitrate desorption curve (Fig.4).

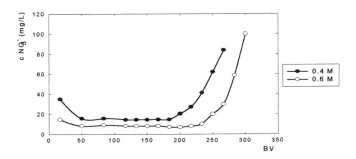

**Figure 8** *Nitrate uptake on Amberlite IRA 996 regenerated by 30 BV (pH 1.4)*

However the regenerant solution will be fully recycled in this process and therefore the solution volume applied is not crucial from the viewpoint of regenerant consumption. This will happen only in the case of the conventional mode of regeneration, where the whole volume of spent solution is immediately wasted.

Results from the sorption runs show that regenerant concentration of 0.6M is the preferred one (Fig.8) but solution concentrations as low as 0.3M can be also successfully used provided the pH value is lowered down to 1.2 (Fig.9).

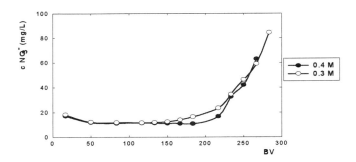

**Figure 9** *Nitrate uptake on Amberlite IRA 996 regenerated by 30 BV (pH 1.2)*

Nitrate leakage level below 20 mg/L can not be constantly achieved with any smaller volume of regenerant solution than 30 BV, as is demonstrated for 22 BV (Fig.10) and for 15 BV (Fig.11), regardless of the solution concentration.

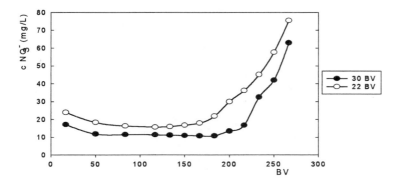

**Figure 10** *Nitrate uptake on Amberlite IRA 996 regenerated by 22 BV (0.4 M; pH 1.2)*

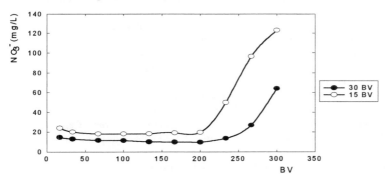

**Figure 11** *Nitrate uptake on Amberlite IRA 996 regenerated by 15 BV (0.6 M; pH 1.4)*

## 2.4 Stoichiometric consumption of regenerant agent

In conventional nitrate removal the consumption of regenerant equals 450 -800% of stoichiometry[7]. In this process the whole volume of regenerant solution is recycled and therefore only the difference in sulfate concentrations between fresh and spent solutions indicates the actual consumption of regenerant ( in this case sulfates ). This decrease in sulfate concentration in the spent regenerant solution is caused by: (a) displacement of nitrates ( and also trace amount of chlorides) from the anion exchanger by sulfate solution according to Eq. 2

$$\overline{NO_3^-} + HSO_4^- \implies \overline{HSO_4^-} + NO_3^- \qquad \{2\}$$

(b) dilution of regenerant solution by rinse water before, and after, the regeneration step (Fig.12).

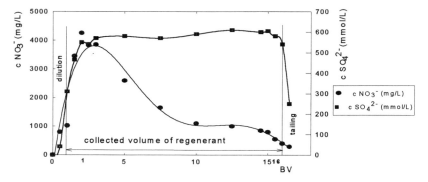

**Figure 12** *The loss of sulfate on regenerant solution reuse*

During repeated operation cycles (sorption - regeneration - electrolytic $NO_3^-$ reduction) the amount of regenerant solution has to be kept at a constant, predetermined ( for instance 15 BV) ,level. On collecting this exact volume of spent regenerant solution a certain loss of sulfate occurs by dilution (at the beginning of regeneration ) and by tailing ( at the end of regeneration ). This is the reason why the actual consumption of regenerant agent (in 0.6 M solution) has been found at 210 % of stoichiometry. Despite this, it is the best result ever achieved for nitrate removal.

The loss of sulfate by tailing and dilution can be minimized by
(a) using low concentration of regenerant solution (0.3 M ; 30 BV; pH 1.2 ) ;
(b) starting and ending the regeneration step with a column totally free of any solution (and by filling the column by solution again with an upflow-mode to displace air-bubbles).

## 2.5 The composition of water treated by $HSO_4^-$ anion exchanger

The concentration of $Cl^-$ in the treated water does not exceed its loading solution level, which is the advantage of this process over the operation of $Cl^-$ form resin. The amount of sulfates in the filtrate is elevated, however, in relation to the total concentration of anions in the loading solution (Fig. 13). The small amount ( 1 BV ) of 5% NaCl solution used as a *secondary regenerant* will eliminate - after mixing the resin in a column - the unwanted-high concentration of sulfates in the nitrate-free water (Fig.14). The careful separation of this secondary regenerant (NaCl) from *primary regenerant* (NaHSO₄) is required and its reuse is possible.

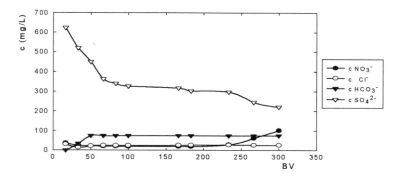

**Figure 13** *The history of nitrate removal run using only sulfate as primary regenerant*

The pH value of the filtrate at the very beginning of the sorption run (approx. 25 BV) was found to be within the slightly acidic range of 2.5 - 6.0 when only primary regenerant was applied (Fig.15). This low pH - which is unacceptable for drinking water - is a consequence of the hydrolytic reaction

$$\overline{2\ HSO_4^-} + H_2O \implies \overline{SO_4^{2-}} + H_2SO_4 + H_2O$$

A small mixing tank should be installed to improve this situation.
When a secondary regenerant solution is used, the pH value is kept above 7.0 throughout the whole sorption run (Fig. 15).

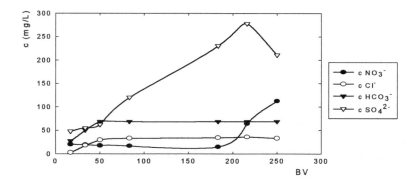

**Figure 14** *The history of nitrate removal run using chloride as secondary regenerant*

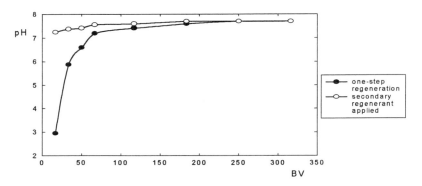

**Figure 15** *The history of pH value on nitrate removal run*

3 CONCLUSION

Nitrate removal from drinking water by $-N^+Et_3$ anion exchanger in $HSO_4^-$ form and the reuse of regenerant solution were investigated.

Sulfate solution at pH lower than 1.4 - which is virtually $HSO_4^-$ solution already - was selected as the primary regenerant and its optimum operating conditions , necessary to achieve low nitrate leakage (below 20 mg/L) , low concentration of chlorides in the treated water and high breakthrough capacity toward nitrates were estimated ( solution amount 30 BV and concentration of $HSO_4^-$ in the range of 0.3 -0.4 M).

The electrolytic reduction of nitrates from the spent acidic sulfate solution proceeds efficiently at low pH on a copper cathode. Residual concentrations of nitrate lower than 50 mg/L after electrolytic process were achieved, which enables regenerant solution reuse. Regenerant agent (sulfate) consumption as low as 200% of stoichiometry and virtually no nitrates in waste water are achieved.

**References**

1. G.A.Guter , US Patent  4,479 877,( 1983)
2. R.E.Barron and J.Fritz, *J.Chromatography*, 1984,**316**, 201.
3. P. Ambrus, J.G.Grantham and D.Grasser, *in " Ion Exchange for Industry",M.Streat (Ed.),Ellis Horwood Ltd.,Chichester,U.K..,1988 (Appendix).*
4. F.X.McGarvey, B.Bachs and  E.W.Hauser, *Reactive Polymers* ,1990, **13**, 324.
5. B.T.Croll, *in "Ion Exchange Processes: Advances and Applications"*
*A.Dyer.M.J.Hudson.P.A.Williams (eds),Royal Soc.Chem., Cambridge,U.K.,1993,p.141*
6. O.Samuelson, *Ion exchange  in analytical chemistry*, John Wiley and Sons, New York, 1963,p.69.
7. Product Data Sheets for Amberlite IRA 996 (Rohm  and Haas) and for Purolite A 520E (Purolite Ltd.).

# TRACE METAL ANALYSIS IN WATER QUALITY MONITORING: SEPARATION OF Zn(II) FROM Cu(II), Pb(II) Ni(II) AND Cd(II) USING IMPREGNATED RESINS

E.Castillo[2], M. Granados[2], M.D. Prat[2] , and J.L. Cortina[1].
1) Departament d'Enginyeria Química, Universitat Politècnica de Catalunya, Diagonal 647, 08028 Barcelona (Spain).
2) Departament de Química Analítica, Universitat de Barcelona, Martí Franqués 1, 08028 Barcelona (Spain).

## 1 INTRODUCTION

There has been a growing demand in analytical chemistry for highly sensitive methods for trace components in complex matrices. Recent advances in new reagents, chelating ligands and instrumentation have improved both detection and specificity (1-2). In the field of metal ion trace analysis ion exchange resins are well known as useful preconcentration and separation tools. However they have some drawbacks: slow adsorption and desorption rates, poor selectivity and requirement of concentrated solution of electrolytes for recovery. Impregnated Resins, which are easily prepared from inert supports and extractant solutions, are an excellent alternative to ion exchange resins. This approach combines many of the advantages of both liquid-liquid extraction and ion-exchange separation. Thus, the use of impregnated resins for preconcentration and clean-up in metal ions analysis in complex matrices offers the following benefits: a) the active part of the resin, i.e. the metal complexing, can be selected depending on the nature of the metal ion, the sample matrix and the analytical procedure which is going to be applied; b) the impregnation procedures of the complexing molecules are very simple and c) their structure and composition can be designed to be compatible with integrated detection systems, when used in solid phase spectroscopic measurements. Two different approaches have been tried for the preparation of these supports: the direct adsorption of the reagent onto high surface area macroporous adsorbents to produce "Solvent Impregnated Resins" (SIR)(1-3) and a polymerisation process by mixing the extractant with a mixture of monomers (styrene and divinylbenzene) to produce Levextrel Resins (4-5). Whereas numerous extractants are available for SIR, the extractants available for Levextrel resins are, at present, limited to phosphoryl compounds: tributylphosphate, di(2-ethylhexyl) phosphoric acid, and di(2,4,4-trimethyl pentyl)phosphinic acid (6).

In the determination of trace metals in surface waters by means of fully-automated analytical methodologies based on UV-VIS spetrophotometry, separation is an important issue. Some non toxic metal ions, such as Fe(III) and Zn(II), can be in surface waters at relatively high concentration levels, which may become a serious interference problem when the aim of the analysis is the determination of toxic metal ions (e.g. Cd(II), Pb(II), Ni(II)) . Under this situation the best approach to solve this problem is a separation process.

The present work is devoted to the evaluation of impregnated resins in on-line schemes for the separation of Zn(II) from other divalent metal ions (Cu(II), Cd(II), Pb(II) and Ni(II)), as a pre-treatment step before the spectrophotometric determination of these metal ions. Impregnated resins based on the immobilisation of di(2,4,4-tri-methylpentyl) phosphinic acid onto high surface macroporous polymeric supports were evaluated. Two different types of impregnated resins were assayed: a) those prepared by direct adsorption of the extractant onto Amberlite XAD2 macroporous polymeric supports (SIR) and b) those prepared by polymerisation of the support in presence of the active component.

## 2 EXPERIMENTAL

### 2.1 Reagents

**Levextrel Resins**. The Levextrel resins Lewatit TP807'84, containing di(2,4,4-tri-methylpentyl)phosphinic acid (DTMPPA) (7), the active component of Cyanex 272, were conditioned before use by treatment with aqueous hydrochloric acid solutions several times, and finally with deionized water. The amount of DTMPPA impregnated was evaluated after washing a proper amount of resin with ethanol, which completely elutes the ligand, and subsequent potentiometric titration with NaOH. The total extractant capacity of the resin was 1.22 mol.kg$^{-1}$ dry resin.

**Solvent Impregnated Resins**. Amberlite XAD-2 resin supplied by Rohm and Haas, size 0.3-0.9 mm, was used as inert support. Before impregnation, XAD-2 was kept in contact for 12 h with a 50% methanol-water solution containing 4 M HCl, in order to eliminate inorganic impurities as well as monomeric material.

XAD2-DTMPPA resins were prepared according to a modified version (7) of the dry impregnation method. The amount of DTMPPA impregnated was evaluated by washing a known amount of resin with ethanol, which completely elutes the extractant, and for subsequent determination by  potentiometric titration with NaOH. The total extractant capacity of the resin was 0.60 mol.kg$^{-1}$ dry resin.

**Solutions**. Stock solutions of Zn(II), Cu(II), Pb(II), Ni(II), Cd(II) and Fe(III) (1 g.dm$^{-3}$) were prepared by dissolving the corresponding salts (nitrate or chloride) ( Merck a.r. grade) in  water.  Sodium  nitrate,  sodium  chloride,  sodium  hydroxide,  sodium dihydrogenphosphate, nitric, hydrochloric and formic acids (Merck a.r. grade), were used for the preparation of the different solutions.

### 2.2 Procedures
### Batch extraction experiments

The extraction of Zn(II), Cu(II) and Cd(II) was carried out with batch experiments at 25°C. Samples of 0.2 g of Lewatit TP807'84Oc resin were mixed mechanically in glass stoppered tubes with an aqueous solution (20 mL) having composition 0.1 M (M$^{2+}$, H$^+$, Na$^+$)NO$_3^-$ or (M$^{2+}$, H$^+$, Na$^+$)Cl$^-$ until equilibrium was achieved. After phase separation with a high-speed centrifuge the equilibrium pH was measured using a Metrhom AG 9100 combined electrode connected to a CRISON digital pH meter, model Digilab 517. Metal content in both phases was determined by Atomic Absorption Spectrophotometry. A Perkin-Elmer 2380 Spectrophotometer with air-acetylene flame was used.

## Column extraction experiments

Known amounts (about 0.13 g) of resin were slurry-packed in Omnifit borosilicate glass columns fitted with porous 25 micron polyethylene frits and teflon end pieces. A peristaltic pump at the column entrance delivered solution at constant flow rate of 0.5-1 ml/min. Metal ions were determined in the fractions collected at the exit of the columns to follow the extraction histories.

After each metal extraction experiment, the column was washed successively with water and 0.4 M hydrochloric acid solution at flow rate of 1 ml/min through the resin bed in the column.

## 3 RESULTS AND DISCUSSION.

### 3.1 Optimisation of Zn(II) separation of Cu(II) and Cd(II)

Adsorption isotherms of Zn(II) as a function of pH were determined in order to obtain the experimental conditions where the interfering metal Zn(II) could be efficiently separated from other metal ions. Special attention was paid to the separation of Zn(II) from Cu(II) and Cd(II) because they are extracted in a narrow pH range. Figures 1 and 2 report the results as relative percentage of recovery for Zn(II), Cu and Cd(II) as a function of pH for both: (a) a SIR type resin prepared by impregnation of DMTPPA onto Amberlite XAD2 and b) the Levextrel type resin Lewatit TP807'84.

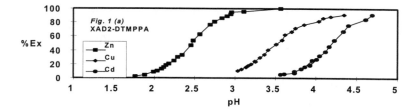

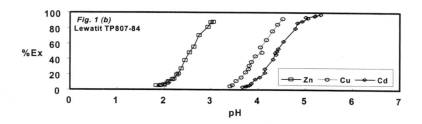

*Figure 1 (a-b). pH extraction isotherms of Zn(II), Cu(II) and Cd(II) with (a) Amberlite XAD2-DTMPPA resin ([HL]=0.6 mol.kg⁻¹) and b) Levextrel resin Lewatit TP807'84 ([HL]=1.22 mol.kg⁻¹).*

Table 1 collects differences in so-called separation factors by using $pH_{50}$, that is the pH at which 50% of the metal is extracted, calculated from the pH extraction isotherms

*Table 1.(a) $pH_{50}$ and (b) $\Delta$ $pH_{50}$ values for Zn(II), Cu(II), Cd(II), Co(II) and Ni(II) with XAD2-DTMPPA and Lewatit TP807'84 resins.*

(a)

| Resin | $pH_{50}$ (Zn) | $PH_{50}$ (Cu) | $pH_{50}$ (Cd) | $pH_{50}$ (Co) | $pH_{50}$ (Ni) | Sample Matrix |
|---|---|---|---|---|---|---|
| XAD2-DTMPPA | 2.3 | 3.6 | 4.3 | ----- | ----- | $NaNO_3$ |
| Lewatit TP807'84 | 2.3 | 3.6 | 4.3 | ----- | ----- | $NaNO_3$ |
| Lewatit TP807'84 | 2.3 | 3.7 | 4.6 | ----- | ----- | NaCl |
| Lewatit TP807'84* | ----- | ----- | ----- | 3.8 | 4.5 | $NH_4NO_3$ |
| Lewatit TP807'84* | ----- | ----- | ----- | 4.0 | 5.0 | $(NH_4)_2SO_4$ |

Experimental data obtained from reference 10.

(b)

| Resin | $\Delta pH_{50}$ (Zn/Cu) | $\Delta pH_{50}$ (Zn/Cd) | $\Delta pH_{50}$ (Zn/Co) | $\Delta pH_{50}$ (Zn/Ni) |
|---|---|---|---|---|
| XAD2-DTMPPA | 1.3 | 2.0 | ----- | ----- |
| Lewatit TP807'84 | 1.4 | 2.1 | ----- | ----- |
| Lewatit TP807'84 | 1.4 | 2.3 | ----- | ----- |
| Lewatit TP807'84* | ----- | ----- | 1.5 | 2.2 |
| Lewatit TP807'84* | ----- | ----- | 1.7 | 2.7 |

Differences in $\Delta pH_{50}$ (Zn/$M_i$) higher than 2 indicates that is possible to obtain the quantitative extraction of Zn(II) with a contamination of metal ($M_i$) lower than 1%, while the $\Delta pH_{50}$ (Zn/$M_i$) between 1.4 and 2 indicates that it is possible to achieve the quantitative separation of Zn(II) with a contamination of metal $M_i$ lower than 7%. From these results it can be observed that the less favourable results are obtained for the Zn(II)/Cu(II) pair, for which $\Delta pH_{50}$ is about 1.4. This means that it is possible to find a pH range where Zn(II) will be removed between 95-100% while other metals ions, such as Cd(II), Cu(II) and Ni(II) will not be retained by the resin.

## 3.2 On Column Zn(II) separation

In order to explore the possibility of selective separation of Zn(II) from divalent metal ions (Cu(II), Cd(II), Ni(II), Pb(II)) in a column, experiments to study the effect of different parameters were carried out.

Taking into account results shown in Figures 1-2, pH values around 3 could be suitable to achieve the described purposes. The pH of sample solutions was adjusted between 3 and 4, by using 0.2 M formic acid/ sodium formiate buffers and results of the Zn(II) column retention are shown in Table 2. The results indicate that Zn(II) retention on the column is efficient enough at pH values higher than 3.2

*Table 2. Effect of pH on Zn(II) retention. Lewatit TP807'84 resin, [Zn(II)] 1 mg/L. Sample flow-rate 1 mL/min.*

| pH | % Retention |
|---|---|
| 3.0 | 74±3 |
| 3.2 | 97±4 |
| 3.7 | 94±2 |
| 4.0 | 96±4 |

Samples containing variable concentrations of Zn(II), from 1 mg/L to 50 mg/L, were adjusted to pH 3.2 and pumped to the column at 1,0 ml.min$^{-1}$. Zn(II) retention efficiencies, collected in table 3, indicate that Zn(II) is adsorbed quantitatively in all the concentration range assayed.

Taking into account that Fe(III) is a common metal ion present in surface natural waters its retention on the column has been evaluated. Samples containing variable concentrations of Fe(III) from 15 mg/L to 30 mg/L were adjusted to pH 3.2 with a formic acid/sodium formiate buffer solution and pumped through the column. Fe(III) retention efficiencies, collected in table 3, indicates that in these conditions Fe(III) is quantitatively adsorbed in the concentration range evaluated.

*Table 3. Zn(II) and Fe(III) retention at pH 3.2. SIR resin XAD2-DTMPPA (0.60 mol/Kg).*

| [Zn(II)](mg/L) | % Retention | [Fe(III)](mg/L) | % Retention |
|---|---|---|---|
| 1.0 | 99±4 | 15.0 | 98±3 |
| 5.0 | 97±2 | 20.0 | 97±4 |
| 10.0 | 94±3 | 25.0 | 96±2 |
| 25.0 | 97±4 | 30.0 | 96±4 |
| 50.0 | 96±4 | | |

Samples containing mixtures of Cu(II), Cd(II), Ni(II) and Pb(II) and different amounts of Zn(II), from 5 to 50 mg/L, were pumped through a column containing Lewatit TP807'84. Evaluation of metal retention indicates that target analytes (Cu(II), Cd(II), Ni(II) and Pb(II)) passed the column with minimal adsorption (lower than 1.5%) and low Zn(II) contamination (between 2 and 10%). Results obtained are shown in table 4.

*Table 4. Zn(II) removal from multi metal samples at pH 3.2. Lewatit TP807'84 resin.*

| Metal Ion | [Zn(II)]$_o$ (mg/L) | %Retention (Zn) | [M(II)]o (mg/L) | [M(II)]f (mg/L) | %Retention (M(II)) |
|---|---|---|---|---|---|
| Cd(II) | 5.0 | 98 | 0.600 | 0.60 | 0.0 |
| | 10.0 | 97 | 0.580 | 0.576 | 0.7 |
| | 25.0 | 90 | 0.611 | 0.608 | 1.5 |
| | 50.0 | 92 | 0.601 | 0.598 | 0.3 |
| Cu(II) | 5.0 | 98 | 0.093 | 0.0928 | 0.3 |
| | 10.0 | 97 | 0.101 | 0.099 | 0.8 |
| | 25.0 | 90 | 0.202 | 0.199 | 0.5 |
| | 50.0 | 92 | 0.100 | 0.099 | 0.1 |
| Ni(II) | 5.0 | 98 | 0.200 | 0.197 | 1.5 |
| | 10.0 | 97 | 0.201 | 0.200 | 0.1 |
| | 25.0 | 90 | 0.221 | 0.219 | 0.5 |
| | 50.0 | 92 | 0.151 | 0.150 | 0.1 |
| Pb(II) | 5.0 | 98 | 0.501 | 0.501 | 0.0 |
| | 10.0 | 97 | 0.420 | 0.419 | 0.1 |
| | 25.0 | 90 | 0.602 | 0.601 | 0.0 |
| | 50.0 | 92 | 0.561 | 0.559 | 0.2 |

Finally, spiked tap water samples were prepared to evaluate the performance of the resins under real conditions. Pre-treatment of the water samples was reduced to a filtering step with a 0.45 μm filter. Results, given in table 5, point out that Zn(II) removal can be achieved, after adjusting the pH of the sample to 3.2, with no significant losses of the target metal ions.

*Table 5. Metal removal from spiked tap water samples by using SIR resin XAD2-DTMPPA (0.60 mol/Kg).*

| [Zn(II)] | %Re | [Cd(II)] | %Re | [Cu(II)] | %Re | [Ni(II)] | %Re | [Pb(II)] | %Re |
|---|---|---|---|---|---|---|---|---|---|
| 5mg/L | 98 | 0.1mg/L | 0.0 | 0.1 mg/L | 0.3 | 0.3 mg/L | 0.1 | 0.5 | 0.1 |
| 10 mg/L | 97 | 0.1 mg/L | 0.1 | 0.2 mg/L | 0.1 | 0.3 mg/L | 0.3 | 0.6 | 0.1 |
| 25 mg/L | 100 | 0.1 mg/L | 0.1 | 0.2 mg/L | 0.2 | 0.2 mg/L | 0.0 | 0.6 | 0.2 |
| 50 mg/L | 93 | 0.1 mg/L | 0.1 | 0.2 mg/L | 0.1 | 0.2 mg/L | 0.1 | 0.6 | 0.1 |

## 4 CONCLUSIONS

The results obtained in the extraction of Zn(II), Cu(II), Cd(II), Ni(II), Pb(II) and Fe(III) with both SIR and Levextrel resin Lewattit TP807'84 show a suitable pH window to allow the separation of Zn(II) and Fe(III)) from the target analytes (Cu(II), Cd(II), Ni(II) and Pb(II)). The fact that both types of impregnated resins, i.e.SIR and Levextrel, provide the same results, indicates that the different nature of the resin preparation does not have a significative influence on the extraction behaviour of the metal extractant. Therefore direct impregnation procedure to prepare SIR can be very useful, because the active part could be easily varied, without recourse to the sometimes difficult and time-consuming procedures needed for covalenty linking the reagent to the skeleton of the resin.

This system will be used on-line in a multicomponent spectrophotometric method for the determination of low levels of toxic metal ions in surface waters.

## 5 ACKNOWLEDGEMENTS

We wish to acknowledge Dr. Rickelton (Cytec) for their support in Cyanex 272 sample supply and to Bayer Hispania for Levextrel resin supply. Finally, this work was supported by CICYT Projects QUI96C02-002 and AMB97-0387 (Ministerio de Educación y Ciencia de España) .

## 6 REFERENCES

1 M. Torre, M.L.Marina, Critical Reviews in *Anal. Chem.*, 1994, **24**, 327.
2 Abollino, E. Mentasti, V. Porta and C. Sarzanini, *Anal. Chem.*, 1990, **62, ** 21.
3 S.K. Menon and Y.K. Agrawal, Reviews in Analytical Chemistry, 1992, **XI**(3-4), 1509
.
4 A. Warshawsky, in Solvent Extraction and Ion Exchange, J.A. Marinsky, Y. Marcus, (Eds.)., Marcel Dekker, New York, 1981, **8**, 229.
5 J.L. Cortina, A. Warshawsky, in Solvent Extraction and Ion Exchange, J.A. Marinsky, Y. Marcus, (Eds.)., Marcel Dekker, New York, 1996, **13**, 209.
6 H.W. Kauczor and A.Meyer, *Hydrometallurgy.*, 1975, **3**, 65.
7 J.L. Cortina, A.M. Sastre, N. Miralles, M. Aguilar, A. Profumo, and M. Pesavento, *React. Polym.*, 1992, **18**, 67.
8 J.L. Cortina, N. Miralles, A. Sastre, M. Aguilar, A. Profumo and M. Pesavento, *Reactive Polymers*, (a) 1993, **23**, 81 and (b) 1993, **23**, 103.
9 K. Inoue, Y. Baba, Y. Sakamoto and H.Egawa, Sep. Sci. and Technol., 1987, **24,**1349.
10 K. Yoshizuka, Y. Sakamoto, Y. Baba, K. Inoue, *Hydrometallurgy.*, 1990, **23**, 309.

# Theoretical Aspects

# DESCRIPTION OF MULTISPECIES ION EXCHANGE EQUILIBRIA USING THE SURFACE COMPLEXATION THEORY - FUNDAMENTALS AND APPLICATIONS

Wolfgang H. Höll

Forschungszentrum Karlsruhe, Institute for Technical Chemistry, Section WGT
D-76021 Karlsruhe, Germany

## 1 INTRODUCTION.

Inorganic and organic ion exchangers consist of either a crystalline or polymeric matrix and functional groups. Depending on the pH value of the liquid phase these groups can either be protonated or dissociated. By this means exchangers are able to interact with ions in the liquid phase. There is a large variety of such interactions: they may be due to electrostatic and van der Waals- forces, heteropolar and covalent binding, or coordination forces. The resulting sorption phenomena take place at the inner and/or surface of the exchangers[1].

The surface charge generates an electric potential normal to the surface. Consideration of the resulting electrostatic interactions has led to several models, e.g. the Helmholtz, Gouy-Chapman and Stern models. Another group of theoretical approaches considers the adsorption of counter-ions as a result of chemical interactions between the surface groups and dissolved species. Sorption of protons or of any other kind of ions is treated as a local equilibrium reaction. Specific (for protons and hydroxyl ions) and non-specific interactions lead to the formation of ion pairs at the surface which are designated as surface complexes. Different kinds of surface complexes can be discriminated[1].

Spectroscopic investigations of surfaces gave rise to the assumption that more than one single layer has to be assumed to account for the uptake of counterions. A typical approach used by many authors is the triple layer model with the surface, the inner and outer layers. Adsorption data of one kind of counterions can well be correlated[2].

These two main difficulties of the triple layer have been overcome by a modification of the existing approaches which has originally been developed by Horst[3]. In his model individual sorption layers are credited to each kind of counter-ion. The respective theoretical approach allows the description of pure cation or anion exchange equilibria as well as of amphoteric equilibria encountered with activated alumina or activated carbon.

## 2 THEORY

### 2.1 Amphoteric reactions at a charged surface

Derivation of the mathematical relationships uses a generalised exchanger with a surface containing OH-groups as functional sites. In aqueous systems these surface groups can be protolysed in two different ways:

In acid media the surface may be protonated according to

$$S - OH + H^+ \quad \Leftrightarrow \quad S - OH_2^+ \tag{1}$$

In order to maintain the condition of electroneutrality the charge of the surface has to be balanced by the negative charge of an anion, e.g.

$$S - OH_2^+ \quad + \quad Cl^- \quad \Leftrightarrow \quad S - OH_2Cl \tag{2}$$

Therefore, in sufficiently acid media sorption of acid takes place similar to the uptake by weakly basic exchange resins.

In alkaline solutions the surface hydroxyl groups can dissociate and release protons to the liquid phase:

$$S - OH \quad \Leftrightarrow \quad S - O^- \quad + \quad H^+ \tag{3}$$

The negative surface charge has to be balanced by cations, e.g. by sodium ions:

$$S - O^- \quad + \quad Na^+ \quad \Leftrightarrow \quad S - ONa \tag{4}$$

Together the formal equations (3) and (4) represent a cation exchange on an arbitrary cation exchanger.

Each of the above models holds for the uptake of acids or the exchange of protons for other monovalent cations in arbitrary pure anion or cation or anion exchange processes. If the state of equilibrium is considered for an amphoteric exchanger and a liquid phase containing different cations and anions the uptake of anions will decrease with increasing pH value until the sorption is completely suppressed. In the same direction the cation exchange of protons for cations increases. The pH regions of almost exclusive cation and anion sorption are separated by a maximum of non-charged surface groups. In this transition region a sorption of both, cations and anions occurs. At the point of zero charge, therefore, the surface is covered with equal amounts of equivalents of positively and negatively charged counter-ions.

### 2.2 Surface model assumptions

During the protonation or dissocation of the surface OH groups either a negative or a positive surface potential is generated which depends on the pH and on the ionic strength of the liquid phase. Therefore, the respective counterions are submitted to an attraction while the co-ions are rejected. Due to the sorption of counter-ions at functional groups the surface potential is partly decreased. Unlike in the description widely used in the literature it is assumed that each kind of counter-ion is located at a characteristic distance from the surface. Thus, ordered double layers or STERN layers are formed. Ion pairs between surface groups and counter-ions in the ordered layer are designated as surface complexes. Excess charges of the surface are balanced by counterions in the diffuse layer which also contains co-ions. As a consequence, the surface potential continuously decreases normal to the surface to zero in the liquid phase. The equilibrium between solid and liquid phases is mainly dominated by protolytic reactions. Therefore, the protons are the potential-determining species.

For a system with $Cl^-$, $NO_3^-$, and $Na^+$ ions the schematical arrangement of counter-ions and co-ions and the corresponding development of the potential are plotted in figure 1. With respect to the radii of the hydrated species the anion layers are assumed closer to the surface than those of the cations. ($Cl^-$, $NO_3^-$, and $Na^+$ ions are used as model counter-ions. Individual properties are not taken into account. Furthermore, the absolute values of the potential in the different layers are of no importance for the further mathematical derivations).

## 2.3 General sorption equilibrium relationships for an amphoteric surface

*2.3.1 General assumptions.* The derivation of equilibrium relationships makes use of the following assumptions:

1. The functional groups are uniformly distributed across a plane surface.
2. Activity coefficients in the exchanger phase are assumed to be 1.
3. Any ion exchange develops as the replacement of one surface complex by a new one. As a consequence, an oxide valence is defined which is equal to the smallest common multiple of the valences of the counter-ions.

*2.3.2 Surface reactions.* For the derivation of the equilibrium relationships a simple system with $Cl^-$ and $Na^+$ ions is considered. The protolytic reactions at the surface are considered as local equilibria which can be described by the mass action law. The formation of the two surface complexes can be described by the respective formation constants.

a) Protonation:

$$S - OH \ + \ H_s^+ \ \Leftrightarrow \ S - OH_2^+ \tag{5}$$

$$K_H^+ \ = \ \frac{c(S - OH_2^+)}{c(H^+)_s \cdot c(S - OH)} \tag{6}$$

b) Dissociation/neutralisation:

$$S - O^- + H_s^+ \ \Leftrightarrow \ S - OH \tag{7}$$

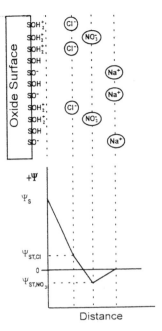

**Figure 1**  *Schematical arrangement of ions at the amphoteric surface of activated alumina and the respective development of electrical potential* (by courtesy of Academic Press)

$$K_H^- = \frac{c(S-OH)}{c(S-O^-) \cdot c(H^+)_s} \tag{8}$$

The sorption of counter-ions leading to two further surface complexes is considered in an analogous way:

c) Sorption of anions:

$$S-OH_2^+ + Cl^- \Leftrightarrow S-OH_2Cl \tag{9}$$

$$K_{Cl}^- = \frac{c(S-OH_2Cl)}{c(S-OH_2^+) \cdot c(Cl^-)_s} \tag{10}$$

d) Sorption of cations:

$$S-O^- + Na^+ \Leftrightarrow S-ONa \tag{11}$$

$$K_{Na}^+ = \frac{c(S-ONa)}{c(S-O^-) \cdot c(Na^+)_s} \tag{12}$$

Symbols with index s designate concentrations in the respective STERN layer. These unknown quantities can be expressed in terms of the concentrations in the liquid phase by means of the POISSON-BOLTZMANN relationship:

$$c(i)_x = c(i) \cdot \exp\left\{-\frac{z(i) F}{R T} \cdot \Psi_x\right\} \tag{13}$$

The uptake of anions develops as the simultaneous sorption of protons and chloride ions. Since we have a sequence of reactions, the equilibrium of this common sorption can be expressed by the product of the respective formation constants:

$$K_{Cl}^H = K_H^+ \cdot K_{Cl}^- \tag{14}$$

The uptake of cations, however, develops as the competitive sorption of protons and sodium ions. As a consequence the equilibrium of the cation exchange is expressed by the ratio of the corresponding formation constants:

$$K_{Na}^H = \frac{K_H^-}{K_{Na}^+} \tag{15}$$

By means of eqns (6) to (15) one obtains:

$$K_{Cl}^H = \frac{c(S-OH_2Cl)}{c(S-OH) \cdot c(Cl^-) \cdot c(H^+)} \cdot \exp\left\{+\frac{F}{R T} \cdot (\Psi_S - \Psi_{Cl})\right\} \tag{16}$$

$$K_{Na}^H = \frac{c(S-OH) \cdot c(Na^+)}{c(S-ONa) \cdot c(H^+)} \cdot \exp\left\{+\frac{F}{R T} \cdot (\Psi_S - \Psi_{Na})\right\} \tag{17}$$

As shown in the literature[5] the concentrations of surface complexes are converted to exchanger loadings q(OH), q(Cl⁻), and q(Na⁺), respectively (q(OH) and y(OH) denote surface groups which are neither dissociated nor protonated).

As a consequence, the first factor on the right hand sides exclusively contains quantities which can be derived from experiments. Both expressions are designated as *generalised separation factors*[5]. After introducing dimensionless loadings y(i) = q(i)/$q_{max}$ the following expressions are obtained:

$$Q_{Cl}^H = \frac{q(Cl)}{q(OH) \cdot c(H^+) \cdot c(Cl^-)} = \frac{y(Cl)}{y(OH) \cdot c(H^+) \cdot c(Cl^-)} \tag{18}$$

$$Q_{Na}^H = \frac{q(OH) \cdot c(Na^+)}{q(Na) \cdot c(H^+)} = \frac{y(OH) \cdot c(Na^+)}{y(Na) \cdot c(H^+)} \tag{19}$$

Similar relationships can be derived for any monovalent counter-ions.

    *2.3.3 Equilibrium relationships.* The difference of electrical potentials in the exponential terms of eqns. (16) and (17) need considerations about the surface charge densities in the series of electric capacitors formed by the surface, STERN and diffuse layers. As has been shown in earlier publications[5] the unknown differences of the electrical potential can be expressed in terms of the loading of the amphoteric surface with Cl⁻ and

$Na^+$ ions.                         After resolving the equations for the generalised separation factor one obtains:

$$\log Q^H_{Cl} = \log K^H_{Cl} + m(H,Cl)\cdot\left[-y(Cl)+y(Na)\right] \tag{20}$$

$$\log Q^H_{Na} = \log K^H_{Na} - m(H,Cl)\cdot y(Cl) + m(H,Na)\cdot y(Na) \tag{21}$$

The quantities m(H,i) are abbreviations of

$$m(H,i) = \frac{1}{\ln 10}\cdot\frac{q_{max}\cdot F^2}{A_0\cdot C(S,i)\cdot R\cdot T} \tag{22}$$

The derivation has been given in previous papers[5]. For a system with an arbitrary number of monovalent counter-ions the following relationship can be deduced from similar considerations:

$$\log Q^H_i = \log K^H_i + \sum_{j=2}^{i-1}\left\{m(H,j)\cdot V(j)\cdot y(j)\right\} + m(H,i)\cdot\sum_i^n\left\{V(i)\,y(i)\right\} \tag{23}$$

$i = 2,3,...,\,n$    index of counterions
$j = 2,3,...,\,i\text{-}1$   running index

Hydrogen ions are always taken as component „1". The factors V(j) and V(i) are the signs of the charge of the ions having the values of -1 for anions and +1 for cations. The first summation comprises counterions from index „2" to „i-1" (closer to the surface than species „i"). The second summation considers the counterions with indices running from „i" to „n". For a system with n counterions there are n-1 equations (23). By subsequent evaluation of data all constants $\log K^H_i$ and m(H,i) can be derived from the multicomponent system.

   *2.3.4 Extension to systems with mixed monovalent/divalent counterions.* During the exchange of one divalent counterion for two monovalent ions the divalent one replaces two adjacent monovalent species at the surface. In the case of the presence of at least one divalent kind of counterion  the sorbent has to be assumed to have divalent functional sites (assumption #3). As a consequence, a surface complex consisting of two sites and two monovalent counterions is defined which is replaced by a suface complex which consists of two sites and a divalent counterion. For a system with $H^+$, $Na^+$, $Cl^-$, and $SO_4^{2-}$ the generalised separation factors are then given by the following expressions:

$$Q^H_{IISO_4} = \frac{c\left\{(S-OH_2)_2 = SO_4\right\}}{c\left\{S-OH,S-OH\right\}c(H^+)^2\,c(SO_4^{2-})} \tag{24}$$

$$Q^H_{IICl} = \frac{c\left\{S-OH_2Cl, S-OH_2Cl\right\}}{c\left\{S-OH,S-OH\right\}c(H^+)^2\,c(Cl)^2} \tag{25}$$

$$Q_{II\,Na}^{H} = \frac{c\{S-OH, S-OH\}}{c\{S-ONa, S-ONa\}\, c(H^+)^2}$$

(26)

(Index „II" refers to the calculation under the assumption of divalent functional sites). The probability of two adjacent monovalent counterions i,i at the surface is given by[5]:

$$y(i,i) = \left(\frac{y(i)}{\sum y(j)}\right)^2 \sum y(j)$$

(27)

with $\sum y(j)$ as the sum of dimensionless loadings of all monovalent counterions with valencies +1 and -1.

By this means the above separation factors can be expressed as:

$$Q_{II\,SO_4}^{H} = \frac{y(SO_4)\{y(H)+y(Cl)+y(Na)\}}{y(OH)^2\, c(H^+)^2\, c(SO_4^{2-})}$$

(28)

$$Q_{II\,Cl}^{H} = \frac{y(Cl)^2}{y(OH)^2\, c(H^+)^2\, c(Cl)^2} \equiv \left[Q_{I\,Cl}^{H}\right]^2$$

(29)

$$Q_{II\,Na}^{H} = \frac{y(OH)^2\, c(Na^+)^2}{y(Na)^2\, c(H^+)^2} \equiv \left[Q_{I\,Na}^{H}\right]^2$$

(30)

*2.3.5 Transformation of equilibrium parameters for different valencies of counterions.* If divalent functional sites are assumed for systems with exclusively monovalent counterions the equilibrium parameters $\log K_{II\,i}^{H}$ and $m_{II}(H,i)$ can be transformed to the parameters for monovalent functional sites according to

$$\log K_{I\,i}^{H} = \frac{\log K_{II\,i}^{H}}{2}$$

(31)

$$m_I(H,i) = \frac{m_{II}(H,i)}{2}$$

(32)

The indices „I" and „II" refer to monovalent or divalent functional sites, respectively. If a systems contains monovalent and divalent counterions the entire calculation has to assume divalent functional sites. Equilibrium parameters derived fom systems with only monovalent counterions have to be converted corresponding to eqns. (31)and (32).

## 2.4 Relationships for pure cation or anion exchange on weak electrolyte resins

*2.4.1 Weakly acidic exchangers.* In the case of weakly acidic exchange resins there is no uptake of anions. In the general equation (24) the terms for anions vanish. Correspondingly, equation (21) simplifies to

$$\log Q_{Na}^{H} \quad = \quad \log K_{Na}^{H} \quad + \quad m(H, Na)\cdot y(Na) \tag{33}$$

*2.4.2 Weakly basic exchangers.* In a similar way, there is no uptake of cations by weakly basic exchangers. Therefore, eq.(20) yields

$$\log Q_{Cl}^{H} \quad = \quad \log K_{Cl}^{H} \quad - \quad m(H, Cl)\cdot y(Cl) \tag{34}$$

The uptake of anions by protonated functional groups is equivalent to the exchange of anions for hydroxyl groups:

$$S-OH\cdot H_2O \quad + \quad Cl^- \quad \Leftrightarrow \quad S-O_2Cl \quad + \quad OH^- \tag{35}$$

A theoretical development similar to that in section 2.3 leads to the equilibrium relationship

$$\log Q_{Cl}^{OH} \quad = \quad \log K_{Cl}^{OH} \quad + \quad m(OH, Cl)\cdot y(Cl) \tag{36}$$

### 2.5 Strongly electrolyte exchangers

Relationships for the description of equilibria with strongly acidic and strongly basic resins can also be derived from the basic equation (23). The difference, however, is that hydrogen ions cannot be assumed to be adsorbed in layer one at the surface. In the case of the cation exchangers this can be explained by the fact the exchanger is completely dissociated even in the $H^+$ form. Since hydrated protons have a rather large diameter, they are unable to come close to the surface. Similar considerations hold for strongly basic anion exchangers and hydroxyl ions. For both cases, therefore, the sequence of layers starts with the kind of counterion which is located the closest to the surface.

### 2.6 Liquid phase equilibrium relationships

The liquid phase equilibrium is calculated by means of
- the dissociation of water,
- the condition of electroneutrality, and
- mass balances for each component.

Activity coefficients in the liquid phase can be calculated from the ionic strength using the extended DEBYE-HÜCKEL relationship:

$$\log f_i(T, p, I) \quad = \quad -\left[\frac{A\cdot z(i)\cdot I^{0.5}}{I + a_i\cdot B\cdot I^{0.5}}\right] \quad + \quad \beta_i\cdot I \tag{37}$$

with
$$A = 1.823 \cdot 10^6 \cdot \frac{1}{(T \cdot DK)^{1.5}}$$

and
$$B = 50.3 \cdot \frac{1}{(T \cdot DK)^{1.5}}$$

In systems with trivalent species like e.g. aluminium, hydrolysis of water in the presence of $Al^{3+}$ species has to be taken into account. Constants for the formation of positively and negatively charged as well as of neutral species have to be taken from literature[6,7].

## 4 EXPERIMENTAL EVALUATION OF EQUILIBRIUM DATA

### 4.1 General

Exchange equilibria have been determined for pure cation and anion exchangers as well as for amphoteric materials like activated alumina and activated carbon. The pretreatment of these materials and the experimental conditions needed for obtaining proper equilibrium data have been described in detail in several previous publications[4,5,8-10].

Cation concentrations were measured by titration, photometric methods or by atomic absorption spectroscopy. Nitrate concentrations were measured photometrically, chloride and sulphate concentrations by ion chromatography.

In the case of pure anion or pure cation exchange equilibria samples with increasing amounts of exchanger material were contacted with a fixed volume of the solution at constant temperature until equilibrium had been adjusted. Experiments with amphoteric equilibria were carried out at constant concentrations in the liquid phase and constant quantities of alumina or activated carbon in each of the samples. The volume/mass ratio was adjusted in a way that there was a substantial degree of sorption as well as an equilibrium concentration which could easily be measured. Since the state of equilibrium is mainly dominated by the pH of the liquid phase the pH value was varied in each series of samples. Details are described in previous publications[4].

### 4.2 Evaluation of data

For evaluation of equilibrium parameters the generalised separation factors have to be determined from equilibrium concentrations and resin loadings and have to be plotted vs. the respective dimensionless resin loading of (algebraic) sum of resin loadings. In the case of binary systems and pure cation or anion exchange, the respective graphical representations directly yield the equilibrium parameters: The intersection of the linear relationship leads to log K whereas the slope of the straight line yields m(i,j). Figs. 2 - 5 show examples for different types of exchangers.

The results demonstrate that linear relationships are obtained in most cases. Systematic deviations are found for small loadings. This has to be credited to the neglection of counterions in the diffuse layer. As has been demonstrated by HORST[3] for simple systems this can be considered in addition yielding an excellent agreement. For strongly acidic exchangers this region is comparatively larger than for other resin types. Evaluation of data for the exchange of sulphate for chloride on strongly basic exchangers clearly exhibits two different straight lines. Obviously, there are two different layers which are subsequently filled.

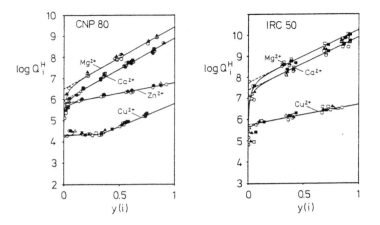

**Figure 2** *Development of the generalised separation factor for two weakly acidic ion exchange resins[8] (by courtesy of Marcel Dekker Inc.).*

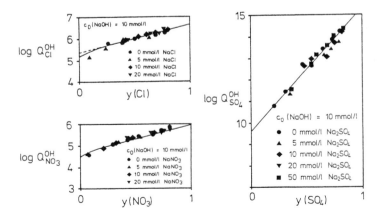

**Figure 3** *Development of the generalised separation factor for a weakly basic ion exchange resin[10] (by courtesy of Marcel Dekker Inc.).*

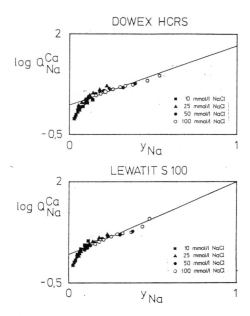

**Figure 4** *Development of the generalised separation factor for the exchange of calcium for sodium ions on two different strongly acidic resins*

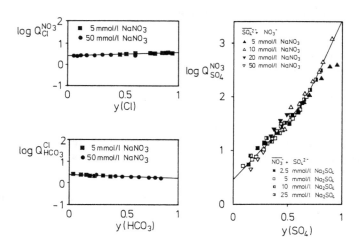

**Figure 5** *Development of the generalised separation factor for a strongly basic ion exchange resin[5] (by courtesy of Marcel Dekker Inc.).*

Evaluation of amphoteric data requires a more sophisticated method: The first step is the assumption of a certain sequence of layers. Evaluation of data starts with the evaluation of the equilibrium for the uptake of protons and ions in the first layer. The respective parameters are required for evaluation of the sorption of species in the next layer. Determination of the parameters for the third layer requires both sets of parameters of the inner layers. This kind of evaluation has to be continued until all sets of binary parameters are derived. The sequence of layers was assumed correctly if both the log K and m(H,i) values show a steady increase with increasing distance. If this condition is not fulfilled the calculation has to be repeated with a modified sequence. Considering a system of layers $H^+$, $Cl^-$, $NO_3^-$, $Na^+$ on activated alumina as an example the equilibrium parameters can be evaluated from the following relationships:

$$logQ_{Cl}^H = logK_{Cl}^H + m(H,Cl) \cdot \{y(Na) - y(Cl) - y(NO_3)\} \qquad (38)$$

$$logQ_{NO_3}^H - m(H,Cl) \cdot \{-y(Cl)\} = logK_{NO_3}^H + m(H,NO_3) \cdot \{y(Na) - y(NO_3)\} \quad (39)$$

$$logQ_{Na}^H - m(H,Cl) \cdot \{-y(Cl)\} - m(H,NO_3) \cdot \{-y(NO_3)\} = logK_K^H + m(H,K) \cdot y(Na)$$
$$(40)$$

Results of the evaluation of equilibrium data for the sorption of $Cl^-$, $NO_3^-$, and $SO_4^{2-}$ on COMPALOX AN7V800 plotted in the figures 6 and 7. Although the results for the uptake of sodium ions are not satisfactory the data demonstrate that linear relationships are obtained.

## 5 PREDICTION OF MULTICOMPONENT EQUILIBRIA

### 5.1 General

Prediction of arbitrary exchange equilibria requires a system of conditions to be fulfilled:

- conservation of mass
- electroneutrality on the liquid phase
- electroneutrality on the resin phase
- adjustment of equilibrium parameters for the exchange of ions
- dissociation equilibria in the liquid phase

As can easily be shown the number of equations always outnumbers those of unknown quantities. Therefore, some conditions are fulfilled automatically. The remaining set of coupled non-linear equations has to be solved numerically by appropriate numerical methods.

Respective results for the uptake of inorganic ions by an activated carbon are plotted in Figs. 8 and 9. With few exceptions the equilibrium data satisfactorily fall on linear relationships as predicted by the theoretical approach.

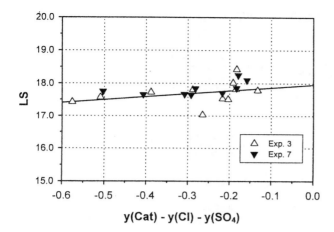

**Figure 6** *Development of the generalised separation factor for the uptake of nitrate by activated alumina from a multicomponent system[4].Cat = Na[+] or K[+] (by courtesy of Academic Press).*

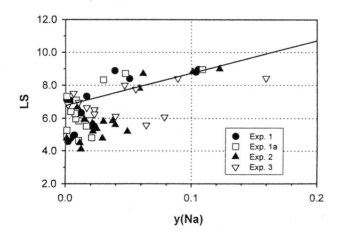

**Figure 7** *Development of the generalised separation factor for the sorption of sodium ions by activated alumina[4] . LS = log Q(H,Na) + m(H,Cl) y(Cl) + m(H,NO₃) y(NO₃) (by courtesy of Academic Press).*

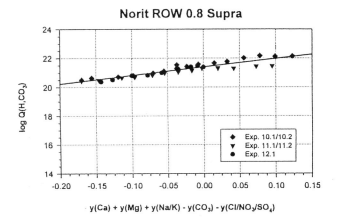

**Figure 8** *Development of the generalised separation factor for the sorption of carbonate species by activated carbon from a multicomponent system.*

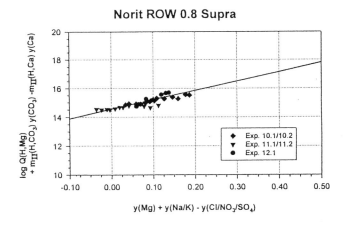

**Figure 9** *Development of the generalised separation factor for the sorption of magnesium ions by activated carbon from a multicomponent system*

### 5.2 Pure cation or anion exchange

For prediction of multicomponent equilibria it is convenient if the binary parameters for the exchange of different counterions for the same reference ion have been deduced. For pure cation or anion exchange systems consideration of a three-component equilibrium with ions A(reference ion), B, and C leads to the relationships:

$$\log Q_B^A \;=\; \log K_B^A \;+\; m(A,B)\cdot\big[y(B)+y(C)\big] \tag{41}$$

$$\log Q_C^B \;=\; \log K_C^B \;+\; m(B,C)\cdot y(C) \tag{42}$$

As has been demonstrated earlier the binary parameters for the exchange of „A" vs. „B" are the same as in the pure binary case. Also, the binary parameters for the exchange of „A" for „C" ions have been evaluated. In the above system, therefore, the parameters for the exchange of „B" for „C" are unknown. However, they can easily be calculated: the unknown parameter $\log K_C^B$ results from the definition yielding

$$\log K_C^B \;=\; \log K_C^A - \log K_B^A \tag{43}$$

For derivation of the unknown slope we make use of the fact that the sequence of layers is considered as a series of layers of an electric capacitor. Since the definition of $m(i,j)$ contains the electric capacitance we can make use of the well-known relationship

$$\frac{1}{C(A,C)} \;=\; \frac{1}{C(A,B)} + \frac{1}{C(B,C)} \tag{44}$$

Thus, one obtains:

$$m(B,C) \;=\; m(A,C) - m(A,B) \tag{45}$$

Similar relationships are obtained for systems with more than three counterions.

Examples of a comparison between experimental results and equilibria predicted from binary data are given in figs. 10 - 13.

For amphoteric sorbents the individual equilibrium parameters were already determined during evaluation of data. Therefore, no additional calculations are required. Figure 14 shows the comparison between experimentally determined oxide loadings and developments calculated by means of the set of equilibrium parameters determined before. The data are plotted as relative loadings $(q(i)/q_{max})$ as a function of pH. The system contained the five components $H^+$, $Cl^-$, $NO_3^-$, $SO_4^{2-}$ and $Na^+$. Obviously, sulphate ions are preferred over nitrate and chloride species. The constant sulphate loadings for pH values below 5 are due to the constant initial sulphate concentration in the system. Only minor loadings with monovalent cations are observed. There is an excellent agreement with the predicted loadings.

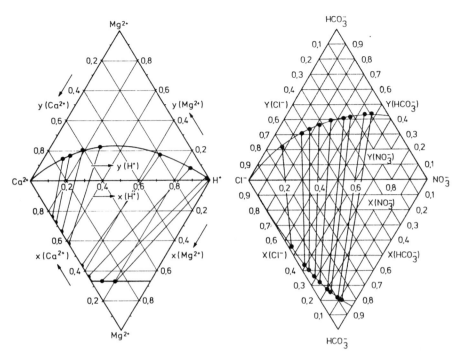

**Figure 10(left)** *Comparison of experimental data and predicted equilibria for the ternary system $H^+/Mg^{2+}/Ca^{2+}$ on the weakly acidic resin LEWATIT OC 1046.*

**Figure 11(right)** *Comparison of experimental data and predicted equilibria for the ternary system $Cl^-$, $HCO_3^-$, $NO_3^-$ on the strongly basic resin AMBERLITE IRA 410.*

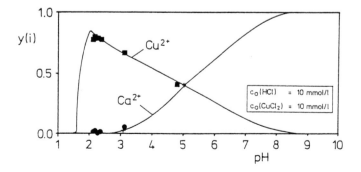

**Figure 12** *Comparison of experimental data and predicted equilibria for the ternary system $H^+$, $Ca^{2+}$, $Cu^{2+}$ on the chelating resin LEWATIT TP 207.*

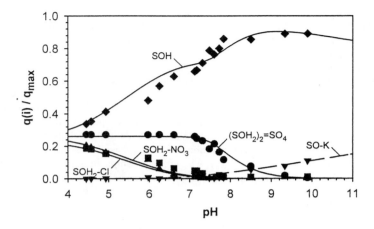

**Figure 13** *Comparison of experimental data and predicted loadings of activated alumina for a multicomponent system. Activated alumina: Compalox AN7V800.*

Figure 14 shows the results from one experiment with activated carbon and protons, carbonate, nitrate, calcium and potassium as counterions. As can be deduced the relative loadings with both, carbonate and nitrate ions decrease with increasing pH of the solution. The loadings of calcium and potassium slightly increase. Nevertheless their uptake remains small indicating that there is only y small capacity for the uptake of cations. Again, there is good agreement with the predicted developments.

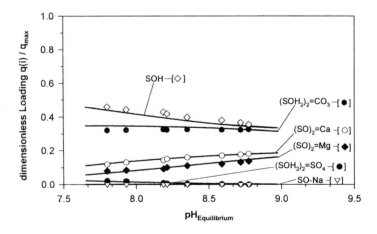

**Figure 14** *Comparison of experimental data and predicted loadings of activated carbon with inorganic species. Activated carbon: Norit ROW 0.8 Supra.*

*Advances in Ion Exchange for Industry and Research*

## 6 FURTHER APPLICATIONS

The excellent prediction of multicomponent exchange equilibria can well be applied for the prediction either multispecies exchange kinetics or the filter breakthrough performance, particularly for reaction-coupled exchange processes. Application for multispecies kinetics has been demonstrated by Franzreb[11,12] who included the description into the solution of the Nernst-Planck equations for several ternary and quaternary cases. Prediction of the filter performance has so far been based on an equilibrium approach in which the filter is divided into a series of equilibrium stages. By means of the porosity of the filter section the quantities of resin material and liquid phase are known. Consequently, the exchange equilibrium in each stage can be calculated. The column calculation develops as follows: Raw water enters the first stage, after equilibration the liquid phase leaves for the second stage whereas further raw water enters stage 1. The effluent composition of the last stage is given by the equilibrium in the last stage. By this means and by an appropriate assumption of the number of equilibrium stages a fairly good agreement between experimental and predicted concentration histories can be obtained. Figure 15 shows an example for the respective comparison for a weakly acidic resin filter in the hydrogen form which is applied for softening/dealkalisation of tap water[11]. As becomes obvious a very satisfactory agreement is obtained.

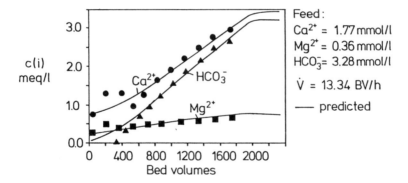

**Figure 15** *Comparison of experimental and predicted breakthrough curves during softening/dealkalisation of tap water. Resin: AMBERLITE IRC 50. Volume of resin: 1L. Throughput: 13.34 BV/h.*

## 7 LIST OF SYMBOLS

| | | |
|---|---|---|
| $a_i$ | - | ion size parameter for species „i" |
| A | - | constant in DEBYE-HÜCKEL relationship |
| Ao | $m^2/g$ | specific surface area |
| B | - | constant in DEBYE-HÜCKEL relationship |
| $c(i)$ | mol/L | concentration of species „i" |
| $c(i)_s$ | mol/L | concentration of species (i) in STERN layer |
| $c(i)_x$ | mol/L | concentration of species „i" at position „x" in an electrical field |
| $C(i,j)$ | $F/m^2$ | electric capacitance of capacitor formed by the layers of species „i" and „j" |
| $C(S,i)$ | $F/m^2$ | electric capacitance of capacitor formed by the surface and the layer of species „i" |
| DK | C/vm | permittivity |
| $f_i$ | - | activity coefficient of species „i" in the liquid phase |
| F | A s | FARADAY constant |
| I | mol/L | ionic strength |
| $K_H^+, K_H^-, K_{Cl}^-, K_{Na}^+$ | L/mol | formation constants, defined in 2.3.2 |
| $K_j^i$ | | equilibrium constant of surface reaction, |
| $m(H,j)$ | - | abbreviation, defined by eq. (22) |
| $q(i)$ | eq/g | oxide loading with species „i" |
| $q_{max}$ | eq/g | maximum exchanger loading |
| $Q_j^i$ | - | generalised separation factor |
| R | J/mol K | gas constant |
| S- | | abbreviation used to designate the surface |
| T | K | temperature |
| V | - | sign of charge |
| $y(i)$ | - | dimensionless loading with species „i" |
| $y(i,i)$ | - | probability of the presence of two adjacent ions „i" at the surface |
| $z(i)$ | - | valency of species „i" |
| $\beta_i$ | - | factor |
| $\sigma_i$ | $As/m^2$ | surface charge density of layer with ions „i" |
| $\Psi_x$ | V | electrical potential at position „x" |

## 8 REFERENCES

1. W. Stumm, 'Chemistry of the solid-water interface', Wiley and Sons, New York, 1992.
2. J. A. Davis, R. O. James and J. O. Leckie, *J. Colloid Interface Sci, 1978.* **63**, 480.
3. J. Horst, PhD Thesis, University and Research Centre of Karlsruhe, 1988.
4. J. Horst, W. H. Höll, *J. Colloid Interface Sci, 1997,* **195**, 250.
5. W. H. Höll, J. Horst and M. Franzreb, S. H. Eberle, Solv. Ext. Ion Ex. **11** (1993), 151.

6.      R. M. Smith and A. E. Martell, 'Critical stability constants', Plenum Press, New York 1976.
7.      S. H. Eberle, Karlsruhe Nuclear Research Center Report KfK 3930UF, 1986.
8.      J. Horst, W. H. Höll and S. H. Eberle, *Reactive Polymers, 1990,*13, 209.
9.      W. H. Höll, J. Horst and M. Wernet, *Reactive Polymers,1991,* 14, 251.
10.     W. H. Höll, J. Horst and M. Franzreb, *Reactive Polymers,1993,* 19, 123.
11.     M. Franzreb, W. H. Höll, H. Sontheimer, *Reactive Polymers,1993,* 21, 117.
12.     M. Franzreb, W. H. Höll, S. H. Eberle, *Ind. Eng. Chem. Res.* 1995, 34, 2670.
13.     W. H. Höll, Proc. Int. Conf. ICIE'91, M. Abe, T. Kataoka, and T. Suzuki, Eds., KODANSHA LTD., Tokyo, 1991, 277.

# DETERMINATION OF ION EXCHANGE CONSTANTS FOR CHROMIUM-RESIN SYSTEMS

Kamel Kebdani Moufdi [2], Federico Mijangos [1], Jorge Díaz[1] and María Puy Elizalde[2]
Departments of Chemical Engineering [1] and Analytical Chemistry [2]
Universidad del País Vasco, Apdo. 644, 48080 Bilbao, Spain

## SUMMARY

An iterative method for the study of the chromium (III)-resin (carboxylic Lewatit CNP 80) system using potentiometric titrations has been investigated in order to determine both the capacity of the exchanger and the dissociation constant of the resin. Deprotonation of the resin and complexation of chromium (III) were studied using two-phase potentiometry. From these experiments apparent formation constants could be calculated and distribution curves obtained. A two-site model was applied for the experimental results analysis, based on the observation of the titration curves.

## 1 INTRODUCTION

Chromium exists in oxidation states ranging from minus one to plus six [1]. In natural waters the two main oxidation states are (III) and (VI), characterized by a markedly different chemical behaviour. The divalent state is unstable with respect to evolution of oxygen, the trivalent state has broad stability, and hexavalent chromium occurs under strongly oxidizing conditions [2]. The oxidation state of an element can have an important effect on bioavailability and toxicity. Cr(III) low levels are essential for the maintenance of glucose, lipid and protein metabolism in mammals, whereas Cr(VI) is considered a toxic material for animals [3]. The presence of Cr(III) and Cr(VI) in the environment is a result of effluent discharge from tanning industries [4], metal processing activities such as electroplating and other metal-related industries. Tanning waste liquors contain several ionic compounds as well as small amounts of organic species which include fats, proteins and enzymes.

Chromic salts are used to tan a treated animal hide to produce leather. Significant quantities of chromium are applied in tanning ($20,000$ mg $dm^{-3}$) and approximately 60% of this amount is taken up by the leather in the tanning process. The remaining chromium ($8,000$ mg $dm^{-3}$) is discharged in the waste liquors. For this reason, the recovery of the chromium from these waste liquors is necessary for environmental and economic reasons.

Ion exchange is one of the most effective techniques for the pretreatment of liquid tannery wastes [4,5]. In addition, several recent publications concerning the separation of Fe(III), Al(III) and Cr(III) in various aqueous solutions generally showed that carboxylic resins offer a higher apparent capacity and slow sorption kinetics.

The ion exchange behaviour of ion exchange resins is chiefly determined by the fixed ionic groups, and the nature of the groups affects the selectivity of the resins. Carboxyl groups are only fully ionized at pH 9. At pH 2 the COOH groups are almost completely undissociated. In contrast, strong acid groups, such as sulfonic acid remain ionized at pH values lower than 2.

A series for ion equilibrium data for strong acid and weak acid ion exchange resins is usually determined by equilibrating the ion exchanger with solutions containing various competing counter-ion species and analysing the ionic composition of the ion exchanger after separation from solution. Equilibration can be carried out by using the batch technique. In contrast for weak acid ion exchange resins, ion exchange equilibrium data must be determined by using the progressive neutralization technique. Because of the progressive neutralization of fixed ionic groups with decreasing or increasing pH, the ion exchange reaction for weakly dissociating ion exchange resins is progressively driven to completion [6].

The ion exchange reaction for a weak acid cation exchange resin in the H form with solutions of metal cation $M^{n+}$ may be simplified by the following equilibrium:

$$nRH + M^{n+} \Leftrightarrow R_nM + nH^+ \tag{1}$$

where R is a fixed ionic group in matrix of the resin, and $M^{n+}$ is a metal cation.

In the present study, ion exchange equilibria in the chromium (III)-carboxylic system have been determined by potentiometric titrations. The ion exchanger used in the experiments was Lewatit CNP 80. The properties of this resin have been investigated by examination of equilibrium between metal complexes (initially in the aqueous phase) and the resin. For this purpose, the resin was first partially converted to the metal form by equilibration of the acid form of the resin with a metal chlorate solution. We propose the potentiometric method for evaluating the equilibrium constant (in chromium- resin system) for weakly dissociating ion exchange resin.

## 2 EXPERIMENTAL METHODS

The standard titration is carried out with a weighed amount (1.0 g) of the $H^+$ form of wet cation exchangers. The macroreticular acrylic acid cation resin Lewatit (CNP 80) supplied by Bayer was employed in the experimental program.

### 2.1 *Chemicals and solutions.*

Chromium chloride, sodium hydroxide and hydrochloric acid from Fluka, p.a. were used without further purification. All solutions were prepared freshly to avoid the aging process of chromium [7]. The NaOH solutions were standardized with a standard hydrochloric acid solution (Panreac) using phenolphthalein as indicator.

### 2.2 *Pretreatment of the resin.*

To remove any water-soluble residues, or undesired cations remaining on the resin after the manufacturing process, the resin samples were subjected to a pre-conditioning process. For this purpose the resin was converted by three successive cycles changing from the Na to H-form and finally equilibrated with 1M hydrochloric acid and washed with deionized water until free from imbibed electrolyte.

### 2.3 *Determination of the resin moisture content.*

All calculations were based on the resin dry weight. Precisely weighed portions of

0.5 g of ion exchanger in the acid form were dried at 60°C. During the drying, the resin was cooled in a desiccator and weighed periodically to assure the attainment of a constant weight

## 2.4 *Methods of analysis.*

The liquid phase of each run at equilibrium was analysed for chromium and H+. Total chromium metal concentration was determined by atomic absorption spectroscopy (Perkin-Elmer, model 1100B) and free hydrogen ion concentration was determined using a combination glass electrode with encapsulated Ag/AgCl reference electrode in 3M KCl (Ref Electrode 6.0233.100 Metrohm Company)

## 2.5 *Potentiometric measurements.*

Prior to the potentiometric measurements, the glass combined electrode was calibrated with a standard buffer solution over the experimental pH range. The electrode calibration was also checked at the end of each experiment in order to assure electrode stability over the course of the measurements.

Potentiometric titrations were carried out using an automated system and a thermostatic bath at $25.0 \pm 0.1°C$, using $N_2$ free from oxygen and presaturated with solvent by being bubbled through a sample of the ionic medium [8]. The titrant was added using a Metrohm 702 automatic burette connected to a computer via a RS-232c interface. Each titration took between 12 and 72 hours to complete, depending on the number of chemical phases. The purge of $CO_2$ and $O_2$ was effected by covering the solution with $N_2$. The additions were managed with a potentiometric titrator. The e.m.f of a combination glass electrode was meausred with 0.1 mV resolution. For each part of the experiments, a data file was created on a PC.

Titrations of the resin in the presence and absence of chromium were carried out by a Metrohm 702 Titroprocessor. The compositions of the experimental solutions are given in Table 1(see later). The ionic strength was maintained constant by using NaCl. The resin was ground to a fine powder to minimise possible problems from slow kinetics [9]. Titration starts from low to high pH by the addition of 0.2 mL aliquots of 0.2 M sodium hydroxide, and the stabilized potential measurement after 25 minutes needed to attain equilibrium was recorded.

The titroprocessor was programmed not to add titrant solution until the electrode drift was less than 0.5 mV.min$^{-1}$, or when twenty five minutes had elapsed from the previous addition.

The free hydrogen ion concentration (h) was determined by the Nernst equation:

$$E(mV) = E^o + 59.16 \log h + E_j(h) \tag{2}$$

where $E^o$ is the standard potential of the glass electrode determined for each titration, and $E_j(h)$ is the liquid junction potential expressed as:

$$E_j(h) = j_{ac} \, h + j_{alk} \, K_w \, h^{-1} \tag{3}$$

$j_{ac}$ is the acid junction liquid potential coefficient, $j_{alk}$ is the alkaline liquid junction potential coefficient and $K_w$ is the autoprotolysis constant of water.

The value of h cannot be analytically solved, so an iterative procedure, such as the Newton method, is used. The equivalence point of each titration was calculated using Gran's method [10].

A computer program, MINEQL$^+$, from Environmental Research Software, was used for metal ion speciation computations in the liquid phase.

## 3 RESULTS AND DISCUSSION

### Chromium speciation in aqueous phase

Acid base reaction of Cr(III) in aqueous medium has been studied by several authors [1, 11, 12]. In order to evaluate the contribution of different species of Cr(III) in aqueous phase, it has been necessary to make use of MINEQL$^+$ program. The logarithms of concentration of the various protonated forms are plotted against the pH in Figure 1.

Chromium (III) speciation in aqueous solution is dominated by chromium(III) hydrolysis products. Chromium(III) hydroxo complexes are expected to be the dominant species of Cr(III) in natural waters. In the pH range of 5 to 12.5, $Cr(OH)_3$ is the dominant species. $Cr(OH)_4^-$ becomes dominant above pH 12.5 and $Cr(OH)^{2+}$ is dominant in the pH range 3.8-6.3 according to Rai et al [11]. Figure 1 shows the speciation diagram of chromium(III) 0.01M indicating that Cr(III) has an amphoteric behaviour, in acidic solution it is found as a cation but in basic solution the formation of $Cr(OH)_4^-$ and probably the anionic species $CrO_2^-$, is thought to occur.

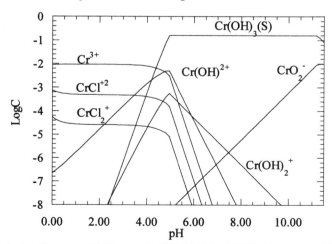

**Figure 1.** LogC versus pH diagram Cr(III) 0.01M. $Cr(OH)_3(s)$ is the solid species

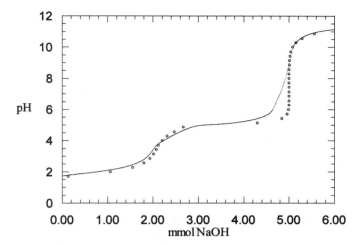

**Figure 2.** Experimental (dash line) and simulated (MINEQL) (∘) titration curves of Cr(III) 0.01M in 1.0M NaCl.

From Figure 2, can be appreciated the effect of polynuclear species, especially at pH around 5, where the solid species would be expected and responsible for the very slow kinetics of the process.

## *Effect of ionic strength in system H/Na*

The dissociation of the carboxylic resin in NaCl media was studied. The experiments were performed at 25°C and at three ionic strengths: 0.1, 0.5 and 1.0 M NaCl.

Neutralization of weak-acid resin in the free acid form with sodium hydroxide yields titration curves in which the resin loading can be plotted as a function of the pH of the solution. This provides one possible representation of the exchange equilibrium. Figure 3 shows the titration curves for Lewatit CNP 80.

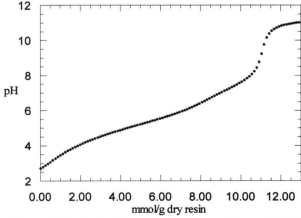

**Figure 3.** Typical acid base titration curve of Lewatit CNP 80 resin.

Figure 4 shows that the sensitive response, during neutralization, to changes in salt concentration levels is immediately apparent from the separation of the curves associated with such change.

On the other hand, a plot of $\log\dfrac{1-\alpha}{\alpha}$ vs pH should yield a straight line of slope 1.0 if the mass-action law holds. The degree of dissociation ($\alpha'$) is calculated taking into account the mass-balance equation for the exchange site:

$$\frac{[\text{RCOOH}]}{Q} + \frac{\left[\text{RCOO}^-\right]}{Q} = \alpha + \alpha' = 1 \tag{4}$$

where Q is the total exchange capacity of the resin, stated to be 11.1 mmol/g dry resin for this sample (see Figure 3) and ($\alpha$) is the fraction of non dissociated groups. This value is in agreement with those obtained by Tamura et al [13], and is summarized in Table 1.

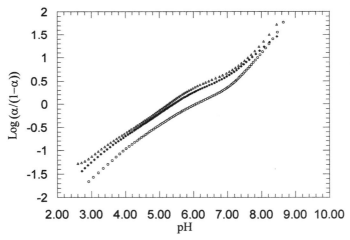

**Figure 4.** Fitting of the experimental results to the one site equilibrium Equation. (o) 0.1M NaCl, (•) 0.5M NaCl (▵) 1.0M.

However, the function $\log\dfrac{1-\alpha}{\alpha} = f(\text{pH})$ plotted in Figure 4 has a slope less than 1, which seems to indicate that the acidity constant K is not unique. In this sense, it can be assumed that the resin has two types of carboxyl groups (I and II). The reactions at type I and II sites can be expressed as:

$$\text{RCOO}_\text{I}\text{H} \Leftrightarrow \text{RCOO}_\text{I}^- + \text{H}^+ \tag{5}$$

$$\text{RCOO}_\text{II}\text{H} \Leftrightarrow \text{RCOO}_\text{II}^- + \text{H}^+ \tag{6}$$

The equilibrium constants are respectively

$$K_{a_1} = \frac{\left[RCOO_I^-\right]\left[H^+\right]}{\left[RCOO_IH\right]} \qquad (7)$$

$$K_{a_2} = \frac{\left[RCOO_{II}^-\right]\left[H^+\right]}{\left[RCOO_{II}H\right]} \qquad (8)$$

and the mass-balance equation for sites I and II are:

$$\left[RCOO_IH\right] + \left[RCOO_I^-\right] = \left[RCOO_IH\right]_o \qquad (9)$$

$$\left[RCOO_{II}H\right] + \left[RCOO_{II}^-\right] = \left[RCOO_{II}H\right]_o \qquad (10)$$

where $\left[RCOO_IH\right]_o$ and $\left[RCOO_{II}H\right]_o$ are the initial concentrations of the respective sites. The sum of these initial concentrations is equal to the total exchange capacity of the resin Q, as expressed in Equation (4), with the fraction of site I to the total sites, r:

$$\left[RCOO_IH\right]_o = rQ \qquad (11)$$

$$\left[RCOO_{II}H\right]_o = (1-r)Q \qquad (12)$$

$$\alpha_{Tot} = \frac{\left[RCOO_I^-\right] + \left[RCOO_{II}^-\right]}{\left[RCOO_IH\right] + \left[RCOO_I^-\right] + \left[RCOO_{II}H\right] + \left[RCOO_{II}^-\right]} \qquad (13)$$

$\alpha_I$ and $\alpha_{II}$ being respectively, the dissociation coefficients for sites I and II. The total dissociation coefficient can be expressed as:

$$\alpha_{Tot} = \frac{\alpha_1 R + \alpha_2}{R+1} \qquad (14)$$

where

$$R = \frac{r}{1-r}, \quad \alpha_1 = \frac{1}{1 + \left[H^+\right]/K_{a_1}} \quad \text{and} \quad \alpha_2 = \frac{1}{1 + \left[H^+\right]/K_{a_2}}$$

By fitting Equation (14) to the experimental data, the values $K_I$, $K_{II}$ and R were determined by minimizing the sum of square errors U, defined as:

$$U = \sum_{N_p} (\alpha_{calc} - \alpha_{exp})^2 \qquad (15)$$

where $N_p$ is the number of experimental points.

**Table 1.** Values of calculated parameters.

| I (M) | g dry resin | Q (mmol/g dry resin) | Log $K_{a1}$ | Log $K_{a2}$ |
|-------|-------------|----------------------|--------------|--------------|
| 0.1 | 0.5327 | 10.740 | -3.86 | -6.12 |
| 0.5 | 0.4803 | 11.168 | -4.27 | -6.56 |
| 1.0 | 0.4533 | 11.003 | -4.50 | -6.74 |

When a charged polymer is dispersed in aqueous media the concentration of counter ions associated with it will always be a sensitive function of the ionic strength of the solution.

The different reactivities of the resin carboxyl sites may correspond to different chemical states of the carboxyl groups. The causes of the polyfunctionality are explained in terms of the position isomerism of the carboxylic groups. An investigation of chemical structure will be necessary and may be the microenvironment of the sites could be a good way to explain this phenomenon.

## Chromium-resin system

The ion exchange reaction of Cr(III) from chloride media by the resin can be described by:

$$Cr^{3+} + iCl^- + (3-i)\overline{HR_{I,II}} \Leftrightarrow \overline{CrCl_iR_{I,II(3-i)}} + (3-i)H^+ \qquad (16)$$

where the bar refers to species into the resin phase and $X^-$ represents co-ions $\left(OH^-, Cl^-\right)$ in the liquid phase and i is the stoichiometric coefficient of the chloride ions. The equilibrium constants for metal retention at different sites can be defined:

$$K_I = \frac{\left[\overline{CrCl_iR_{3-i}}\right]\left[H^{3-i}\right]}{\left[\overline{HR_I}\right]^{3-i}\left[Cl^-\right]^i\left[Cr^{3+}\right]} \qquad (17)$$

$$K_{II} = \frac{\left[\overline{CrCl_iR_{3-i}}\right]\left[H^{3-i}\right]}{\left[\overline{HR_{II}}\right]^{3-i}\left[Cl^-\right]^i\left[Cr^{3+}\right]} \qquad (18)$$

were $\left[\overline{HR_I}\right] = rQ$ and $\left[\overline{HR_{II}}\right] = (1-r)Q$. The mass of the resin takes into account that it can be found fully dissociated, protonated, or complexing. The metallic cation concentration in the resin can be expressed as a function of $K_a$:

$$\left[\overline{HR_I}\right] = \frac{rQ - (3-i)\left[\overline{CrClR_{I3-i}}\right]}{1 + K_{a_1}/[H]} \tag{19}$$

$$\left[\overline{HR_{II}}\right] = \frac{(1-r)Q - (3-i)\left[\overline{CrClR_{II3-i}}\right]}{1 + K_{a_2}/[H]} \tag{20}$$

and the distribution coefficient (D) can be expressed as:

$$D = \frac{\left[\overline{CrCl_iR_{I3-i}}\right] + \left[\overline{CrCl_iR_{II3-i}}\right]}{[Cr]_{aq}} \tag{21}$$

The mass balance of Cr(III) in solution can be expressed as:

$$[Cr]_{aq} = \left[Cr^{3+}\right] + \left[Cr(OH)^{2+}\right] + \left[Cr(OH)_2^+\right] + \left[Cr(Cl)^{2+}\right] \tag{22}$$

$$[Cr]_{aq} = \left[Cr^{3+}\right]\alpha_{Cr} \tag{23}$$

with $\alpha_{Cr}$ being the side reaction coefficient due to the formation of both hydroxo and chloride complexes in the aqueous solution. This can be determined by taking into account the formation constant values of the complexes as well as the concentration of $\left[OH^-\right]$ and $\left[Cl^-\right]$.

Taking into account equations (17)-(19), (20) and (23), equation (21) can be written as:

$$D_{Cr} = \left\{K_I\left(\frac{rQ - (3-i)\overline{CrCl_iR_{I3-i}}}{[H] + K_{a_1}}\right)^{3-i} + K_{II}\left(\frac{(1-r)Q - (3-i)\overline{CrCl_iR_{II3-i}}}{[H] + K_{a_2}}\right)^{3-i}\right\}\left[Cl^-\right]^i \tag{24}$$

If $\left[H^+\right] > Ka_1$, Equation (24) can be simplified as:

$$Log(D\alpha_{Cr}) = log\left\{K_I\left(rQ - \overline{CrCl_iR_{I3-i}}\right) + K_{II}\left((1-r)Q - \overline{CrCl_iR_{II3-i}}\right)\right\}\left[Cl^-\right]^i + (3-i)pH \tag{25}$$

and if $K_{a_1} \triangleright \left|H^+\right| \triangleright K_{a_2}$ , it can be rearranged to give

$$\frac{D\alpha K_{a_1}{}^{3-i}}{Cl^{-i}K_I\left(rQ-(3-i)\overline{CrCl_iR_{I3-i}}\right)^{3-i}}=1+\frac{K_{II}K_{a_1}{}^{3-i}\left((1-r)Q-(3-i)\overline{CrCl_iR_{II3-i}}\right)^{3-i}}{K_I\left(rQ-(3-i)\overline{CrCl_iR_{I3-i}}\right)^{3-i}\left[H^+\right]^{3-i}}$$

(26)

and assuming that the unity term in the equation could be neglected, the equation above can be rewritten as

$$LogD+Log\alpha_{Cr}=LogK_{II}\left((1-r)Q-(3-i)\overline{CrCl_iR_{II3-i}}\right)^{3-i}\left[Cl^-\right]^{3-i}+(3-i)pH$$

(27)

The variation of the distribution coefficient $\left(D=\overline{Cr\left[Cl^-\right]R_{3-i}}\Big/Cr_{aq}\right)$ with pH is shown in Figure 5.

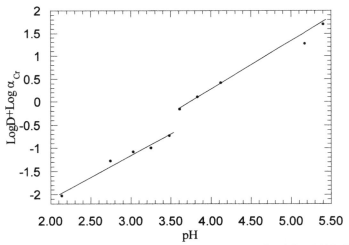

**Figure 5.** Variation of Log (D) + Log ($\alpha_{Cr}$) *versus* f(pH) in 1M NaCl.

It can be appreciated from Figure 5 that two straight lines occur of the same slope (close to 1) from which the value of (3-i)=1 can be deduced, and the species $CrCl_2R$ seems to be responsible for the retention of the metal ion.

The plot of $\log D\alpha_{Cr}$ versus pH is linear for $2 \leq pH \leq 4$, hence the Y axis intercept is $\log\left\{K_I\left(rQ - \overline{CrCl_iR_{I3-i}}\right) + K_{II}\left((1-r)Q - \overline{CrCl_iR_{II3-i}}\right)\right\}\left[Cl^-\right]^i$. On the other hand, the plot of $\log D\alpha_{Cr}$ versus $[pH]$ at $pH \geq 4$ yields a straight line with intersect $\log\left\{K_{II}\left((1-r)Q - \overline{CrCl_iR_{II3-i}}\right)\right\}\left[Cl^-\right]^i$.

The equilibrium constant values $K_I$ and $K_{II}$ graphically obtained have been optimized by minimization of the deviation between the experimental and calculated distribution coefficients. The results are given in the Figure 6, and the values of $LogK_I = -5.0$ and $LogK_{II} = -4.3$.

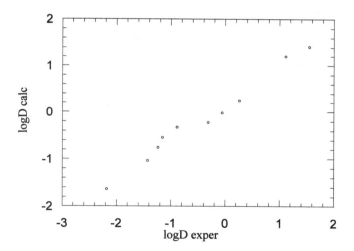

**Figure 6.** Experimental and calculated Log D.

## References

1   H.M.N.H. Irving, The XVth Procter memorial lecture fact or fiction? How much do we really know about the chemistry of chromium today?, Journal of the Society of Leather Technologists and Chemists, 1974, 58, 51.

2   C.F. Baes, R.E. Mesmer, The hydrolysis of cations, John Wiley and sons, New York, 1976, pp. 211

3    R. Escobar, Q.Lin and A. Guiraum, Int. J. Environ. Anal. Chem., 1995, <u>61</u>, 169.

4    T. F. O'Dwyer B. K. Hodnett, J. Chem. Tech. Biotechnol, 1995, <u>62</u>, 30.

5    M. Loizidou, K. J. Haralambous, A. Loukatos, D. Dimitrakopoulou, J. Environ. Sci. Health, 1992, <u>A27</u>, 1759.

6    Tao Zuyi and Wang Changshou, Solvent extraction and ion exchange, 1993, <u>11</u>, 171.

7    M. R. Baloga, J. E. Earley, J. Am. Chem. Soc, 1961, <u>83</u>, 4906.

8    H. S. Rossotti, 1974, Talanta, <u>21</u>, 809.

9    N. V. Jarvis and J. M. Wagener, Solvent extraction and ion exchange, 1995, <u>13</u>, 591.

10   G. Gran., Analyst, 1952, <u>77</u>, 661.

11   D. Rai, b. M. Sass, D. A. Moore, Inorg. Chem. 1987, <u>26</u>, 345.

12   H. Stunzi, L. Spiccia, F. P. Rotzinger, W. Marty, Inorg. Chem. 1989, <u>28</u>, 66.

13   H. Tamura, T. Oda, R. Furuichi, Anal. Chim. Acta. 1991, <u>244</u>, 275.

# MICROKINETIC ANALYSIS OF HEAVY METAL ION EXCHANGE ONTO CHELATING RESINS

*F. Mijangos, M. Ortueta, L. Bilbao
*Department of Chemical Engineering, Science Faculty, University of the Basque Country, Apdo. 644, Bilbao 48080, Spain.*

## ABSTRACT

This work considers the kinetics of heavy metal uptake onto chelating resins on the basis of the internal concentration profiles inside one bead, swelling-shrinking measurements are also reported, and results are compared with those derived from external solution analysis.

Metal concentration profiles inside the particles at different reaction times were measured by the Scanning Electron Microscopy-Energy Dispersive X-Ray (SEM-EDX) technique, which allows one to obtain a linescan along diametrical positions. This study was complemented with measurements of particle size during the ion exchange process: metal uptake, elution and regeneration. Experiments on batch contact for metal uptake and sodium/potassium interdiffusion have been carried out to help in the understanding of the ions' intraparticular motion. In the experiments operated in a batch-wise procedure, alkaline metals were analysed by ion chromatography and heavy metal by Atomic Absorption Spectrophotometry.

The experimental results are discussed in terms of pseudosteady state kinetic models and the effective diffusion coefficients evaluated were compared against the corresponding values reported in the literature. Swelling measurements were taken into account in these models. On the other hand, interdiffusion kinetics for sodium/potassium were fitted to the Nernst-Plank model, and the effective diffusion coefficient was used to estimate the microporosity of the iminodiacetic-type ion exchanger.

## 1 INTRODUCTION

The recovery of heavy metals from wastewaters is discussed in this work because of hazards in the environmental field and their high economical value. This application of ion exchangers requires studies of metal distribution between phases and also kinetic experiments of metal uptake to evaluate its industrial feasibility. There are many reports in the literature about these topics, but more recently a deeper knowledge of the intraparticular motion and distribution is required in order to optimise the exchanger behaviour.

Several authors have studied the swelling behaviour of ion exchangers that usually accompanies the exchange process, but the influence of this behaviour on the overall process has not been well quantified. Hariharan and Peppas[1] studied the dynamic swelling behaviour in cationic networks and applied a semi-empirical equation in which the main parameter was an effective diffusion coefficient. Chu et al.[2] combined quantitatively the theory of equilibrium swelling into the model of swelling kinetics. By doing so, the advantage is taken of applying relatively more mature knowledge of gel swelling thermodynamics to predict less knowledgeable dynamic behaviour of gels. The diffusion of hydrogen ions through polyelectrolyte matrices is concluded to be an important factor in gel swelling processes.

This effect is more noticeable for chelating resins, like the iminodiacetic type, where the interaction between the cation and functional group is very strong. For more ionised media, like resins of strong acids or bases, the change in volume is less important and it is basically related to charge and ionic radius of the ions.

Intraparticular concentration profiles of the exchanging species have been measured using different experimental techniques such as optical microscopy, and radioisotopes that allow the researcher an unclear image of the distribution inside the bead. Mijangos and Diaz[3] have reported the temporal evolution of the reaction front appearing inside the bead when a metal is loading. Data analysis of the moving boundaries was based on the pseudosteady state approximation. This has been supported by the observation of sharp reaction fronts under the microscope. They concluded that these experimental results are in general accordance with those results derived from macrokinetic studies.

Mijangos and Bilbao[4] reported the application of SEM-EDX, X-ray microanalysis, Energy Dispersive Spectrometry, for measuring solid phase concentration of the diffusing species and even sulphate and sodium. Radial concentration profiles were strongly influenced by the mobility and weakness of the X-ray signal. For one given time, they reported radial concentration profiles for the simultaneous uptake of copper and cobalt from an equimolar solution. The iminodiacetic type resin used was initially in the sodium form.

We apply this analytical technique and report the temporal evolution of the intraparticular concentration profiles for loading (copper and cobalt) onto an iminodiacetic chelating resin. Experiments were carried out for the monometallic and bimetallic mixtures in order to compare their behaviour and estimate the mass transfer parameters.

## 2 EXPERIMENTAL METHODOLOGY

### 2.1 Batch-wise procedure

The metal loading onto the resin is carried out in a stirred batch reactor. The resin is previously conditioned to assure the functional group is in the Na-form. For this purpose, the resin is washed with sulphuric acid and sodium hydroxide solution three times, alternatively. Once the resin is conditioned, spherical beads and uniform in size are selected under the magnifying glass.

Metal solution (200 mL) are introduced in the batch reactor, pH is adjusted (pH=4.0) in order bring all the assays to the same conditions. Fifty resin beads ranging from 680 to 700 mm in diameter are placed in the reactor. At this moment the exchange reaction starts and at the lapsed time the beads are taken out of the vessel and washed with ultrapure water. This stops the reaction. The loaded metal is then eluted from the resin by keeping the beads in sulphuric acid solution (1 M, 100 mL) for one week, to assure that the metal is totally eluted from the resin. This solution is diluted with ultrapure water and metal concentration measured by Atomic Absorption Spectrometry.

### 2.2 Optical measurement of the internal reaction fronts

For optical observation, the operational procedure is the same as above. Only two beads are selected from the metal load assays, but in this case, the water used to wash the bead must be slightly alkalinised (pH=10) because otherwise the reaction does not stop, and fronts continue to move to the centre of the bead. These beads are cut by scalpel under a magnifying glass, trying to get two similar parts. Under the optical microscope, the radial position of the reaction fronts for each bead and reaction times are measured. This is possible because the chelate formed inside the bead has different colour for the different metal chelates, blue for copper and pale pink for cobalt. Measurements by this method are only possible for a very short period of time, because the cobalt reaction front reaches the centre of the bead very quickly (five minutes) and the whole centre of the bead is pink coloured so only motion of the copper front can be measured.

The measurement of the position of the front are clearly described in a previous paper[3]. The methology is designed to avoid measurement errors. Some are proportional to image size (magnification errors) and others result from incorrect location of the image boundary (criterion errors). Errors of magnification are due to distortion aberration, which should cause a purely radial variation of magnification[5].

## 2.3 Intraparticular Metal Concentration Profiles

Cut bead samples, after measurement by the optical method, are fixed on a special holder to be introduced into the electron microscope. In each holder, six or seven beads are placed, one for each reaction time. These samples have to be treated as we described in a previous paper[4], to avoid changes in composition or microstructure. The drying method must not be aggressive and the sample surface must be conductive.

The microanalysis and sample examinations are carried out on a JEOL 6400 coupled with Energy Dispersive X-ray Spectrophotometer. Using this technique, relative metal concentration profiles are collected from a diametrical linescan plotted over an image of the sample obtained using backscattered electron. This is used to verify the presence of the elements in each zone or layer.

## 2.4 Sodium/potassium Exchange

These experiments are carried out using the batch procedure previously described. The iminodiacetic type resin is initially in the sodium form and potassium is the only cation in the surrounding liquid. Only media that guarantee that no contamination from other alkaline metals are used as sources for resin and solution preparation. All the laboratory material used is inert to this system (PVC, Teflon...) and carefully washed. The resin conditioning to the sodium form tends to eliminate all the impurities that could interfere in this assay. In this case, the conditioning consists of some successive washing of the previously conditioned resin but now with 1M NaOH solution (free of potassium).

The assays are carried out using a phase ratio of 5 g of wet resin to 200 mL solution. Ion chromatography is used for alkaline metals analysis (Dionex 2000i with a cationic column HPIC-CS3). The eluent or mobile phase used is a solution of 25 mM HCl and diaminopropionic acid 0,25 mM and the operating flow 1,7 mL/min. The regenerant for the cations analysis is the tetrabutylamonium 100 mM at a working pressure of 1700 psi.

# 3 EXPERIMENTAL RESULTS

Metal concentration profiles for simultaneous uptake of cobalt and copper from an equimolar solution 0.1 M are show in Figure 1 for 5, 30 and 90 minutes. In this figure is plotted the local attainment of equilibrium because only relative concentrations are

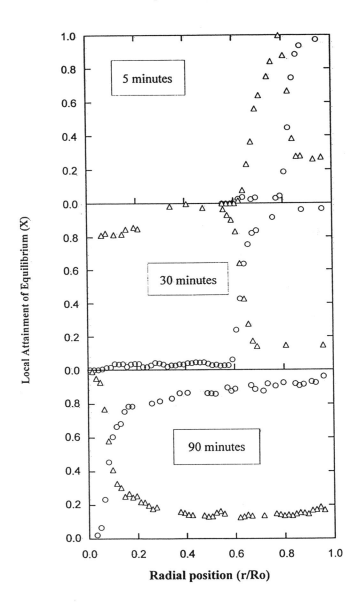

**Figure 1** *Intraparticular concentration profiles of* (O) *copper and* (Δ) *cobalt for different reaction times.*

measured in the experiments. Final values for Figure 1 are calculated assuming that equilibrium has been achieved at the bead surface for any time. It can be seen how cations penetrate inside the bead, and how copper displaces cobalt.

In the outer layer of bead, at radial position r/R near to 1, the solid phase is equilibrated with the external solution and metals concentration is almost constant, but then that copper concentration decreases giving rise to the called reaction front. The position is clearly observed under the optical microscope because of the change in colour, from blue to pink, but the end of this layer coincides with a parallel growing of the cobalt concentration, and consequently to the observable intermediate layer that is pink in colour. The cobalt layer

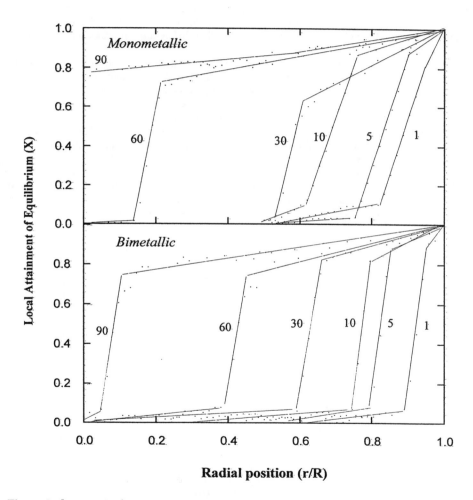

**Figure 2** *Intraparticular concentration profiles of copper for different reaction times (in minutes). Upper figure shows results of Cu/Na interdiffusion and the lower the simultaneous diffusion of copper and cobalt against sodium.*

grows rapidly and moves faster than the copper one. Although the cobalt front achieves the centre of the particle in a few minutes, this does not mean that equilibrium is achieved. In contrast with the assumption made in the optical methods, this equilibrium condition is only achieved near the surface of the bead.

Using the electron microscopy technique (SEM-EDX) for the analysis of the reacted layer of copper in monometallic and bimetallic systems, the evolution of the local attainment of equilibrium in the radial direction inside the resin bead is plotted. For a comparison of the concentration profiles, in Figure 2, only the experimental results are plotted for copper diffusion, from single or monometallic and from bimetallic solutions. If the radial positions of the reaction fronts at the same time are compared, it can be observed that copper diffuses faster alone than in presence of cobalt. Nevertheless, the slope is approximately the same in both cases but for single copper diffusion the slope is even lower. This can be understood if it is considered that different equilibrium values apply in both experiments.

For the bimetallic system, the reaction front advance is summarised in Figure 3, where the copper reaction front motion can be compared with the cobalt front, this moves quicker. However, this advance does not mean that the saturation level is higher. Although copper uptake is slower at each position, the copper reacted layer corresponds to saturated resin and the equilibrium uptake is higher than that for cobalt. Due to the strong bond between copper and the functional group, copper front advance promotes the displacement of cobalt towards the centre of the bead and then the overall cobalt concentration inside the particle starts to decreases.

In the assumption of the intraparticular diffusion coefficient being constant, the slope of the reaction front shows the mass transfer for each species. In Figure 4 can be observed how the slope decreases almost linearly in proportion as reaction fronts advance toward

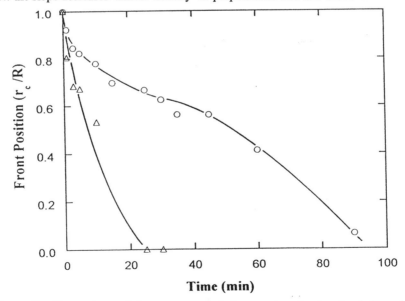

**Figure 3** *Comparison of the movement of the reaction fronts of cobalt and copper observed by SEM-EDX (symbols as in Figure 1).*

the centre of the bead. It is interesting to check that the intraparticular concentration
gradient is equal, within the experimental error (quite important for cobalt), for copper and
for cobalt. Nevertheless, it must be observed that the copper concentration in solid phase
($q_{Cu}^{*}$) is considerably higher than cobalt ($q_{Co}^{*}$). Moreover, slopes calculated for a given
reaction time do not correspond to the same radial position, so the result depicted in Figure
4 should analysed using Equation /1/, which can be derived simply if Fickian diffusion is
assumed.

$$\frac{V_{Co}}{r_{Co}^{2}D_{Co}q_{Co}^{*}} = \frac{V_{Cu}}{r_{Cu}^{2}D_{Cu}q_{Cu}^{*}} \qquad /1/$$

SEM-EDX data give relative local concentrations along a diametrical line, but it is
possible to calculate the average metal concentration ($X$) inside the bead using the Equation
/2/ which has been solved numerically for data reported in Figure 1 or 2.

$$X(t) = -3\int_{0}^{1} X(r,t)\,(r/R)^{2}\,d(r/R) \qquad /2/$$

The average metal concentration versus time is plotted in Figure 5. Here, the cobalt
concentration gains the maximum load at intermediate times and copper continues growing,
displacing the cobalt until the final saturation state is reached.

These experimental results obtained with SEM-EDX are compared with those estimated
with optical microscope and also those uptakes calculated by chemical analysis of the outer
solution. In Figure 6, is depicted the experimental curves of metal uptake for same
conditions as for SEM-EDX. If one compares Figures 6 and 7, the chemically analysed
experimental results are coherent with those obtained with the X-ray microprobe. The
cobalt concentration peak is around 20-30 minutes in both cases.

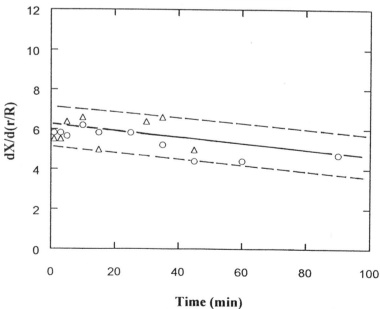

**Figure 4** *Slope of the reaction fronts of copper and cobalt along the reaction (symbols as
in Figure 1).*

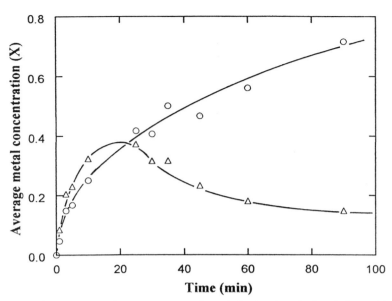

**Figure 5** *Average metal concentration for whole bead calculated from SEM-EDX experimental results (symbols as in Figure 1).*

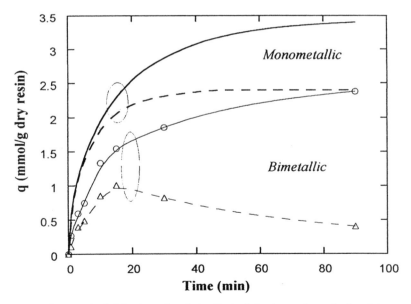

**Figure 6** *Copper (solid line) and cobalt (dashed line) uptake calculated by chemical analysis (symbols as in Figure 1).*

The relationship between cobalt and copper uptake is shown in Figure 7, where $X_{Co}$ versus $X_{Cu}$ for single cations has been also plotted. Due to the faster kinetics of cobalt in solution the relation with uptake of copper is not a linear relation. Nevertheless, for systems with both cations, both metals increase their concentration in parallel, giving rise to a linear relation, but even after the cobalt peak, the linear tendency is kept, but with a negative slope. Another interpretation of metal uptake data is shown in Figure 6 where theoretical values are overplotted, these data have been calculated using the effective intraparticular diffusion coefficient estimated from monometallic solution by fitting experimental results to the unreacted core model,

$$t = \frac{q_i^* \chi R^2}{6 D_i C_{io}} f_2(X)$$

/3/

where

$$f_2(X) = 1 - 3(1-X)^{2/3} + 2(1-X)$$

/4/

The fractional attainment to the equilibrium has been calculated using the metal uptake at infinite ($q^*$)

$$X = q(t)/q*$$

/5/

From these equations the effective diffusion coefficients are calculated for monometallic system:

$$D_{eCu} = 4{,}18 \cdot 10^{-10}\, \text{m}^2/\text{s} \qquad D_{eCo} = 2{,}82 \cdot 10^{-10}\, \text{m}^2/\text{s}$$

Those for simultaneous uptake of copper and cobalt are also estimated, but for the first part of the reaction, when the maximum in cobalt concentration has not been achieved. The

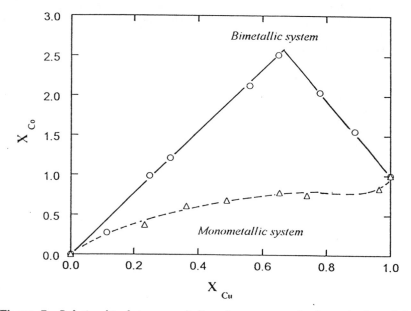

**Figure 7** *Relationship between cobalt and copper uptake from single and bimetallic solutions (symbols as in Figure 1).*

estimated value for this cation contains an important error due to the changing solid phase concentration:

$$D_{eCu} = 4,16 \cdot 10^{-10} \text{ m}^2/\text{s} \qquad\qquad D_{eCo} = 1,02 \cdot 10^{-9} \text{ m}^2/\text{s}$$

The resin capacity values are:

Bication system $\qquad q_{Cu}^* = 2.66$ and $\quad q_{Co}^* = 1.1$ mmol/g dry Na-form

Monocation system $\qquad q_{Cu}^* = 3.23$ and $\quad q_{Co}^* = 2.27$ mmol/g dry Na-form

In order to evaluate the microstructure of the polymeric matrix, the temporal evolution of the potassium uptake has been investigated by sodium/potassium-iminodiacetic resin as the ion exchange system.

The molar fraction of the sodium ion is calculated directly from the area of the peaks of the chromatogram using

$$X_{Na} = \frac{n_{Na}}{n_{Na} + n_K} = \frac{\alpha}{1 + \alpha} \qquad\qquad /6/$$

where

$$\alpha = A_{Na} / A_K \qquad\qquad /7/$$

The concentration of both cations inside the resin has been estimated by the mass balance:

$$X_{Na} + X_K = 1$$
$$Y_{Na} + Y_K = 1 \qquad\qquad /8/$$

$$Y_K = \frac{\left[ Y_K^o - (1 - X_{Na}) \right] / V}{\left[ Q m_R / \left( C_K^o + C_{Na}^o \right) \right]} \qquad\qquad /9/$$

In Figure 8, the molar fraction of potassium inside the resin is plotted. The ion exchange kinetics for these cations is very fast comparing with other processes such as the copper or cobalt ion exchange; in this case 500 seconds are enough to have a good approximation to the achievement of equilibrium state.

Selectivity coefficients are $K_{Na}^K = 4,62$ and $0,55$ for initial potassium concentrations of $0,1$ and $0,01$M, respectively.

Potassium load data have been fitted to the Nersnt-Plank Equation /10/ [6], estimating the interdiffusion effective coefficient, which shows a dependence on the selfdiffusion coefficient as described by Equation /11/.

$$J_i = (J_i)_{dif} + (J_i)_{el} = -D_i \, grad C + \frac{D_i F z_i C_i}{RT} grad \phi \qquad\qquad /10/$$

$$\overline{D}_{12} = \frac{\overline{D}_1 \overline{D}_2 \left( C_1^{z_1} + C_2^{z_2} \right)}{C_1^{z_1} D_1 + C_2^{z_2} D_2} \qquad\qquad /11/$$

Taking into account (in the liquid phase at infinite dilution) that the diffusion coefficients for these cations are:

$$D_K^o = 1.96 \cdot 10^{-9} \text{ m}^2/\text{s} \quad \text{and} \quad D_{Na}^o = 1.33 \cdot 10^{-9} \text{ m}^2/\text{s}$$

and using the Mackie-Meares equation:

$$\frac{\overline{D}}{D^o} = \left( \frac{\varepsilon}{2 - \varepsilon} \right)^2 \qquad\qquad /12/$$

an estimation of the internal porosity can be obtained which is very useful for ion exchange kinetics with heavy metals, where the diffusion and the chelate-forming process are

Table 1 Calculated values for different concentrations of potassium.

| $C_K$ (mol/L) | $D_{Na}^K\, 10^{11}$ (m²/s) | $\varepsilon$ |
|---|---|---|
| 0,01 | 9,0 | 0,385 |
| 0,1 | 7,0 | 0,347 |

accompanied by important changes in the porous microstructure. This estimated value of the "diffusion" porosity for both assays at different concentrations are very similar as can be seen in Table 1.

The values reported in Table 1 can be applied for the analysis of those values obtained for cobalt and cobalt diffusion though the same polymeric matrix. Nevertheless, the values of the internal porosity reported are lower than those estimated from other techniques, such as porosimetry or water content. In our opinion this is due to fact that the overall diffusion process is controlled by the diffusion though the microspheres whose microporosity should be very close to those values.

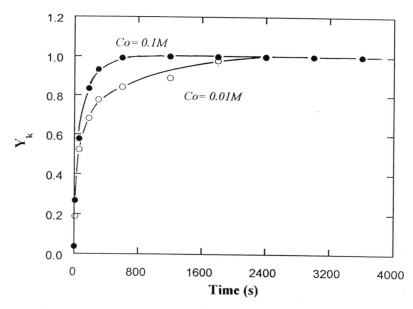

Figure 8 *Experimental data of potassium fractional attainment of equilibrium for the K/Na ion exchange kinetic.*

## 4 NOMENCLATURE

$C_A$      A species concentration (mol/L)

$C_A^o$      initial A species concentration in solution (mol/L)

$\overline{D}$      interdiffusion effective coeficient (m²/s)

| | |
|---|---|
| $D$ | particle diameter at a given time (m) |
| $D_A^o$ | diffusion coeficient at infinite dilution for A species (m$^2$/s) |
| $D_i$ | diffusion coeficient for species i (m$^2$/s) |
| $D_o$ | particle initial diameter (m) |
| $D_\infty$ | particle diameter at infinite time (m) |
| $\overline{D}_{12}$ | diffusion coeficient for both species (m$^2$/s) |
| $F$ | Faraday constant (C/mol) |
| $K_A$ | selectivity coeficient |
| $J_i$ | Flux density of species i (mol/m$^2$ s) |
| $(J_i)_{el}$ | Flux density due to the electric field (mol/m$^2$ s) |
| $(J_i)_{dif}$ | Flux density due to diffusion (mol/m$^2$ s) |
| $m_R$ | humid resin mass (kg) |
| $q^*$ | equilibrium metal uptake (mol/kg dry resin) |
| $Q$ | total capacity of the resin (mol/kg dry resin) |
| $z_n$ | ionic charge |
| $n_A$ | No. moles species A |
| $r$ | particle radius (m) |
| $R$ | universal gas constant (J/K mol) |
| $t$ | time (s) |
| $T$ | temperature (K) |
| $u_i$ | mobility of species i (m/s) |
| $V$ | liquid volume (m$^3$) |
| $X$ | particle swelling |
| $X_A$ | molar fraction for A species outside the particle |
| $Y_A$ | molar fraction for A species inside the particle |

**Greek symbols**

| | |
|---|---|
| $\alpha$ | sodium/potassium ratio |
| $\delta$ | particle relative shrinking |
| $\varepsilon$ | porosity |
| $\phi$ | electric potential gradient (V) |
| $\mu$ | kinetic parameter (min) |
| $\chi$ | structural parameter (kg dry Na-form resin/L wet resin) |

## 5  LITERATURE

1 D. Hariharan, N.A. Peppas, *Polymer*, 1996, **37**, 149.
2 Y. Chu, P.P. Varanasi, M. J. McGlade, S. Varanasi, *J. App. Polymer Sci.*, 1995, **58**, 2161.
3 F. Mijangos, M. Díaz, *J. of Colloid and Inter. Sci.*,1994, 1028.
4 F. Mijangos, L. Bilbao, *Progress in Ion Exchange: Advances and Applications*, ed. A. Dyer, M. J. Hadson and P. A. Willians, 1997.
5 D. H. Freeman, *Ion exchange and Solvent Extractions*, Vol. 1, ed. J.A. Marinsky & Y. Mareus, 1978.
6 F. Helfferich, *Ion Exchange*, McGraw-Hill, New York, 1962.

# TEMPERATURE DEPENDENCE OF THE SELECTIVITY OF AN IMINODIACETIC RESIN TOWARDS TRANSITION METAL IONS

A. Gonzalo, D. Muraviev, M. Valiente

Department of Analytical Chemistry
Autonomous University of Barcelona
08193 Bellaterra (Barcelona), Spain

## 1 INTRODUCTION

Removal of heavy metal ions from industrial effluents and wastewaters by ion exchange[1-3] (IX) has some advantages over other separation methods, such as precipitation or biological processes, because IX allows, not only for the elimination of metals from these effluents, but, also for their recovery directly contributing to resource conservation.[4,5] The main limitation of IX is the generation of chemical wastes produced during conditioning and regeneration steps. This limitation can be overcome by applying reagent-less (and waste-free, as a result) IX technologies such as parametric pumping [6-9], dual-temperature IX processes[10] and thermal IX fractionation[11,12] which exploit the influence of temperature on the affinity of ion exchangers towards certain ions. Chelating resins, which are known to be highly selective towards heavy metal ions, have found little use in studies on temperature-responsive IX separation processes, although they meet the main requirement for their application in dual-temperature IX separation: a high heat of IX reaction.[8, 13-16] The present study was undertaken 1) to confirm a previously developed approach[17,18] which allows for the prediction of temperature dependencies of IX equilibrium on chelating resins and which can be applied for the selection of chelating resins suitable for separation of certain ion mixtures by applying temperature-responsive IX fractionation techniques and also for the preselection of ligands with promising characteristics for use in the synthesis of new chelating resins with temperature-responsive selectivity and 2) to elucidate the influence of complexation conditions in the resin phase on temperature dependence of the cation-exchange equilibrium.

## 2 EXPERIMENTAL

All chemicals of analytical grade were purchased from Panreac (P.A., Spain). Iminodiacetic (IDA) ion exchanger, Lewatit TP-207, was kindly supplied by Bayer Hispania Industrial, S.A. Doubly distilled water was used in all experiments. A Perkin Elmer 16 PC FTIR spectrometer was used to record infrared spectra of the resin phase in the range 4000 - 400 $cm^{-1}$. Solutions of 1:1 mixtures of $Cu^{2+}$ and $Ni^{2+}$ (sulphates or nitrates at pH=1.7) and $Co^{2+}$ and $Zn^{2+}$ (nitrates at pH=3.0) with the total concentration of 0.1M were used in this study. The concentrations of metal ions were determined by the ICP technique using the ARL Model 3410 spectrometer provided with a minitorch. The emission lines used for the spectrochemical analysis were 324.754 nm for $Cu^{2+}$, 227.021 nm for $Ni^{2+}$, 206.200 nm for $Zn^{2+}$ and 228.616 nm for $Co^{2+}$. The uncertainty of metal ions determination was <1.5%. pH was controlled by using a Crison pH meter 507 supplied with a combined glass electrode.

The IX equilibrium at different temperatures was studied under dynamic conditions in thermostatic glass columns by following the procedure described in detail elsewhere[17]. The IX equilibrium in all systems under study has been characterised by the value of equilibrium separation factors, $\alpha$, which can be expressed as follows:

$$\alpha\,^B_{\ A} = \frac{q^*_B}{q^*_A}\frac{C^*_A}{C^*_B} \tag{1}$$

were $q^*$ and $C^*$ are the equilibrium concentrations of metal ions in resin and solution phases respectively. The relative uncertainty on $\alpha$ determination did not exceed 4%.

## 3 RESULTS AND DISCUSSION

### 3.1 Ni-Cu exchange

Temperature dependencies of $\alpha^{Cu}_{Ni}$ for both ionic media studied are shown in Fig. 1. As seen, $\alpha$ vs T dependencies are different for the nitrate and sulphate media. In both systems studied IDA resin preferentially sorbs $Cu^{2+}$ against $Ni^{2+}$ ($\alpha^{Cu}_{Ni} \gg 1$), which correlates with the stability constants of copper and nickel complexes with IDA ligand in aqueous solution ($\log\beta$ IDACu = 10.6 and $\log\beta$ IDANi = 8.2).[19] For the nitrate media, two different zones in $\alpha^{Cu}_{Ni}$ vs T dependency can be distinguished: (1) from 293 to 303 K, where the separation factor diminishes when temperature increases, and (2) from 303 to 353 K where the opposite situation is observed. For sulphate media a smooth decrease of the separation factor with temperature is observed within the whole temperature range studied. For both systems under consideration an increase of the resin capacity towards both nickel and copper with temperature is observed.

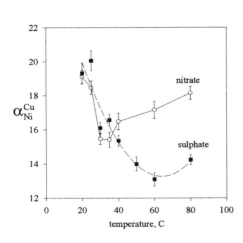

**Figure 1** *Temperature dependency of equilibrium separation factor for Ni-Cu exchange in nitrate and sulphate media*

The capacity of this resin, determined potentiometrically, has been found to be $5.32 \pm 0.23$ meq/g.[20] The maximum resin capacity towards both metal ions ($q_{Cu} + q_{Ni}$) $_{max}$ is observed at T = 353 K and was 4.51 (nitrate medium) and 4.53 (sulphate medium) mequiv/g. This indicates that, even at high temperature, not all functional groups are loaded with metal ions. Carboxylic groups of IDA resin appear partially protonated at all temperatures but their deprotonation increases with temperature resulting in higher metal capacities at high temperatures. This hypothesis is substantiated by IR spectra of the resin phase pre-equilibrated with $Cu^{2+}$ - $Ni^{2+}$ mixture at 293 and 353 K. A comparison of these spectra with those of completely protonated IDA resin testifies to the diminishing of the band at $1716 \text{ cm}^{-1}$, corresponding to the C=O stretching of the carboxylic acid at 293 K, and to the remarkable decrease of this band at 353 K in both systems under study. The difference in the capacity towards proton found in both media can be attributed to the stronger complexation properties of $SO_4^{2-}$ ions with $Cu^{2+}$ and $Ni^{2+}$ in comparison with $NO_3^-$.[21] The presence of stronger sulphate complexes in solution shifts the IX equilibrium in the system and this results to an increase of the content of protonated groups in the resin phase.

The IX reactions, which proceed in these systems can be written as follows:

a) deprotonated resin

$$R\text{-}CH_2\text{-}N(CH_2\text{-}COO^-)_2Ni^{2+} + Cu \Leftrightarrow R\text{-}CH_2\text{-}N(CH_2\text{-}COO^-)_2Cu^{2+} + Ni^{2+} \quad (2)$$

b) partially protonated resin

$$[R\text{-}CH_2\text{-}N(CH_2\text{-}COO^-)_2H^+]_2 \, Ni^{2+} + Cu \Leftrightarrow 2R\text{-}CH_2\text{-}N(CH_2\text{-}COO^-)_2Cu^{2+} + 2H^+ + Ni^{2+} \quad (3)$$

Different $\alpha$ vs T dependencies, clearly distinguishable in the two temperature intervals, in Fig.1 can be attributed to the predominance of different Ni-IDA complexes in different temperature ranges: Ni(IDA)$_2$ at 293 < T < 303 and Ni(IDA at 303 < T 353 K. The probability of preferential formation of Ni-(IDA)$_2$ against Ni-IDA in nitrate medium is higher at low temperatures (T < 303 K), due to the higher content of protons in the resin phase. Under these conditions the interaction of $Ni^{2+}$ (as a minor component) proceeds mainly with semi-protonated IDA functional groups. The same reasoning leads to the conclusion that, in the sulphate medium, formation of Ni(IDA)$_2$ complexes can be observed in a wider temperature range due to the presence of substantial quantities of semi-protonoated IDA groups (uncomplexed) up to the maximal temperatures. Formation of mixed M-IDA and M-(IDA)$_2$ complexes in IDA resin has been recently reported by Waki[22] (for M = $Ni^{2+}$) and Pesavento et al.[23] (for M = $Zn^{2+}$ and $Cd^{2+}$).

Furthermore, the IX reactions (2) and (3) can be considered to be analogous to the following exchange reactions of monomeric IDA complexes:

a) deprotonated IDA ligand

$$HN(CH_2\text{-}COO^-)_2Ni^{2+} + Cu \Leftrightarrow HN(CH_2\text{-}COO^-)_2Cu^{2+} + Ni^{2+} \quad (4)$$

b) partially protonated IDA ligand

$$[HN(CH_2\text{-}COO^-)_2H^+]_2 \, Ni^{2+} + Cu \Leftrightarrow 2HN(CH_2\text{-}COO^-)_2Cu^{2+} + 2H^+ + Ni^{2+} \quad (5)$$

The differential enthalpies of reactions (4) and (5) ($\Delta(\Delta H)$ $_{Ni\text{-}Cu}$) can be easily evaluated from the respective thermodynamic data available in the literature,[21] and compared with the apparent enthalpies, $\Delta H_{ap}$, of the IX reactions (2) and (3)[17] which are obtained in the present study. The $\Delta H_{ap}$ values can be calculated as described before.[17]

Thus, for $Cu^{2+}$- $Ni^{2+}$ couple, a moderate (in comparison with $Cu^{2+}$- $Zn^{2+}$ system, studied previously[17]) temperature dependence of IX equilibrium on IDA resin was predicted from $\Delta(\Delta H)_{CuIDA-NiIDA} = +2.33$ kJ/mol [19] (cf., $\Delta(\Delta H)_{CuIDA-ZnIDA} = -9.4$ kJ/mol[19]), and practically no effect of temperature on IX equilibrium of $Co^{2+}$- $Zn^{2+}$ on IDA resin $(\Delta(\Delta H)_{ZnIDA-CoIDA} = -0.25$ kJ/mol[19]) was expected. A comparison of the results on determination of $\alpha$ vs T dependencies for $Cu^{2+}$- $Ni^{2+}$, and $Co^{2+}$- $Zn^{2+}$ ion couples, shown in Fig. 1 and Table 1, respectively, testifies to the correctness of the predictions made.

Estimation of $\Delta H_{ap}$ values for the systems under study, from the slopes of the respective ln $\alpha$ vs $1/T$ plots assuming the usual Arrhenius dependencies of $\alpha$ on temperature, in case of nitrate medium has been done for both temperature intervals (see above) characterized by different $\alpha$ vs T dependencies (see Fig. 1). The $\Delta H_{ap}$ values, determined for the first and for the second intervals, equal to $\Delta H_{ap}(\Delta T_1) = -15.7$ kJ/mol, and $\Delta H_{ap}(\Delta T_2) = +2.9$ kJ/mol, respectively. The last $\Delta H_{ap}(\Delta T_2)$ value agrees well with $\Delta(\Delta H)_{CuIDA-NiIDA} = +2.33$ kJ/mol (see above), calculated from eq.(4). Calculation of $\Delta(\Delta H)_{CuIDA-Ni(IDA)2}$ using eq.(5) from the data taken from Ref.19, gives a value of -18.8 kJ/mol, which appears to be in a satisfactory agreement with $\Delta H_{ap}(\Delta T_1)$ value obtained. This correlation testifies to the correctness of the above supposition, concerning the formation of $Ni^{2+}$-IDA complexes of different stoichiometry in the resin phase at different temperatures. Note that evaluation of the resin phase composition from $\alpha$ vs T dependencies has been carried out for the first time in the present study and it reveals a clear method to model the resin phase composition. To confirm this conclusion, consider the results obtained in studying the same IX exchange reaction proceeding in sulphate medium within the framework of such an approach.

The $\Delta H_{ap}$ value estimated from the slope of log $\alpha$ vs $1/T$ plot for the temperature range from 293 to 333 K in this medium equals to -8.4 kJ/mol, i.e. it appears to lie in between $\Delta(\Delta H)_{CuIDA-NiIDA}$ and $\Delta(\Delta H)_{CuIDA-Ni(IDA)_2}$ values (see above). In this situation one can expect a coexistence of both Ni-IDA and Ni-(IDA)$_2$ complexes in the resin phase.

### 3.2 Co-Zn exchange

In this system the resin manifests a higher affinity towards $Zn^{2+}$. The resin selectivity in this case is lower than that observed in the previous case, which correlates with the lower difference between stability constants of Co-IDA and Zn-IDA complexes: $log\beta$ IDAZn $= 7.3$ and $log\beta$ IDACo $= 7.0$.[19] As a result, the values of the separation factor, $\alpha_{Co}^{Zn}$, lie within the range $1.32 < \alpha_{Co}^{Zn} < 1.72$, as shown in Table 1. On the other hand, the capacity of the resin for both metals increases with temperature. The increase for $Zn^{2+}$ is more significant than that for $Co^{2+}$. The influence of temperature on $\alpha_{Co}^{Zn}$ is far weaker

**Table 1** *Temperature dependency of equilibrium separation factor for Co-Zn exchange.*

| $T, K$ | 283 | 293 | 298 | 303 | 313 | 323 | 333 | 353 |
|---|---|---|---|---|---|---|---|---|
| $\alpha_{Co}^{Zn}$ | 1.32 ± 0.04 | 1.39 ± 0.04 | 1.34 ± 0.04 | 1.35 ± 0.04 | 1.42 ± 0.04 | 1.51 ± 0.05 | 1.63 ± 0.05 | 1.72 ± 0.05 |

than that observed in $Ni^{2+}$ - $Cu^{2+}$ system, in line with the predictions made previously [17] (see above). For the sake of comparison with literature data, determined at 293 K, values of $\alpha$ ranging between 283 and 313 K have been used to estimate $\Delta H_{ap}$. This estimation gives $\Delta H_{ap}$ = 0.45 kJ/mol, which correlates with $\Delta(\Delta H)$ $_{ZnIDA-CoIDA}$= - 0.25 kJ/mol[19] (this value refers to 293 K) within the uncertainty on $\Delta H_{ap}$ determination (which exceeds 100% in this case). The $\alpha$ values do not vary significantly with temperature which indicates the constancy of stoichiometry of $Zn^{2+}$ and $Co^{2+}$ complexes formed in the resin phase.

### 3.3 Correlation between separation factor and stability constant of a single specie

Any IX reaction proceeding between $M_1$ and $M_2$ metal ions on an IDA resin (see, e.g., eq. 2) can be represented by a simultaneous combination of two independent equilibria describing complexation of a single ionic species by IDA resin. This allows one to connect the $\alpha_{M_1}^{M_2}$ value with the corresponding stability constants of $M_1$ and $M_2$ complexes as follows[18]:

$$\alpha_{M_1}^{M_2} = \frac{\beta_{R-IDAM_2}}{\beta_{R-IDAM_1}} \tag{6}$$

Stability constants of metal complexes in the resin phase, $\beta_{R-IDAM}$ are known to be proportional to $\beta_{IDAM}$ values of IDA ligand complexes in the solution phase.[24] Substantiation of this conclusion is demonstrated in Fig.2, where $\log\alpha$ for different ion couples are plotted against the respective $\Delta\log\beta_{IDAM}$ values taken from Ref.19. As seen, $\log\alpha$ vs $\Delta\log\beta$ dependence is satisfactorily approximated by a straight line. Note, that

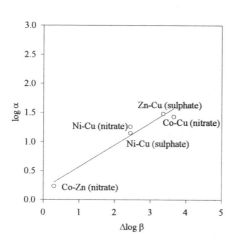

$\log\alpha$ values plotted in Fig.2 refer to T = 353 K, where the formation of 1:1 IDA-M complexes predominates (see above). The corresponding values of $\Delta\log\beta$ refer also to the same type of complexes. Note that plots of $\log\alpha$ determined at T < 353 K vs $\Delta\log\beta$ demonstrate a much higher deviation from the direct proportionality.

**Figure 2** *Correlation between $\log\alpha$ and $\Delta\log\beta$ for different metal ion couples.*

## Acknowledgement

This work was supported by the Research Grant No. EV5V-CT94-556 from the Commission of the European Communities, Programme Environment 1990-1994. Bayer Hispania Industrial, S.A., is gratefully acknowledged for kindly supplying with samples of Lewatit resin. A.G. is a recipient of a fellowship from CIRIT (Commission for Science and Technology of Catalunya). The Catalonian Government is acknowledged for providing financial support to D.M. during his Visiting Professorship at Autonomous University of Barcelona.

## Nomenclature

α   separation factor
β   formation constant
$C^*$   equilibrium concentration in solution phase
$q^*$   equilibrium concentration in resin phase
T   temperature

## References

1.  'Standard Handbook of Hazardous Waste Treatment and Disposal'; H.M. Freeman, Ed., McGraw-Hill, Inc., New York, 1989.
2.  J.E. Etzel, Dyi-Hwa Tseng., 'Metals Speciation, Separation and Recovery', J.W. Patterson, R, Passino, Eds., Lewis Publishers, Inc., Chelsea, 1987, p. 571.
3.  A.K. Sengupta, T. Roy, E. Millan, 'Physicochemical and Biological Detoxification of Hazardous Wastes',Y.C. Wu, Ed., Technomic Publishing Company, Inc., Lancaster, 1989, p.191.
4.  R. Khamizov, D. Muraviev, A. Warshawsky, 'Ion exchange and solvent extraction', J. Marinsky, Y. Marcus, Eds., Marcel Dekker Inc., New York, 1995, Vol.12, p.93.
5.  J.W. Patterson, 'Metals Speciation, Separation and Recovery', J.W. Patterson, R. Passino, Eds., Lewis Publishers, Inc., Chelsea, 1987, p.263.
6.  H.T. Chen, 'Handbook of Separation Technique for Chemical Engineers', P.A. Schweitzer, Ed., McGraw-Hill, New York, 1979, p.467.
7.  G. Grevillot, 'Handbook for Heat and Mass Transfer', N.P. Cheremisinoff, Ed., Gulf Publishing, West Orange, NJ, 1985, Chapter 36.
8.  D. Tondeur, G. Grevillot, 'Ion Exchange: Science and Technology', A.E. Rodrigues, Ed., NATO ASI Series 107, Martinus Nijhoff, Dordrecht, 1986, 369.
9.  P.D. Wankat, 'Percolation Processes, Theory and Applications, A.E. Rodrigues, D. Tondeur, Eds., Sijthoff and Noordhoff, Alphen aan den Rijm, 1978, p. 443.
10. V.I. Gorshkov, M.V. Ivanova, A.M. Kurbanov, V.A. Ivanov, *Vestn. Mosk. Univ., Ser. 2: Khim.*, 1977, **5**, 535 (Russian); *Moscow Univ. Chem. Bull. (Engl. Transl.)*, 1977, **32**, 23.
11. M. Bailly, D. Tondeur, *J. Chromatogr.*, 1980, **201**, 343.
12. M. Bailly, D. Tondeur, *J. Chem. Educ. Symp. Ser.*, 1978, **54**; 111.
13. O.D. Bonner, G. Dickel, H.Z. Brummer, *Phyzik. Chem.*, 1960, **25**, 81.
14. R.A. Glass, *J. Am. Chem. Soc.*, 1955, **77**, 807.
15. K.A. Kraus, R.J. Raridon, *J. Am. Chem. Soc.*, 1960, **82**, 3271.
16. A. Warshawsky, N. Kahana, *J. Am. Chem. Soc.*, 1982, **104**, 2663.

17. D. Muraviev, A. Gonzalo, M. Valiente, *Anal. Chem.*, 1995, **67**, 3028.
18. D. Muraviev, A. Gonzalo, M.J. Gonzalez, M. Valiente, 'Ion Exchange Developments and Applications', J.A. Greig, Ed., The Royal Society of Chemistry, Cambridge, 1996, p. 516.
19. G. Anderegg, *Helv. Chim. Acta,* 1964, **47**, 1801.
20. R. Biesuz, M. Pesavento, A. Gonzalo, M. Valiente, *Talanta,* 1998, accepted.
21. 'Handbook of Metal Ligand Heats and Related Thermodynamic Quantities', 3rd ed., J.J. Christensen, R.M. Izatt, Eds., Marcel Dekker Inc., New York, 1983.
22. H. Waki, 'Ion exchange and solvent extraction', J. Marinsky, Y. Marcus, Eds., Marcel Dekker Inc., New York, 1995, Vol.12, 197.
23. M. Pesavento, R. Biesuz, M. Gallorini, A. Profumo, *Anal. Chem.*, 1993, **65**, 2522.
24. V.I. Gorshkov, D.N. Muraviev, G.A. Medvedev, *Zh. Fiz. Khim.*, 1977, **51**, 2680.

# Analytical Methods

# DEVELOPMENTS IN RETENTION MODELLING IN ION-EXCHANGE AND ION-EXCLUSION CHROMATOGRAPHY

Paul R. Haddad, Kai Ling Ng and John E. Madden

Separation Science Group,
School of Chemistry,
University of Tasmania,
GPO Box 252-75,
Hobart, 7001 Australia

## 1 INTRODUCTION

The study of retention mechanisms in all forms of chromatography is of fundamental importance to the understanding of the chromatographic method and for elucidating the parameters which have most effect on the retention of analytes. When these parameters have been identified they can be employed to develop a desired separation in the most efficient manner. Retention models which provide a mathematical relationship between the retention factor (k') and various properties of the eluent, the stationary phase and the analyte, are of particular importance if the development of a desired separation is to be undertaken using some type of computer optimisation software. One common approach to optimisation, namely interpretive optimisation, uses a retention model to predict retention data over a nominated experimental space and to then use these data to estimate the quality of all possible chromatograms which can be acquired within the experimental space. The optimal conditions can then be identified.

Because computer optimisation is likely to assume increasing importance in the future, sound and reliable retention models are also necessary. In this paper, some retention models for ion-exchange ion chromatography (IC) of inorganic anions and ion-exclusion ion chromatography (IEC) of carboxylic acids are examined.

## 2 RETENTION MODELLING IN ION-EXCHANGE IC

### 2.1 Background

Two types of IC retention models can be identified, namely theoretical and empirical. A theoretical model is derived totally from theory and invariably requires knowledge (or estimation) of a range of parameters relating to the analyte, the stationary phase, and the eluent before calculation of the retention factor of the analyte is possible. These parameters are usually estimated by performing some preliminary experiments in which the retention factors of desired analytes are measured under controlled conditions. Once the relevant parameters are known, a theoretical model enables, in principle, the calculation of retention factors for all possible eluents and stationary phases. However, in practice it is more common for the relevant parameters to be determined only for a single stationary phase and a single type of eluent, and to then use the retention model to predict the effects of varying the composition (typically the concentration of competing ion and the pH) of the eluent. Empirical models concentrate on predicting the manner in which retention changes when the eluent composition is varied between two or more known values. That is, an empirical model is concerned with the observed effects of changes in eluent composition, rather than the underlying theoretical explanation for these changes. Application of an empirical model usually begins with the measurement of retention for the analytes of interest using eluents of known composition, followed by interpolation of retention behaviour at intermediate eluent compositions using the model.

Here, we examine two recent theoretical IC retention models from the literature, and one new empirical model. These models have been evaluated by reference to a set of retention data in order to assess their suitability for inclusion in an interpretive computer-assisted optimisation method.

## 2.2 IC Retention Models

The two theoretical retention models selected for evaluation take into account the fact that the eluent may contain a number of different eluting species each having its own ion-exchange selectivity coefficient, and both models also consider the fact that the form of the analyte might change as the pH of the eluent is varied. In the equations presented below, it is assumed that a phthalate eluent is used. Phthalic acid is denoted by $H_2P$, and the singly and doubly ionised forms by HP and P, respectively.

*2.1.1 Extended Dual Eluent Species Model.* This model is given by the following equation[1]:

$$k_A' = \left(10^{(a(e)+b)}\right)\left(\frac{[H^+]}{4K_{P,HP}K_{a_2}}\right)^e \left(\left(1 + \frac{8QK_{P,HP}K_{a_2}\left(1 + \frac{K_{a_2}}{[H^+]}\right)}{(P_T)[H^+]}\right)^{\frac{1}{2}} - 1\right)^e \tag{1}$$

where, $e = \alpha_1 + 2\alpha_2$ and

$$\alpha_1 = \frac{K_{a_1a}[H^+]}{K_{a_1a}K_{a_2a} + K_{a_1a}[H^+] + [H^+]^2}$$

$$\alpha_1 = \frac{K_{a_1a}K_{a_2a}}{K_{a_1a}K_{a_2a} + K_{a_1a}[H^+] + [H^+]^2}$$

$k_A'$ is the retention factor for analyte A, $K_{P,HP}$ is the ion-exchange selectivity coefficient between the designated species, $P_T$ is the total concentration of phthalate, $K_{a2}$ is the second ionisation constant for phthalic acid, Q is the ion-exchange capacity of the stationary phase, $K_{a1a}$ and $K_{a2a}$ are the first and second acid dissociation constants for the analyte, and a and b are empirical constants.

*2.1.2 Multiple Species Eluent/Analyte Model.* This model takes into consideration the interaction of all eluent species, including hydroxide, with all analyte species, as well as the effects of varying eluent pH. The model[2] is given by the following equation for the case of a triprotic analyte, $H_3A$:

$$k'_{A+HA+H_2A} = \frac{w}{V_m} \left\{ \begin{array}{l} K_{A,HP}\left(\dfrac{\sqrt{p^2+q}-p}{4K_{P,HP}\left[P^{2-}\right]}\right)^3 \Phi_A + K_{HA,HP}\left(\dfrac{\sqrt{p^2+q}-p}{4K_{P,HP}\left[P^{2-}\right]}\right)^2 \Phi_{HA} \\[3ex] + K_{H_2A,HP}\left(\dfrac{\sqrt{p^2+q}-p}{4K_{P,HP}\left[P^{2-}\right]}\right)\Phi_{H_2A} \end{array} \right\} \qquad (2)$$

where, $p = \left[HP^-\right] + K_{OH,HP}\left[OH^-\right]$

$q = 8K_{P,HP}Q\left[P^{2-}\right]$

k' is the retention factor, $\Phi_A$, $\Phi_{HA}$ and $\Phi_{H_2A}$ denote the partial molar fractions of deprotonated and partially protonated forms of the analyte $H_3A$, w is the weight of the stationary phase, $V_m$ is the volume of the mobile phase, and $K_{P,HP}$, $K_{OH,HP}$ and $K_{H2A,HP}$ are the ion-exchange selectivity coefficients between the designated species. For the case of a phthalate eluent at acidic pH, [OH] is so low that any contribution from hydroxide can be neglected, thus the term $K_{OH,HP}$ can be disregarded and p can be simplified to:

$$p = \left[HP^-\right] \qquad (3)$$

*2.1.3 Empirical End Points Model.* This model[3] assumes linearity between log k'$_A$ and log $[P^{2-}]$ and employs purely empirical measurement of the slope of this relationship. The following algorithm gives the model:

$$\log k'_A = \left(f_1 + f_2 P_T\right) + \left(f_3 + f_4 P_T\right)\log\left[P^{2-}\right] \qquad (4)$$

The four chromatographic constants, $f_{1-4}$ can be solved using a set of simultaneous equations which employ experimental retention data for known eluent conditions.

## 2.3 Experimental

The models were compared using a previously published data set[4] in which retention data had been obtained for a series of anions (fluoride, chloride, bromide, iodide, chlorate, bromate, iodate, nitrite, nitrate, sulfite, sulfate, thiosulfate, orthophosphate, acetate and oxalate) using phthalate eluents on three different columns, a Hamilton PRP-X100 polystyrenedivinylbenzene column $150 \times 4.6$ mm ID, a Vydac 302 IC 4.6 silica column $300 \times 4.6$ mm ID, and a Waters IC Pak A polymethacrylate column $50 \times 4.6$ mm ID. The eluents comprised sodium phthalate solutions at concentrations of 1.0, 2.0 and 4.0 mM, and pH values of 4.0, 5.0 and 6.0.

## 2.4 Results

The three models were solved for the experimental conditions used and were then employed to predict the retention factor for each analyte ion on each column using each eluent, for every measured data point in the database[4]. Correlation coefficients were then calculated for each analyte anion by comparing all calculated and observed retention factors. The results are summarised in Figure 1. In a similar manner, correlation coefficients were calculated for each stationary phase by considering the data for all analytes obtained with that stationary phase. These correlation coefficients are given in Table 1.

**Table 1** *Correlation Coefficients for each Stationary Phase Obtained using all Analytes*

| Column | Hamilton PRP-X100 | Vydac 302 IC 4.6 | Waters IC Pak A |
|---|---|---|---|
| Model | Correlation coeff. | Correlation coeff. | Correlation coeff. |
| Extended Dual Eluent Species Model | 0.82179 | 0.86994 | 0.95454 |
| Multiple Eluent-Analyte Species Model | 0.83703 | 0.92570 | 0.84768 |
| End Points Model | 0.99527 | 0.98399 | 0.99190 |

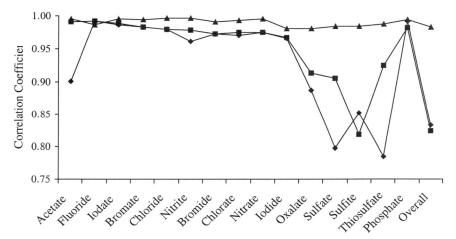

**Figure 1.** *Correlation Coefficients between the predicted and actual retention times given for each analyte separately and combined overall on the Hamilton PRP-X100, Vydac 302 IC 4.6 and Waters IC Pak A columns over the pH range 4.0 to 6.0, and eluent concentration range 1.0 to 4.0 mM.* ◆, *Extended Dual Eluent Species Model;* ■, *Multiple Species Eluent/Analyte Model;* ▲, *End Points Model.*

## 2.5 Discussion

The performance of the models can be summarised as follows:

(i)     The theoretical models gave reasonable performance for singly charged analytes, but showed erratic behaviour for analytes of higher charge. On the other hand, the End-Points model showed consistent performance for all analytes.

(ii)    The ability of a particular model to predict retention factors was found to be dependent to some extent on the type of stationary phase used, and in particular the type of material used to support the functional group. For singly charged analytes, the theoretical models gave negative errors on the polystyrenedivinylbenzene and silica-based stationary phases, whilst for the polymethacrylate stationary phase, positive errors were generally observed. Since the theoretical models consider only electrostatic effects leading to ion-exchange retention, the presence of other retention mechanisms (such as adsorption effects between the analyte and the unfunctionalised portions of the stationary phase) or the occurrence of factors influencing the ion-exchange process (such as steric effects between the analyte and the functional group on the ion-exchanger) will influence the predictive ability of the models.

(iii) Neither of the theoretical models gave reliable predictions of retention factors for all analytes, presumably due to the inability of such models to consider secondary contributions to analyte retention, such as those mentioned in (ii) above. On the other hand, the empirical approach exemplified by the End-points model was applicable to all analytes and stationary phases.

The End-points model offers ease of numerical solution, requires minimal input data and, at the same time, provides the most reliable prediction of retention factors. The only disadvantage of this method is that it requires the greatest number of initial experiments when a two-dimensional optimisation (i.e. variation of both pH and eluent concentration) is to be performed. A further advantage of the End-points model when used as a basis for optimisation is that its accuracy can be improved by iteration during the optimisation process.

## 3  RETENTION MODELLING IN ION-EXCLUSION CHROMATOGRAPHY

### 3.1 Background

Ion-exclusion chromatography (IEC) involves the separation of partially ionised species on strong anion or cation-exchange resins. In the case of carboxylic acids as analytes, separation is achieved on a cation-exchange stationary phase, either silica or polymer based, with chemically bound sulfonate or carboxylate functional groups. Ionic species like strong acids are completely excluded from the interior of the resin by the fixed anionic functional groups, in accordance with the Donnan exclusion effect. Therefore, these species are not retained and pass through the column with the eluent front. Partially ionised species like weak carboxylic acids (pKa = 2.5 - 6.5) permeate selectively into the stationary phase (the occluded liquid trapped within the pores of the resin), resulting in some retention of these species.

Many factors are known to affect retention in IEC[5], but it is generally assumed that the most significant of these is the electrostatic repulsion effect discussed above. A retention model based only on electrostatic effects is simple to derive but has been shown to be deficient for any analyte which exhibits adsorption effects onto the resin backbone of the stationary phase. We have derived a retention model which describes adsorption phenomena in a quantitative manner and shows the influence of two eluent parameters on retention, namely $[H^+]$ and % methanol.

### 3.2 Retention Model for IEC

In IEC, the retention mechanism can be considered to involve three phases: the mobile phase (m), the stationary phase (s), which is the occluded liquid trapped in the pores of the resin, and the resin phase (r). (See Figure 2).

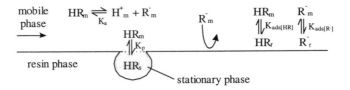

**Figure 2** *Schematic Diagram Showing the Major Equilibria Involved in IEC*

Writing expressions for the four equilibria in Figure 2 and incorporating the solvophobic equation[6] describing the variation of retention in reversed-phase liquid chromatography with changes in the mobile phase composition, we can derive the following retention model for IEC:

$$\ln k' = \ln\left(\frac{V_s + \left(K_{ads[HR]} + \frac{K_a}{[H^+]_m}K_{ads[R^-]}\right)V_r}{\left(1 + \frac{K_a}{[H^+]_m}\right)V_m}\right) - s\Phi \tag{5}$$

where $k'_A$ is the retention factor of the analyte; $V_m$ is the volume of mobile phase, $V_s$ is the volume of the stationary phase, $V_r$ is the volume of the resin phase, $K_a$ is the acid dissociation constant for the analyte, $[H^+]_m$ is the concentration of acid in the eluent, $K_{ads[HR]}$ is the adsorption coefficient for $[HR]_m$, $K_{ads[R^-]}$ is the adsorption coefficient for $[R^-]_m$, $s$ is the slope of the solvophobic equation, and $\Phi$ is the volume percentage of methanol present in eluent.

### 3.3 Experimental

Retention data were obtained for a series of carboxylic acids (formic, acetic, propionic, butyric, valeric, oxalic, malonic, succinic, glutaric, adipic, citric, benzoic, salicylic, phthalic, methanelsulfonic, heptanesulfonic and benzenesulfonic) on the following columns: Tosoh TSKGel SCX (Tokyo, Japan), 300 x 7.8 mm I.D., (a 5 μm polystyrene-divinylbenzene (PS-DVB) co-polymer, functionalised with sulfonate groups, capacity 4.2 mequiv/g), a Tosoh TSKGel SP-5PW (Tokyo, Japan), 300 x 7.8 mm I.D., (a 5 μm polymethacrylate co-polymer, functionalised with sulfonate groups, capacity 0.3 mequiv/mL), and a sulfonated silica column, 300 x 7.8 mm I.D., (packed with 5 μm Develosil silica (Nomura Chemical Co, Japan), home functionalised, capacity 0.275 mequiv/g). The mobile phases consisted of sulfuric acid (5 x $10^{-6}$ M to 5 x $10^{-4}$ M) and HPLC-grade methanol (0% to 20% v/v).

### 3.4 Results

The retention model shown in eqn (5) can be solved by non-linear regression using the retention data obtained experimentally. The adsorption coefficients can be determined in this way and values are shown in Table 2 for some analyte acids which showed adsorption effects. It should be noted here that these analytes showed much longer retention factors than those predicted from electrostatic effects alone, so that retention models which include only electrostatic effects would be inappropriate for these analytes.

**Table 2** *Adsorption Coefficients for some Carboxylic Acids*

| Analyte | TSKGel SCX | | TSKGel SP-5PW | | Sulfonated silica | |
|---|---|---|---|---|---|---|
| | $K_{ads[HR]}$ | $K_{ads[R^-]}$ | $K_{ads[HR]}$ | $K_{ads[R^-]}$ | $K_{ads[HR]}$ | $K_{ads[R^-]}$ |
| Butyric acid | 2.9 | 5.7 | 4.6 | 13.8 | 1.7 | 6.3 |
| Valeric acid | 6.7 | 8.3 | 10.6 | 17.5 | 3.1 | 7.7 |
| Benzoic acid | 106.3 | 24.7 | 121.3 | 28.6 | 4.0 | 3.7 |
| Salicylic acid | 161.2 | 6.5 | 406.1 | 10.5 | 4.2 | 0.9 |
| Phthalic acid | 29.7 | 0.9 | 46.8 | 3.5 | 1.5 | 0.5 |

The retention model can be used to predict the retention factors for each analyte under each of the measured mobile phase conditions, and the predicted retention factors can be compared to the observed values. Figure 3 shows the results obtained.

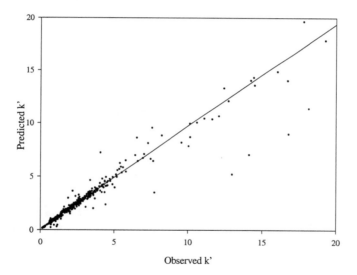

**Figure 3** *Correlation between predicted and observed retention factors for 13 carboxylic acids at 14 eluent conditions for TSKGel SCX and TSKGel SP-5PW columns and for 13 carboxylic acids at 9 eluent conditions for the sulfonated silica column. The correlation coefficient ($r^2$) is 0.9645, slope = 0.9673, intercept = 0.0631, and n = 479*

Finally, the retention model can be used to predict the retention factors for a selected group of analytes and to model their retention over a desired range of eluent compositions. Retention plots for acetic, succinic and benzoic acids are shown in Figure 4. From these plots, a response surface showing the quality of the separation (in this case, expressed using the normalised resolution product, r, for the chromatogram[7]) can be constructed and the maximum value of r used to find the optimal separation. Such a response surface is shown below in Figure 5 for a mixture of acetic, succinic and benzoic acids.

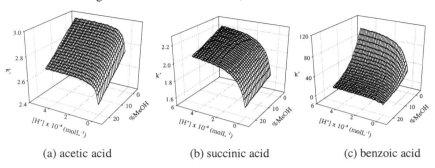

(a) acetic acid                    (b) succinic acid                    (c) benzoic acid

**Figure 4** *Predicted retention factor (k') responses of carboxylic acids on TSKGel SCX column as a function of % methanol and concentration of acid eluent.*

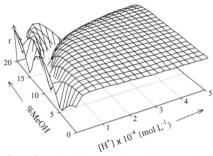

**Figure 5** *Response surface showing the variation of r over the parameter space of the eluent conditions. The parameters are the concentration of the acid eluent over the range $5 \times 10^{-4}$ to $5 \times 10^{-6}$ M and the amount of methanol present over the range 0 to 20%.*

### 3.5 Discussion

Changes in the eluent pH affect the degree of ionisation of the analytes, which in turn governs their electrostatic interaction with the ionic functional groups on the stationary phase. That is, eluent pH controls the electrostatic component of analyte retention. On the other hand, the amount of methanol present in the eluent governs the degree of hydrophobic adsorption of the analyte onto the unfunctionalised portion of the stationary phase material.

The values of the adsorption coefficients, $K_{ads[HR]}$ and $K_{ads[R^-]}$, for aliphatic carboxylic acids increased as the chain length of the acid increased from C1 to C5 and were larger for the monocarboxylic acids than for the corresponding dicarboxylic acid. For aromatic acids (see Table 2), the values of the adsorption coefficients were quite large on the polymeric resins due to strong adsorption effects, but were significantly lower on silica. For those analytes showing strong adsorption effects, the adsorption coefficient for the neutral analyte was much larger than that for the ionised eluent, which was expected due to the lower solvophobic effects for the ionised analyte. Finally, it was observed that the adsorption coefficients were generally highest on the polymethacrylate resin, followed by the PS-DVB resin and were lowest on silica.

## 4 CONCLUSIONS

Retention models in IC and IEC have been developed and have been shown to give reliable predictions of retention factors for analytes under a wide range of chromatographic conditions. In the case of IC, the complexity of the technique decrees that an empirical model gives the best performance. On the other hand, IEC can be modelled satisfactorily using a theoretical model since the two main influences on retention, namely electrostatic and adsorption effects, can be described mathematically in a relatively simple manner. The retention models discussed in this report have the potential to be incorporated into interpretive optimisation software for computer-assisted selection of the optimal separation conditions for a desired analysis. The development of such optimisation software is currently in progress.

## 5 ACKNOWLEDGEMENTS

We gratefully acknowledge financial support from the Dionex Corporation, Sunnyvale, USA, and the donation of columns for the IEC work by Dr. K. Tanaka, National Industrial Research Institute of Nagoya, Japan.

### References

1. D.R. Jenke, *Anal. Chem.*, 1994, **66**, 4476.
2. P. Hajos, O. Horvath and V. Denke, *Anal. Chem.*, 1995, **67**, 434.
3. P.R. Haddad and P.E. Jackson, *Ion Chromatography - Principles and Applications*; Elsevier: Amsterdam, 1991.
4. A.D. Sosimenko and P.R. Haddad, *J. Chromatogr.*, 1991, **546**, 37.
5. B.K. Glod, *Chemia Analityczna.* 1994, **39**, 399.
6. C. Horvath, W. Melander and I Molnar, *J. Chromatogr.* 1976, **125**, 129.
7. P.R. Haddad, A.C.J.H. Drouen, H.A.H. Billiet, and L de Galan, *J. Chromatogr.* 1983, **282**, 71.

# PHYSICOCHEMICAL CHARACTERIZATION OF METAL-ALGINATE GELS

**Jodra, Y. and \*Mijangos, F.**
*Department of Chemical Engineering, Science Faculty, University of the Basque Country, Apdo. 644, Bilbao 48080, Spain.*

## ABSTRACT

The gelling and physical properties of alginate gels depend on the conditions under which alginate gels are formed. A wide range of sodium alginate and calcium nitrate concentrations were used to study how some physicochemical properties of calcium alginate gels depend on the alginate and calcium concentration, and also on the ionic strength of the gelling solution. The kinetic of sol/gel transformation was a process controlled by the diffusion of metal ions into alginate spheres. It was described by the unreacted core model assuming an average particle radius, and by the modified unreacted core model allowing for the variation of the bead size.

## 1 INTRODUCTION

Alginates are linear polymers of 1,4-linked β-D-mannuronate (M) and α-L-guluronate (G) residues, which are arranged in homopolymeric blocks of each monomer, together with mixed blocks[1-4]. They are mainly extracted from the intercellular material of various brown algae, but also from the extracellular material of some bacteria. The composition and sequential arrangement of two uronic acids depends on the source from which they were extracted, so that different alginates might exhibit different physical and chemical properties.

Sodium alginate and other alginates of monovalent metals form highly viscous solutions in water, and these are used in various fields, such as textile industries and food industry[3]. One of the most important properties of these polysaccharides is the ability to form water insoluble gels. The interaction of divalent metal ions, like calcium or copper, with carboxyl groups of monomers, results in a formation of ionotrophic metal alginate gels, setting up the classical "egg-box" structure[1-4]. This property has led alginates to play an important commercial role, i.e., in biotechnology and biomedicine as an important medium for immobilizating enzymes and cells[5,6], in the pharmaceutical industry, as phase-stabilizers in foods and beverages[3], in agriculture and recently as immobilization matrices of some adsorbents[7]. The ion exchange properties of alginates have also attracted considerable attention[8-16].

The gelling and physical properties of alginate gel beads depend on their composition (M/G ratio), sequential structure and molecular size[1-3,5,17,18], as well as on the alginate and divalent metal concentration used [5,17-21]. The conditions under which alginate beads are produced (pH and temperature) also have influence on the physical properties[18]. Careful selection of the starting material and experimental conditions is necessary to obtain the best results. In the present work, a wide range of concentrations of sodium alginate and calcium nitrate solutions were used to study how the some physicochemical properties of calcium alginate gel beads depend on the concentration of the polymer and calcium ions, and the ionic strength of the gelling solution, which was controlled adding sodium nitrate in the calcium solution. From the experimental results obtained in this analysis, the "optimum" conditions for preparing spherical calcium alginate gels were chosen in order to use these gels as ion exchanger materials.

For practical interest, we have also studied the gelation rate by following the diffusion process of calcium and copper ions (separately) into sodium alginate spheres. A physical model was applied for the estimation of diffusion coefficients of each divalent metal ion.

## 2 EXPERIMENTAL

### 2.1 Materials

Sodium alginate solutions (2, 3, and 4% w/v) were prepared by dissolving a known mass of finely powdered sodium alginate (Carlo Erba, pharmaceutical grade) in a known volume of ultrapure water at room temperature. The solutions were homogenised to ensure good mixing. Calcium nitrate (0.02, 0.04-0.05, and 0.08 M) and sodium nitrate (0.5 and 1.0 M) solutions were also prepared by dissolving the commercial salts in ultrapure water.

### 2.2 Preparation of calcium alginate gels

Calcium alginate gels were prepared by using the device described in a previous work[11,16]. Sodium alginate solution is delivered by a peristaltic pump and extruded through a plastic pipette, dropping it into the magnetically stirred calcium nitrate solution at room temperature. The drop of sodium alginate sol becomes immediately a translucent, semi-rigid calcium alginate gel upon contacting the calcium solution.

### 2.3 Characterization of calcium alginate gels

After 24 hours, time enough for complete penetration of calcium ions [16], calcium alginate gels were taken out of the gelling solution and washed with ultrapure water. The following parameters were determined:

*a) Weight reduction.* The weight reduction was calculated as the difference in weight of sodium alginate drops before gelling and after overnight in calcium solution [17].

*b) Equilibrium water fraction (EWF).* The equilibrium water fraction of each calcium alginate gel refers to the weight of water in gel per total weight of gel. It was calculated after drying a known number of gels at 60°C [22].

*c) External appearance.* Calcium alginate gels were observed by microscopy and some photographs were also taken using a microscope (Nikon SE) and a magnifying glass (Nikon SMZ-2T), both provided with a photographic camera (Nikon FX-35A) and an automatic control mechanism (Nikon HFX-II).

## 2.4  Gelation experiments

The gelation rate of sodium alginate by calcium ions was studied by a discontinuous batch experiment at room temperature and neutral pH. Using the experimental setup used for the preparation of calcium alginate gels, five drops of sodium alginate sol 3% (w/v) were poured into 50 ml (or 100 ml) of 0.05M calcium nitrate solution. Time zero was taken as that when the third drop fell into calcium solution. The beads remained suspended in the stirred gelling solution for a known period of time. They were then removed from the calcium solution, washed with ultrapure water and slightly dried with filter paper. The gel diameter at each time was measured by a digital micrometer.

To determine the amount of calcium in the gels after a period of time, the beads were dissolved in concentrated boiling nitric acid solution and calcium concentration was measured by atomic absorption spectrophotometry (Perkin Elmer 1100B) at 422.7nm and air-acetylene flame. These steps were repeated for different periods of time in order to estimate the diffusion of calcium into the gels.

The gelation of sodium alginate by copper ions was also studied, using the same device, conditions and experimental method. The sodium alginate solution drops were directly dispersed in a 0.05M copper sulphate solution.

## 3  RESULTS AND DISCUSSION

### 3.1  Weight reduction in calcium alginate gels

As Figure 1 shows, at any sodium nitrate concentration the weight reduction of calcium alginate gels goes up with increasing concentration difference between calcium solution and the gel[17]. This weight reduction is due to a variation of the ionic charge within the gel. As the crosslinking increases, the number of ionized groups converted to the unionized form in the polymer also increases. The electrostatic repulsions between charged groups are thereby reduced and also the free volume available for  diffusion[17,23-25]. It can be seen in Figure 2, for the gelation of 3% (w/v) sodium alginate in 0.05M calcium solution, that the increase of calcium ions in the gels with time results in a shrinkage of the beads that could be interpreted as weight reduction. The shrinking proceeds rapidly during the first few seconds and reaches an equilibrium state in approximately 5 seconds. After that period of time the size does not change.

For higher calcium concentrations than 0.08M, the weight reduction seems to be approximately constant as Martinsen observed[17]. Calcium ions will be present in sufficient amount to counterbalance the osmotic pressure in the gel and to saturate the binding sites and, as a result, physical changes in the gel system will be smaller.

### 3.2  Equilibrium water fraction in calcium alginate gels

The shrinkage of alginate gels produces a water transference from the gel to the external solution to achieve the osmotic equilibrium, so that the water content inside the gels decreases[17]. Figure 3 shows the equilibrium water fraction (EWF) with respect to the initial water fraction of the sodium alginate drop (EWF$_o$), i.e., 0.98, 0.97 and 0.96 for 2%, 3% and 4% (w/v) sodium alginate solutions, respectively. As expected, it decreases with the increasing of the calcium solution concentration, being a little higher for the smaller sodium alginate concentration. In this figure it can also be noted that all the values are higher than 0.89, which means that the water content is higher than 89%.

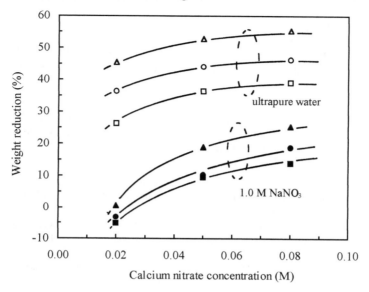

**Figure 1** *Weight reduction versus the calcium nitrate concentration for gelation of (▲,△) 2%, (●,○) 3%, and (■,□) 4% (w/v) sodium alginate. Open symbols: gelification without $NaNO_3$ in calcium solution, solid symbols: gelification with 1.0M $NaNO_3$.*

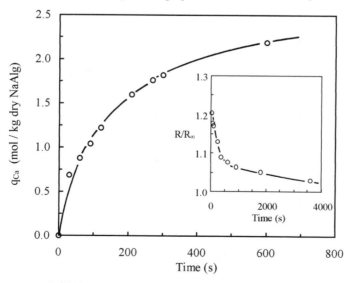

**Figure 2** *Gelation of 3% (w/v) sodium alginate by 0.05M calcium ions at room temperature.*

The volume fraction of polymer ($\phi$) in the beads is also affected by the shrinking of alginate gels as Figure 4 shows[17,25]. It can be related to EWF by the following Equation[22],

$$\phi = \frac{1 - EWF}{1 + EWF(\rho_p - 1)} \tag{1}$$

where $\rho_p$ is the polymer density relative to water. Taking into account the elevated water content of gels, it could be assumed that $\rho_p = 1$, and hence Equation (1) becomes Equation (2),

$$\phi = 1 - EWF \tag{2}$$

which was used to estimate the values of $\phi$ in Figure 4. This parameter reflects the weight of completely dried gels per the total mass of those gels, and differs from the fraction of dry sodium alginate ($\phi_{NaAlg}$) obtained in Equation (3).

$$\phi_{NaAlg} = M_{NaAlg} / w_g \tag{3}$$

The relationship between $\phi$ and $\phi_{NaAlg}$ illustrated in Figure 5, gives a straight line for each sodium alginate concentration. In absence of sodium nitrate in the gelling solution, the slope of the lines is near to one and, so, $\phi$ and $\phi_{NaAlg}$ agree. But at 1M sodium nitrate concentration $\phi > \phi_{NaAlg}$. The difference beween these parameters has been related to the amount of free electrolyte (ionic fraction) that has penetrated into the gel during the sol-gel transformation process, and is represented in Figure 6 *versus* the ionic strength of the gelling solution calculated by the Debye-Hückel equation. In this figure, the ionic fraction is nearly zero when the sodium nitrate concentration in the gelling solution is zero, but increases with the increase in the sodium nitrate concentration.

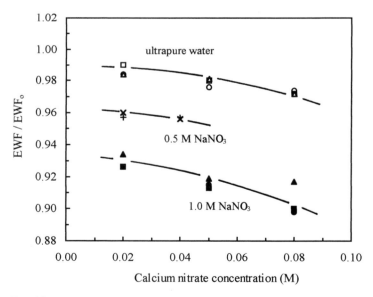

**Figure 3** *Equilibrium water fraction respect to the initial water fraction of alginate drops versus the calcium concentration for gelation of ($\triangle$,+,$\triangle$) 2%, ($\bullet$,×,○) 3%, and ($\blacksquare$,□) 4% (w/v) sodium alginate.*

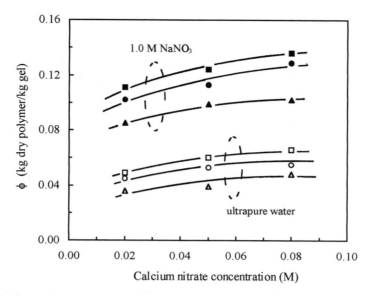

**Figure 4** *Volume fraction of polymer versus the calcium nitrate concentration (symbols as in Figure 1).*

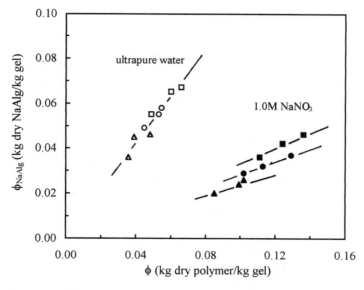

**Figure 5** *Fraction of dry sodium alginate versus volume fraction of polymer (symbols as in Figure 1).*

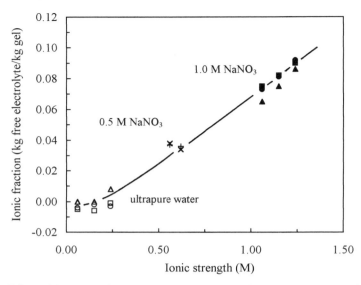

**Figure 6** *Effect of the external solution ionic strength on the ionic fraction (symbols as in Figure 3).*

### 3.3 Effect of the ionic strength on physicochemical properties of calcium alginate gels

In previous figures it was observed that the physicochemical properties of calcium alginate depend also on the sodium nitrate concentration, i.e., the ionic strength. As Figures 7 and 8 show, with the increasing of the ionic strength of the gelling solution, the water content decreases as the volume fraction of the polymer increases. This behaviour at high ionic strengths is due to the water transference that takes place from the gel to the external solution in order to achieve an osmotic equilibrium.

### 3.4 External appearance of calcium alginate gels

All the calcium alginate gels formed under different experimental conditions were observed by microscopy and with a magnifying glass. Ionic strength also affected the shape of calcium alginate gels. The results indicate that, in calcium solution with 1M NaNO$_3$, only spherical gels were prepared by dropping 4% (w/v) sodium alginate in 0.05 and 0.08M calcium nitrate solutions. The photographs in Figure 9 illustrate calcium alginate beads prepared from 3% (w/v) sodium alginate solution. They were translucent, quite soft, and, with increasing calcium concentration, they become more rigid and white opaque due to the penetration of calcium ions.

The gels prepared by 2% (w/v) sodium alginate in calcium solution, with 0.5M sodium nitrate, were not spherical either, whereas the ones prepared by 3% (w/v) sodium alginate were. However, all the gels formed in absence of sodium nitrate in the calcium solution were spherical, white/opaque and rigid, for any sodium alginate and calcium nitrate concentration. Figure 10 shows a photograph of a calcium alginate bead formed by 3% (w/v) sodium alginate in 0.05M calcium nitrate solution.

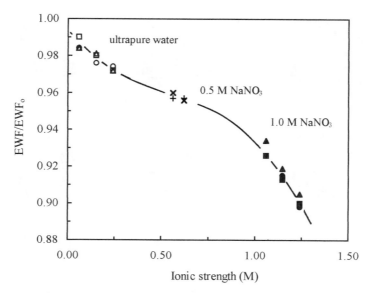

**Figure 7**  *Effect of the ionic strength of the gelling solution on the equilibrium water fraction (symbols as Figure 3).*

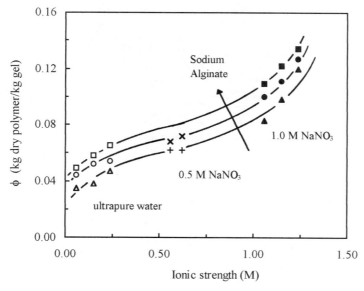

**Figure 8**  *Effect of the ionic strength of the gelling solution on the volume fraction of polymer (symbols as in Figure 3).*

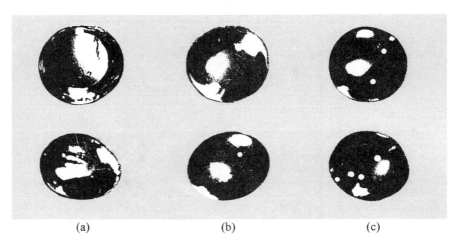

(a) (b) (c)

**Figure 9** *Photographs of calcium alginate gels prepared by 3% (w/v) sodium alginate in (a) 0.02M, (b) 0.05M and (c) 0.08M calcium nitrate solution with 1M NaNO₃.*

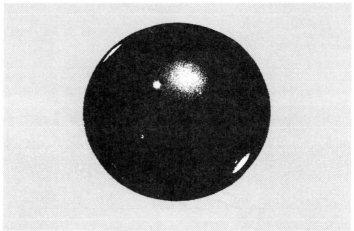

**Figure 10** *A spherical calcium alginate gel prepared by 3% (w/v) sodium alginate in 0.05M calcium nitrate solution (without NaNO₃).*

### 3.5 Gelation of sodium alginate by calcium and copper ions

As soon as 3% (w/v) sodium alginate sol was dispersed into the 0.05M calcium nitrate solution (without NaNO₃) spherical beads were formed. A primary membrane will be formed on the surface of the sol which will separate the sol from the surrounding electrolyte. The metal ions diffuse through this membrane into the alginate sol and, simultaneously, sodium ions start to diffuse through the membrane out of the alginate[19]. In Figure 3 we have already shown the increase of the amount of calcium ions in gels ($q_{Ca}$) and the shrinking of beads during the gelation process. After 24 hours, the calcium concentration in the beads was 2.347 mol Ca/kg dry sodium alginate. The change of the gel volume with time is represented in the same figure. As the amount of calcium ions increases inside the gels, the degree of crosslinking will increase, and shrinkage of the beads will occur[17,23]. The diameter of calcium alginate spheres was 3.62 mm at t=30 s and 2.97 mm at the end of the experiment, so the the volume reduction was 42%.

When sodium alginate drops came in contact with copper ions in the bulk solution copper alginate gels are formed. At the end of the experiment 3.393 mol Cu/kg dry sodium alginate were diffused into the gels. Figure 11 shows the increase of the metal concentration in the gels during the sol/gel transformation. The maximum amount of copper exchanged is greater than the amount of calcium.

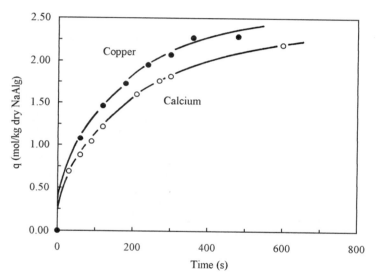

**Figure 11** *Variation of calcium and copper ions concentration in 3% (w/v) sodium alginate drops during the sol/gel transformation.*

### 3.6 Estimation of effective diffusion coefficients

The *unreacted core model* (UCM) was used for describing sol/gel transformation. This model has been used for ion exchange processes accompanied by reactions[27] and some ion exchange processes[11,27,28]. The application of this model for gelation kinetic analysis conforms to the results of Potter et al.[21] who visualized the displacement of the reaction front (sol/gel interface) towards the centre of the gel using magnetic resonance imaging. Lin[19] also considered a moving gelling front which moved towards the centre of a sphere.

Sodium alginate gelation involves simultaneously exchange diffusion of metals ions with chemical reaction at the fixed ionic groups. The reaction appears to proceed faster than the diffusion of divalent ions into the gel (calcium ions react rapidly with alginate to form a gel state) and hence the gelation seems to be controlled by the rate of diffusion of the metal ions through the gel network[19,21]. For particle diffusion control, experimental data can be correlated by the following expression:

$$f(X_i) = 1 - 3(1 - X_i)^{2/3} + 2(1 - X_i) = \frac{6C_i D_{eff,i}}{q_i^* \rho_i R^2} t \qquad (4)$$

where $X_i$ is the fractional attainment of equilibrium of $i$ species at any instant of time during the experiment, and $C_i$ is the concentration of $i$ species in the gelation solution which remains constant during gelation (=0.05M). A plot of $F(X_i)$ versus $t$ will give a straight-line relationship and the effective diffusivity of $i$ species ($D_{eff,i}$) could be obtained from the slope:

$$D_{eff,i} = \frac{[slope]\, q_i^* \rho R^2}{6C_i} \qquad (5)$$

Here the alginate gel density $\rho$ was calculated according to the Equation (6),

$$\rho = M_{NaAlg} \left/ \left( \frac{4}{3} \pi R^3 N \right) \right. \qquad (6)$$

where $M_{NaAlg}$ is the total mass of dry Na-alginate and $N$ is the number of beads[11,26]. The values of the parameters in Equation (4) and the effective diffusivities determined are listed in Table 1, and Figure 12 shows how the experimental points fit to the theoretical curve by the assumption of an averaged radius in the Equation (3) and (5)[11].

However, it was experimentally observed that alginate gels shrink throughtout the gelation. This increases the polymer concentration in the gel and this will affect the diffusional properties of the polymer[17,30,31]. To take into account this change, a *modified unreacted core model* was considered allowing for the variation of the bead size with reaction[29,32]. According to this model, when diffusion through the reacted layer is controlling, the corresponding Equation for $X_i$ vs $t$ is

$$f(X_i) = 3\left\{ \frac{z_v - [z_v + (1 - z_v)(1 - X_i)]^{2/3}}{z_v - 1} - (1 - X_i)^{2/3} \right\} = \frac{6C_i D_{eff,i}}{q_i^* \rho R_o^2} t \qquad (7)$$

where $z_v = (R_\infty / R_o)^3$. It is seen that if $z_v$ is lower than one, the beads shrink with time, whereas if $z_v$ is higher than one, there is an increase in particle size during the course of reaction. When $z_v$ is equal to one, the Equation (7) becomes Equation (4). The values of the effective diffusion coefficients obtained by applying this model, which are also included in Table 1, are smaller than the ones estimated considering an average radius. The good fit to this model is shown in Figure 12.

The rate process can be described with a constant diffusion coefficient for an averaged particle radius or by considering the variation of $R$ in time. Both models fit the experimental data correctly but the values of $D_{eff,i}$ obtained by Equation (7) are smaller than those obtained by Equation (4). The first will be more suitable according to the characteristics of the sol/gel transformation. It is noticed that the effective diffusivity for calcium is greater than for copper.

**Table 1** *Effective diffusion coefficients of calcium and copper in alginate beads from UCM and modified UCM (particle diffusion control).*

| Unreacted Core Model | Calcium | Copper |
|---|---|---|
| Slope (s⁻¹) | $1.085\times10^{-3}$ | $6.098\times10^{-4}$ |
| $q^*$ (mol/kg dry NaAlg) | 2.347 | 3.393 |
| $R$ (m) [a] | $1.65\times10^{-3}$ | $1.65\times10^{-3}$ |
| $\rho$ (kg dry NaAlg/L wet bead) [b] | $3.895\times10^{-2}$ | $3.387\times10^{-2}$ |
| $D_{eff}$ (m²/s) | $8.98\times10^{-10}$ | $6.35\times10^{-10}$ |
| Modified Unreacted Core Model | Calcium | Copper |
| Slope (s⁻¹) | $7.749\times10^{-4}$ | $3.986\times10^{-4}$ |
| $z_v$ | 0.55 | 0.55 |
| $D_{eff}$ (m²/s) | $7.75\times10^{-10}$ | $5.01\times10^{-10}$ |

[a] Average bead radius. [b] Average polymer density.

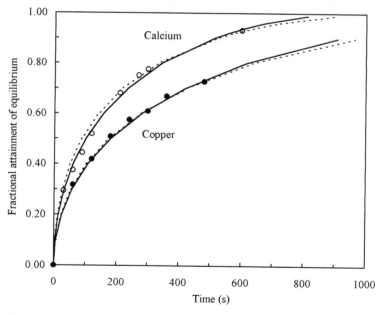

**Figure 12** *Gelation rate of calcium alginate and copper alginate beads. The fitting of the unreacted core model with particle diffusion control for an average radius (solid line), and for considering size variation (dashed line).*

## 4 CONCLUSIONS

Calcium alginate gels shrink during the sol/gel transformation. This yields changes in bead weight, water and polymer volume fraction. This behaviour is due to a variation of the ionic charge within the gel, the weight being a function of the ratio of calcium to sodium alginate concentrations.

At high ionic strength, sodium and calcium ions compete for the polymer binding sites. In this salt medium the weight reduction becomes higher and, hence, the volume fraction of polymer increases. In such ionic conditions the external appearance of the gels were non-spherical in shape.

The gelation kinetics of sodium alginate (3% w/v) with calcium or copper solutions (0.05M) are both particle diffusion controlled processes. Experimental data were fitted to the *Unreacted Core Model* assuming an averaged bead radius, and to the *Modified Unreacted Core Model*, allowing for variation of the bead size with reaction. The effective diffusion coefficients determined applying these models were $8.98 \times 10^{-10}$ and $7.75 \times 10^{-10}$ m$^2$/s for calcium, and $6.35 \times 10^{-10}$ and $5.01 \times 10^{-10}$ m$^2$/s for copper, respectively.

## 5 NOMENCLATURE

| | |
|---|---|
| $C_i$ | $i$ species concentration in solution (M) |
| $D_{eff,i}$ | $i$ species effective diffusion coefficient (m$^2$/s) |
| $f(X_i)$ | function defined in Equations (4) and (7) |
| $M_{NaAlg}$ | the amount of dry sodium alginate per gel (kg dry NaAlg/gel) |
| $N$ | number of beads |
| $q_i$ | $i$ species load (mol/kg dry NaAlg) |
| $q_i^*$ | $i$ species equilibrium load (mol/kg dry NaAlg) |
| $R$ | particle radius (m) |
| $R_o$ | initial particle radius (m) |
| $R_\infty$ | final particle radius (m) |
| $t$ | time (s) |
| $w_g$ | weight of wet calcium alginate gels (kg) |
| $X_i$ | $i$ species fractional attainment of equilibrium ($= q_i / q_i^*$) |
| $z_v$ | molar volume ratio of product to reactant ($= R_\infty^3 / R_o^3$) |
| $\rho$ | particle density (kg dry NaAlg/L wet bead) |
| $\rho_p$ | polymer density relative to water |
| $\phi$ | fraction of polymer in gel |
| $\phi_{NaAlg}$ | fraction of dry sodium alginate in gel |

6 REFERENCES

1   R. Kohn, *Pure and Appl. Chem.*, 1975, **42**, 371.
2   D.A. Rees, E.J. Welsh, *Angew. Chem. Int. Ed. Engl.*, 1977, **16**, 214.
3   M. Hardman, *Water and Food Quality*, Elsevier, London, 1989.
4   R. J. Petroski, S. P. McCormick, *Secondary-Metabolite Biosynthesis and Metabolism*, Plenum Press, New York, 1992.
5   M. Kierstan, J. Reilly, *Biotechnol. Bioeng.*, 1982, **XXIV**, 1507.
6   B. Ruggeri, A. Gianetto, S. Sicardi, V. Specchia, *Chem. Eng. J.*, 1991, **46**, B21.
7   Yong-xiang Gu, Zhong-cheng Hu, R. A. Korus, *Chem. Eng. J.*, 1994, **54**, B1.
8   O. Smidsrod, A. Haug, *Acta Chem. Scand.*, 1968, **22**, 1989.
9   A. Haug, O. Smidsrod, *Acta Chem. Scand.*, 1970, **24**, 843.
10  O. Smidsrod, A. Haug, *Acta Chem. Scand.*, 1972, **26**, 2063.
11  L. K. Jang, W. Brand, M. Resong, W. Mainieri, *Environ. Prog.*, 1990, **9**, 269.
12  L.K.Jang, S.L.Lopez, S.L.Eastman, P.Pryfogle, *Biotechnol. Bioeng.*, 1991, **37**, 266.
13  W. Hartmeier, R. Schumacher, W. Gloy, R. Lassak, *Med. Fac. Landbouww. Univ. Gent.*, 1992, **57**, 1713.
14  S. A. El-Shatoury, R. M. Hassan, A. A. Said, *High Performance Polymers*, 1992, **4**, 173.
15  Y. Konishi, S. Asai, Y. Midoh, M. Oku, *Sep. Sci. and Tech.*, 1993, **28**, 1691.
16  F. Mijangos, Y. Jodra, *Afinidad*, 1995, **459**, 313.
17  A. Martinsen, G. Skjåk-Kræk, O. Smidsrod, *Biotechnol. Bioeng.*, 1989, **33**, 79.
18  A. Martinsen, I. Storrø, G. Skjåk-Kræk, *Biotechnol. Bioeng.*, 1992, **39**, 186.
19  S. H. Lin, *Chem. Eng. Sci.*, 1991, **46**, 651.
20  K.I. Draget, M.K. Simensen, E. Onsoyen, O. Smidsrod, *Hydrobiologia*, 1993, **260/261**, 563.
21  K. Potter, B.J. Balcom, T.A. Carpenter, L.D. Hall, *Carbohydr. Res.*, 1994, **257**, 117.
22  R.A. Siegel, B.A. Firestone, J. Cornejo, B. Schwarz, *Polymer Gels*, Plenum Press, New York, 1991, 309.
23  R.M. Hassan, A.M. Summan, M.K. Hassan, S.A. El-Shatoury, *Eur. Polym. J.*, 1989, **25**, 1209.
24  K. Dusek, *Advances in Polymer Science*, No. 110, Springer-Verlag, Berlin, 1993.
25  H. Deepak, N. A. Peppas, *Polymer*, 1996, **37**, 149.
26  L.K. Jang, N. Harpt, D. Grasmick, T. Uyen, G. G. Geesey, *J. Polym. Sci.*, 1989b, **27**, 1301.
27  M. G. Rao, A. K. Gupta, *Chem. Eng. J.*, 1982, **24**, 181.
28  D. Chen, Z. Lewandowski, F. Roe, P. Surapaneni, *Biotechnol. Bioeng.*, 1993, **41**, 755.
29  F. Mijangos, M. Diaz, *Can. J. of Chem. Engin.*, 1994, **72**, 1028.
30  A. H. Muhr, J. M. V. Blanshard, *Polymer*, 1982, **23**, 1012.
31  B.A. Westrin, A. Axelsson, *Biotechnol. Bioeng.*, 1991, **38**, 439.
32  L. K. Doraiswamy, M. M. Sharma, *Hetereogeneous Reactions*, Vol. 1, Wiley-Interscience, New York, 1984.

*E-mail iqpmianf@lgdx04.lg.ehu.es*

# NEW ASPECTS OF THE INTERACTION BETWEEN POLYCATION/DYE COMPLEXES AND POLYANIONS

Stela Dragan, Mariana Cristea, Demetra Dragan, A. Airinei and Luminita Ghimici

"Petru Poni" Institute of Macromolecular Chemistry, 6600 Iasi, Romania

## 1 INTRODUCTION

Specific and practical aspects related to dye-binding on polyions relate to the quantitative determination of small amounts of polyelectrolytes in water,[1,2] removal of the dyes from water,[3,4] improving the dyeability of fibers[5] and so on. The removal of the dyes from waste waters is still a difficult problem.

Polyion-dye-polyion assemblies could be a way to increase the efficiency of dye removal from waste waters especially in the case of dyes with a high solubility in water. Interpolyelectrolyte complexes (PECs) have been extensively studied,[6-10] but the parameters which influence the formation of the polyelectrolyte/small organic molecules/polyelectrolyte complexes have not been so well explored. In previous papers we reported on water insoluble PECs containing poly(N,N-dimethyl-2-hydroxypropylene ammonium chloride) (PC) with different degree of branching, as a cationic component, and poly(sodium acrylate) (NaPA)[11] or poly(2-acrylamido-2-methylpropane sulfonate) (NaPAMPS)[12] as an anionic component. We have also pursued the stoichiometry of complexes formed by the interaction of the above polycations with anionic dyes having two or three sulfonic groups.[13]

The purpose of this study is to investigate the stoichiometry and the stability of the polycation/dye/polyanion complexes obtained by the interaction between non stoichiometric PC/dye complexes and two polyanions with sulfonic groups, differing both in their structure and molecular weight.

## 2 EXPERIMENTAL

### 2.1 Materials

Polycations (PC$_1$ and PC$_2$, Scheme 1) were synthesized by condensation polymerization of epichlorohydrin with dimethylamine and N,N-dimethyl-1,3-diaminopropane and purified as shown before.[14] The intrinsic viscosities of these polymers in 1M NaCl at 25°C were measured: $[\eta]_{PC1} = 0.420$; $[\eta]_{PC2} = 0.355$ (dL/g). Poly(sodium styrenesulfonate) (NaPSS) and NaPAMPS, synthesized by the polymerization of monomers in aqueous solution in the presence of $(NH_4)_2S_2O_8/Na_2S_2O_5$ as initiator-activator, were used as polyanions. Polyanions were purified by dialysis against distilled water, concentrated in vacuum, precipitated with methanol and finally purified by reprecipitation from aqueous solution with methanol. The viscosity average molecular weights evaluated in 1M NaCl at 25°C, were: $M_v$ = 140,000 for NaPSS, $M_v$ = 168,000 for $NaPAMPS_{II}$ and $M_v$ = 1,400,000 for $NaPAMPS_{67}$. The concentration of all polyelectrolytes was expressed as moles of charged sites per litre. The concentration of polycations was confirmed by the potentiometric titration of Cl⁻ ions.

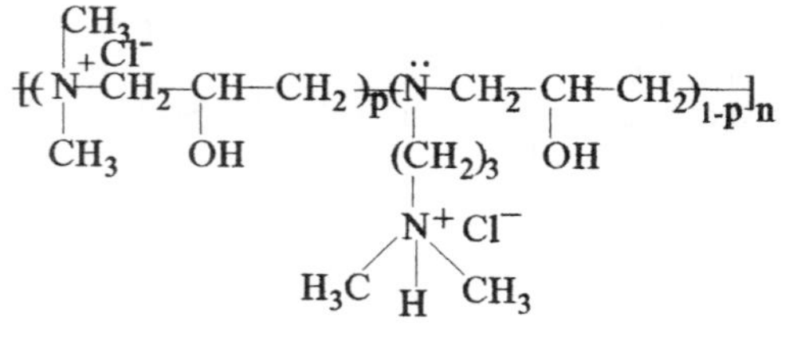

where: p = 0.95, polycation $PC_1$; p = 0.80, polycation $PC_2$

Scheme 1

The structures of the dyes used in this work are included in Scheme 2.

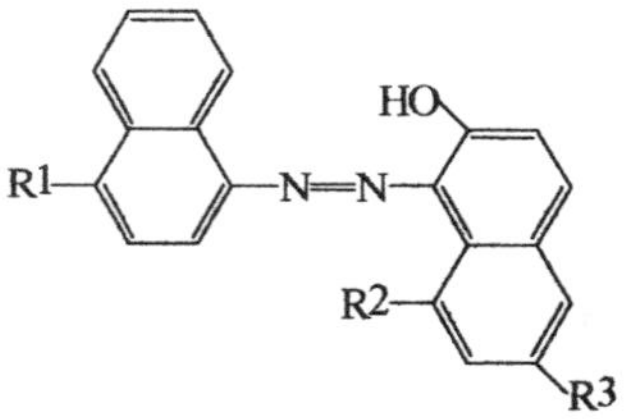

where:   $R^1$ = H, $R^2$ and $R^3$ = $SO_3Na$ -- Crystal Ponceau 6R (CP6R):
$R^1$, $R^2$ and $R^3$ = $SO_3Na$ -- Ponceau 4R (P4R)

Scheme 2

The dyes were dried at 70°C for 48 hours before preparation of the aqueous solutions.

## 2.2 Methods

Non stoichiometric polycation/dye complexes were prepared first, as follows: 25 mL of the dye aqueous solution was slowly dropped into 25 mL of the polycation aqueous solution having a concentration of $2 \times 10^{-3}$M, with magnetic stirring, at room temperature. The concentration of the dye solution was chosen to achieve a P/D molar ratio higher than 2.0 for PC/CP6R and higher than 2.5 for PC/P4R as already shown.[13] After 30 min of mixing, the aqueous solution of a polyanion was added dropwise with stirring, to give a final molar ratio [PA]/[PC] of 1.5. The mixing was continued for 2 hours, then the precipitates which formed were filtered off. The homogeneous solutions were analysed by UV-Vis spectrophotometry, viscometry and conductometry. All spectra were recorded using a SPECORD M 42 spectrophotometer (standard 1cm quartz cell). Viscosities were measured at 25°C using an Ubbelohde viscometer with internal dilution. The polymer concentration in a certain system, at a given moment, was calculated according to the following equation:

$$C = [(C_{PC} \times V_{PC} + C_{PA} \times V_{PA}) / (V_{PC} + V_D + V_{PA})] \times 10^2, \text{ g/dL}$$

where: $C_{PC}$ - polycation concentration, g/mL; $V_{PC}$ - volume of the polycation solution, mL; $C_{PA}$ - polyanion concentration, g/mL; $V_{PA}$ - volume of the polyanion solution, mL; $V_D$ - volume of the dye solution, mL.

Conductivities were measured in a specific cell with platinum/platinum electrodes (Radiometer Copenhagen, type CDM 2d).

### 3 RESULTS AND DISCUSSION

First information on the formation of the tricomponent complexes was obtained from the variation of the relative absorbance ($A_t/A_d$, where $A_t$ was the absorbance at 497 nm for P4R and at 513 nm for CP6R respectively, at different unit molar ratios [PA]/[PC] and $A_d$ was the absorbance at the same wavelengths, before the addition of polyanion) vs. [PA]/[PC]. The formation of $PC_1$/CP6R/PA complex when NaPAMPS was used as polyanion is shown in Figure 1.

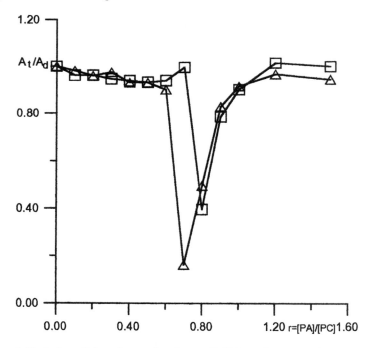

Figure 1 *Variation of the relative absorbance ($A_t/A_d$) vs. the unit molar ratio[PA]/[PC] in the formation of the $PC_1$/CP6R/NaPAMPS complexes, P/D=7.7: (□) NaPAMPS$_{II}$; (Δ) NaPAMPS$_{67}$.*

The increase of amount of polyanion brought about a very small decrease of the $A_t/A_d$, until the stoichiometric point was reached. This point was emphasized by an abrupt decrease of the $A_t/A_d$ at about 0.7 unit molar ratio for NaPAMPS$_{67}$ and at about 0.8 for NaPAMPS$_{II}$. It was accompanied by a phase separation. Further amounts of NaPAMPS increased the $A_t/A_d$ which indicated an increase in the free dye content in solution, caused by the interaction between PC/D/PA complex and the excess of polyanion. The $A_t/A_d$ was not much influenced by the molecular weight of NaPAMPS until near the endpoint, when a small increase was evident in the case of the polyanion with the low molecular weight; also after the endpoint the variation of the $A_t/A_d$ did not depend so much on the molecular weight of the polyanion, which indicated that the tricomponent complexes were stable enough in both systems.

The formation of the tricomponent complexes was also reflected in the decrease of the reduced viscosity until about zero, near the stoichiometric point, followed by an increase due to the excess of the polyanion present in the aqueous solution after the endpoint (Figure 2, NaPAMPS$_{67}$). As can be seen in Figure 2, the viscometric endpoint for NaPAMPS$_{67}$ was at about 0.7 unit molar ratio [PA]/[PC], as in the case of the spectroscopic endpoint (Figure 1). The conductometric endpoint was evident at about 0.9 unit molar ratio [PA]/[PC], irrespective of the molecular weight of NaPAMPS, which meant the migration of the low molecular weight species went on after the phase separation had taken place. The clear inflection point for both systems was also evidence for the stability of the tricomponent complexes.

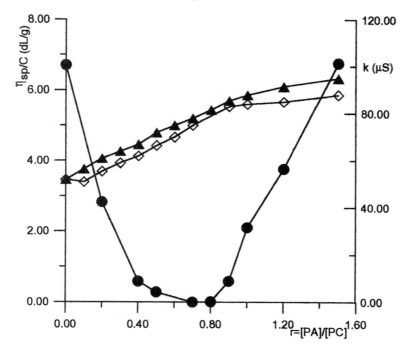

Figure 2 *Variation of the reduced viscosity ($\eta_{sp}/C$) and conductivity (k) vs. the unit molar ratio[PA]/[PC] in the formation of the PC$_1$/CP6R/NaPAMPS complexes for P/D=7.7: (●) $\eta_{sp}/C$, NaPAMPS$_{67}$; (▲) k, NaPAMPS$_{II}$; (◇) k, NaPAMPS$_{67}$.*

If we take into account the value of the $PC_1$/CP6R molar ratio used in the preparation of these complexes was 7.7, the composition of the polycation/dye complexes would be *1 mole charged sites of polycation for 0.13 moles of CP6R*. It was already mentioned that in the case of this dye both sulfonic groups were involved in the formation of the complex and this led to the following composition of the $PC_1$/CP6R complex: *1 mole charged sites of polycation for 0.26 moles $SO_3^-$ groups*. From this assumption the fraction of free dye which could exist in the aqueous solution of the $PC_1$/CP6R complex could be neglected. This ratio suggested a theoretical composition of the tricomponent complex of about *1 mole charged sites of polycation / 0.26 moles $SO_3^-$ groups / 0.74 moles of anionic sites of polyanion*. That the apparent stoichiometry of the $PC_1$/CP6R/NaPAMPS complex was close to that calculated one can be explained by the presence of the strong sulfonic groups and of the spacer with 4 atoms between $SO_3^-$ group and the main chain which leads to a fast, and total, compensation of the opposite charges.

Although the molecular weight of NaPSS was close to that corresponding to $NaPAMPS_{II}$ and the charged group was $SO_3^-$, the behaviour of this polyanion as a complementary polymer for the formation of tricomponent complexes with $PC_1$/CP6R showed some distinct features (Figures 3 and 4). As in the case of $NaPAMPS_{II}$ the $A_t/A_d$ decreased slowly by the addition of NaPSS and showed a small increase just before the minimum, the value of the unit molar ratio [PA]/[PC] corresponding to this moment being 0.7. The addition of the polyanion was followed by slow increases of the free dye content (Figure 3).

The variation of the $A_t/A_d$ by the interaction of NaPSS with a nonstoichiometric $PC_2$/CP6R complex was also included in Figure 3.

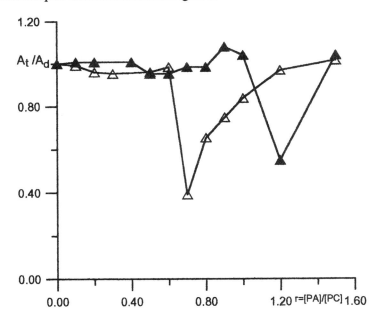

Figure 3 *Variation of the relative absorbance ($A_t/A_d$) vs. the unit molar ratio[PA]/[PC] in the formation of the PC/CP6R/NaPSS complexes, P/D=7.7: ($\triangle$) $PC_1$; ($\blacktriangle$) $PC_2$.*

The PC/D molar ratio used in the preparation of this dicomponent complex was about 7.7, as in the case of $PC_1$/CP6R. The influence of the branched structure of this polycation was noted first by the high increase of the $A_r/A_d$ just before the endpoint and then by the very high content of polyanion needed to achieve the minimum of the $A_r/A_d$ (unit molar ratio [PA]/[PC] at the endpoint was about 1.2).

The variation of the reduced viscosity, and of the conductivity, with the increase of NaPSS content, in the $PC_1$/CP6R/NaPSS system, confirmed a structure close to that predicted above, for the tricomponent complex (Figure 4). The variation of the reduced viscosity *vs.* the unit molar ratio [PA]/[PC] for the $PC_2$/CP6R/NaPSS system was different, suggesting an inclusion of an excess of polyanion in the complex, as was observed before for the formation of the interpolyelectrolyte complex of $PC_2$ with NaPA.[11] The slow decrease of the reduced viscosity until about 1.0 unit molar ratio [PA]/[PC] indicated the presence of soluble macromolecular species which could be mainly soluble nonstoichiometric tricomponent complexes (Figure 4). The continuous increase of the conductivity of this system with the increase of NaPSS content was also an argument for a non cooperative mechanism of the reaction in this system (Figure 4).

Formation of $PC_1$/CP6R/PA complexes with NaPAMPS and NaPSS suggested that the main driving force was the electrostatic interaction between the polyanion and the free positive charges of the nonstoichiometric $PC_1$/CP6R complex. The presence of the free dye in solution, only after the endpoint, showed the cooperative binding of this dye and the binding of the polyanion in an ordered structure. The bound dye was not disturbed until the stoichiometric point was reached.

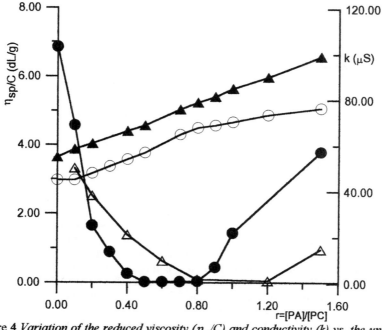

Figure 4 *Variation of the reduced viscosity ($\eta_{sp}/C$) and conductivity (k) vs. the unit molar ratio[PA]/[PC] in the formation of the PC/CP6R/NaPSS complexes for P/D=7.7: (●) $\eta_{sp}/C$, $PC_1$; (△) $\eta_{sp}/C$, $PC_2$; (○) k, $PC_1$; (▲) k, $PC_2$.*

As regard the influence of the dye structure on the formation and stability of the tricomponent complexes Figure 5 is very suggestive. The use of NaPAMPS$_{67}$ as polyanion in the formation of PC$_1$/P4R/PA led to a great instability of the tricomponent complex, just before the apparent stoichiometric point. The P/D molar ratio used in the preparation of the nonstoichiometric PC$_1$/P4R complex was about 4.0, that means a higher concentration of the dye was present in the initial aqueous solution. The apparent stoichiometry of these complexes was experimentally found to be 0.7 unit molar ratio [PA]/[PC] for NaPAMPS$_{67}$ (Figure 5). These results suggested that a smaller amount of the dye was included in the tricomponent complex. This situation could be explained by the exclusion of a dye fraction from the PC$_1$/P4R complex before the formation of the tricomponent complex, since P4R has a higher solubility in water compared with CP6R (10.0 wt. % compared with 1.9 wt. % in water, at 15°C).

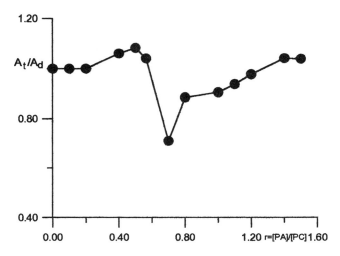

Figure 5 *Variation of the relative absorbance ($A_t/A_d$) vs. the unit molar ratio[PA]/[PC] in the formation of the PC$_1$/P4R/NaPAMPS$_{67}$ complex.*

## 4 CONCLUSIONS

Formation of PC$_1$/CP6R/PA complexes with polyanions with sulfonic groups suggests that the electrostatic interaction between the polyanion and the free positive charges of the nonstoichiometric complex is the main driving force. The presence of the free dye in solution only after the endpoint shows the cooperative binding of this dye and of the polyanion in ordered structure, which does not disturb the bound dye.

The exclusion of a high amount of the dye before the endpoint, when a branched polycation was used (PC$_2$), is the evidence for an irregular distribution of the dye in the PC$_2$/D complex and also for the difficulty in the formation of the tricomponent complex in the case of the polycation with a branched structure.

The solubility of the dye in water is another parameter which influences the formation and the stability of the PC/D/PA complexes. It was more difficult to include P4R in the tricomponent complex than CP6R.

## References

1. H. Terayama, *J. Polym. Sci.* 1952, **8**, 243.
2. K.-H. Wassmer, U. Schroeder, D. Horn, *Makromol. Chem.* 1991, **192**, 553.
3. T. Tashiro, *J. Appl. Polym. Sci.* 1990, **39**, 2279.
4. T. Takagishi, H. Kozuka, N. Kuroki, *J. Polym. Sci., Polym. Chem. Ed.* 1983, **21**, 447.
5. H. Oka, A. Kimura, *JSDC* 1995, **111**, 311.
6. B. Philipp, H. Dautzenberg, K. J. Linow, J. Kötz and W. Dawydoff, *Progr. Polym. Sci.*, 1989, **14**, 91.
7. V. A. Kabanov and A. B. Zezin, *Pure Appl. Chem.*, 1984, **56**, 343.
8. E. Tsuchida, *J. Macromol. Sci.- Pure Appl. Chem.*, 1994, **A31**, 1.
9. H. Dautzenberg, J. Hartmann, S. Grunewald and F. Brand, *Ber. Bunseges. Phys. Chem.*, 1996, **100**, 1024.
10. L. Webster, M. B. Huglin and I. D. Robb, *Polymer*, 1997, **38**, 1373.
11. S. Dragan, M. Cristea, C. Luca and B. C. Simionescu, *J. Polym. Sci., Part A: Polym. Chem.*, 1996, **34**, 3485.
12. M. Cristea, S. Dragan, D. Dragan and B. C. Simionescu, *Macromolecular Symposia*, 1997, **126**, 143.
13. S. Dragan, M. Cristea, A. Airinei and B. C. Simionescu, *Macromol. Rapid Commun.*, 1997, **18**, 541.
14. S. Dragan and L. Ghimici, *Angew. Makromol. Chem.*, 1991, **192**, 199.

# POLYGRAN SORBENTS FOR SEPARATION OF PROTEIN MIXTURES

G. Tishchenko,[1] K. Rozhetsky,[2] M. Bleha,[1] and J. Škvor[3]

[1] Institute of Macromolecular Chemistry, Academy of Sciences of the Czech Republic, Heyrovský Sq. 2, 162 06 Prague 6, Czech Republic
[2] Polygran Ltd., 99 Hahistadrut Ave, Mifraz Haifa, Israel
[3] Seva-Imuno Ltd., 267 18 Karlstejn, Czech Republic

## 1 INTRODUCTION

Among methods used in the separation and fractionation of biologically active compounds, in particular proteins, one of the leading positions is still occupied by ion exchange chromatography.[1] Experimental experience acquired during many years of application of the method in biotechnology allows us to predict optimum conditions of the separation of components of complex biological mixtures and to choose ion exchange supports possessing properties required for the purpose. However, in spite of the success achieved, there is a number of problems due to the slow kinetics of sorption and incomplete desorption of compounds sorbed from polymer supports. Solutions to these problems, and the overall effectiveness of the chromatography, depend predominantly on properties of the support, and this is why the search for new sorbents with improved chromatographic characteristics has to be continued.[2]

There exists a large variety of ion exchange supports for biochromatography, but attention concentrates only on those that are sufficiently hydrophilic, capable of sorbing proteins reversibly, possessing functional groups which have no hydrolytic effects upon biological molecules, which are sufficiently stable from the chemical and microbiological point of view and withstand drastic conditions of sanitation and regeneration with acid and alkali solutions as well as with various solvents.

Carboxylic sorbents produced by Polygran Ltd. (Israel) satisfy all these requirements. These are heterogeneous copolymers synthesized on the basis of methacrylic acid and are named in the paper as Polygran sorbent. The conditions of synthesis result in the preparation of a series of copolymers differing in porosity, pore size distribution as well as in hydrophobic-hydrophilic balance. Among those, the most interesting for the biochromatographic purposes are sufficiently porous copolymers (pore size 100-200 Å) which are "structure stable", i. e. their volume does not vary considerably with pH of the solution (4-9).

Polygran sorbents are a new generation of the well-known carboxylic heterogeneous Biocarb sorbents which have already shown their high capacity and reversibility of sorption in isolation of a number of proteins and enzymes from various biological sources: protease from the culture fluid of *Asp.terricola*,[3] fibrinolytic enzyme urokinase from urine,[4] neuraminidase enzyme from the filtrate of *Cholera vibrions*,[5] microbial enzymes (pectinase,[6] cellulase,[7] polygalacturonase,[8] levansaccharase,[9] protease from the culture fluid of *Bac.subtilis*[10]), enzymes and hormones of animal origin (acetylcholinesterase from the cobra poison,[11] a hormonal substance thymarine from the extract of thymus gland,[12] haemoglobin from the medulla homogenisate[13]).

Polygran sorbents differ from the well-known Biocarb sorbents in their higher porosity and structural stability necessary for the ion exchange chromatography of mouse specific immunoglobulins (Igs), i.e. proteins with molecular weight 180 000-900 000. It should be pointed out that ion exchange supports, such as DEAE and CM Sepharose Fast Flow,

Sartobind, Fracrogel and others are already widely employed in the isolation of specific Igs. On the other hand, processes involving these chromatographic supports have not been optimized (the yield and purity of the target product are not higher than 80-95% and 60-95%, respectively),[1] which stimulates further investigation in the field of synthesis.

The aim of this study consisted in an investigation of chromatographic properties of carboxylic heterogeneous Polygran sorbents in separation of $IgG_1$ possessing specificity with respect to horse-radish peroxidase from mouse ascitic fluid. The experiments were carried out using crude and desalinated ascitic fluid with the aim of evaluating the sorption of the accompanying components with Polygran sorbents and of checking the possibility of separation of specific $IgG_1$ from the crude ascitic fluid. Polygran sorbents obtained while varying the conditions of synthesis differed in their structure characteristics (pore size distribution, porosity, structure stability). The determination of correlation between the structure characteristics of sorbents, on the one hand, and their sorption properties towards the target and accompanying components of ascitic fluid on the other, was the main goal of the investigation, providing information needed for the optimization of the structure of Polygran sorbents.

## EXPERIMENTAL

### 2.1 Polygran Sorbents

The carboxylic heterogeneous Polygran sorbents are the product of POLYGRAN Ltd., Israel. Formation of the structure is defined by the conditions of synthesis. Changing the combinations of these parameters makes it possible to prepare the sorbents with variable hydrophilic-hydrophobic balance as well as structural parameters. These characteristics of Polygran sorbents are not yet published. The swelling of Polygran sorbents in water and phosphate buffers (pH 5-8) are shown in Table 1. The experiments were carried out using the sorbent fraction with particle size 200-300 μm in the swollen state after a preliminary regeneration with 1M HCl and NaOH solutions repeated three times. The sorbents were then equilibrated with 0.1M acetate buffer, pH 5.

**Table 1**  *Dependence of Swelling of Polygran Sorbents on pH of Phosphate Buffer*

| Sorbent No. | Specific Volume of Air-Dried Sorbent | Specific Volume of Swollen Sorbent[a] | Water Content in Swollen Sorbent[a] | Specific Volume of Swollen Sorbent in [H-Na]-Form at pH | | | |
|---|---|---|---|---|---|---|---|
| | | | | 5 | 6 | 7 | 8 |
| | *mL/g Dry* | *mL/g Dry* | *mL/g Dry* | *mL/g Dry* | | | |
| 26 | 3.93 | 11.8 | 7.87 | 12.60 | 27.54 | 29.12 | 29.90 |
| 27 | 2.38 | 7.85 | 5.47 | 9.51 | 14.30 | 17.61 | 17.61 |
| 36 | 4.45 | 9.79 | 5.34 | 14.26 | 16.03 | 18.70 | 18.70 |
| 37 | 3.83 | 9.20 | 5.37 | 12.27 | 18.40 | 22.25 | 20.70 |
| 43 | 3.82 | 8.40 | 4.58 | 10.70 | 12.22 | 18.33 | 18.33 |
| 44 | 3.10 | 4.34 | 1.24 | 4.95 | 6.18 | 6.18 | 6.18 |

[a]*In [H]-form.*

### 2.2 Mouse Ascitic Fluid Containing $IgG_1$ with Specificity against Horse-Radish Peroxidase

The characteristics of crude and desalinated ascitic fluids are summarized in Tables 2 and 3.

**Table 2**  *Characteristics of the Crude Ascites Fluid**

| PI of IgG₁ | pH | Total Proteins mL/g Dry | IgG₂ₐ,₂b,₃ mL/g Dry | IgG₁ mL/g Dry |
|---|---|---|---|---|
| 5.95-6.30 | 8.12 | 21.07 | 1.80 | 2.35 |

*Analysis of the crude ascites fluid was carried out by affinity chromatography on Prosep A (Bioprocessing Technologies Ltd., England)

**Table 3**  *Characteristics of the Desalted Ascites Fluid*

| pH | Colloid Diameter nm | κ mS/cm | Ammonium Sulfate mg/mL (mol/L) | Organic Substances Including Proteins mg/mL | IgG₁ mg/mL |
|---|---|---|---|---|---|
| 6.48-6.58 | $193 \pm 11$ | 4.322-5.924 | 3.40-4.62 (0.026-0.035) | 14.84 | 1.718 |

2.2.1  *Crude Ascitic Fluid.* Mouse ascitic fluid containing IgG₁ with specificity against horse-radish peroxidase was obtained from Seva-Imuno Ltd., Czech Republic. After being collected from the peritoneal region of mice, the ascitic fluid was filtered through special paper filters (597 1/2 and 602h 1/2, Schleicher & Schule, Germany) and Millipore filter, 0.22 μm. To prevent microbial growth, sodium azide (0.1%) and ε-aminocaproic acid (0.03%) were added to the protein solution. The solution was kept in a refrigerator at 4°C. A part of the crude ascitic fluid thus prepared was used in chromatographic experiments.

2.2.2  *Desalinated Ascitic Fluid.* The other part of crude ascitic fluid was free from some accompanying proteins by precipitation with ammonium sulfate followed by removing an excess of salt using dialysis through microporous anion exchange Neosepta membranes. Sodium azide and ε-aminocaproic acid were also added to a desalinated solution of ascitic fluid.

### 2.3 Chromatography of Crude and Desalinated Ascitic Fluid on Polygran Sorbents

Two series of experiments were carried out, one with desalinated and the other with crude ascitic fluid. Chromatographic experiments were performed in a column (0.78 × 3 cm) packed with 2 mL of the sorbent (H-form) and equilibrated with 0.1 M acetate buffer, pH 5. The chromatographic system consisted of a column, UV detector (Pharmacea, Sweden), peristaltic pump (Ismatec, Germany) and fraction collector (Camag, Germany). The loading of protein in the column was 1/5–1/10 of the static sorption capacity of Polygran sorbents to human serum albumin (3.2-1.6 mL). To raise the selectivity of sorption of high-molecular-weight proteins, the ascitic fluid was diluted with the same volume of distilled water and pH of the protein solution was adjusted to 5 prior to its introduction onto the column. Sorption and elution took place at a flow rate of 37 mL/h.cm². On completion, the column was washed with an equilibration buffer (5 column volumes). The proteins were then desorbed at a stepwise pH gradient within the range of 5-7.7 using 0.1 M citrate buffer. The elution was completed by adding 0.2 M NaCl to 0.1 M citrate buffer (pH 7.7) followed by 1/15M phosphate buffer (pH 8.2).

### 2.4 Methods

2.4.1 *Analysis of proteins.* Protein concentration in solution was measured spectro-photometrically. Absorption of eluate fractions was recorded at 260 and 280 nm with a Diode Array Spectrophotometer (Hewlett-Packard, USA), the concentration in solution was calculated using the Warburg and Christian method. Concentration of specific $IgG_1$ in eluate fractions was determined by ELISA titration. The eluate fractions containing the highest protein and $IgG_1$ concentrations were analyzed by SDS-PAGE performed by the discontinuous buffer method with a PhastSystem apparatus (Pharmacea Biotech, Sweden).

## 3 RESULTS AND DISCUSSION

As can be seen from results summarized in Table 1, carboxylic Polygran sorbents are considerably hydrated also in the unionized state (H-form); the water content varies between 1.24 and 7.87 mL/g depending on their structure. Quite usually, the swelling of sorbents in the phosphate buffer accompanied by H-Na exchange increases from 6.18 to 29.9 mL/g with increasing content of more hydrated Na ions in the ion exchangers. Along with the increasing ionization of carboxylic groups, fragments of the polymer network become unfolded or, in other words, development of "latent porosity" takes place. This is the reason why previous attempts at evaluation of the porosity of the first generation of similar sorbents (Biocarb) in the dry state failed.[14] An investigation of swollen sorbents by employing small-angle X-ray scattering and gel chromatography of dextrans of various molecular weights showed that the basic part of porosity is formed by pores having size 100-200 Å while reaching 300 Å for sorbents with a developed heterogeneous structure.[14]

The volume of the water phase in Polygran sorbents-26-43 in ionized and unionized forms is between 1.2-2.3 and 3.2-6.6-times that of the polymeric phase, respectively. Polygran sorbent-44 is an exception to the rule. Its water content is equal to the volume of swollen polymer phase only in the ionized form. Important information about the development of "latent porosity" can be obtained from the dependence of change in water proportion in the swollen polymer phase on relative swelling of the sorbents in phosphate buffers of various pH. The change of water content was calculated as the ratio of the water content in the (H-Na)-form of the sorbent and its specific volume in a given buffer. Relative swelling of the sorbent was found from the ratio of specific volumes of the sorbent in H-Na and H-forms. As can be seen from Figure 1, the water content of Polygran sorbents (H-form) swollen in water changes as follows: 0.7 (sample 27), 0.67 (sample 26), 0.58 (sample-37), 0.545 (sample-36, sample-43) and 0.29 (sample-44).

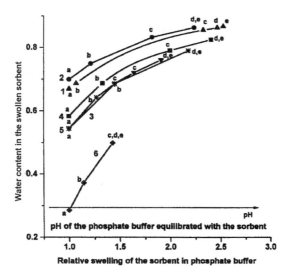

**Figure 1** Dependence of water content in Polygran sorbents on their relative swelling in phosphate buffers.
1 – sample-26, 2 – sample-27, 3 – sample-36, 4 – sample-37, 5 – sample-43, 6 – sample-44. a water, pH: b 5, c 6, d 7, e 8

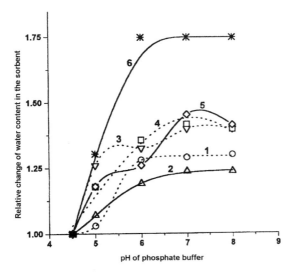

**Figure 2** Relative change of water content in Polygran sorbents on swelling in phosphate buffers.
1 – sample-26. 2 - sample-27. 3 - sample-36, 4 – sample-37, 5 – sample-43, 6 - sample-44

These results show that even in the unionized state, the sorbents have developed systems of pores. Ionization of carboxylic groups in Polygran sorbents takes place differently. For example, the highest and the lowest change of water contents were observed for sample-36 and sample-26, respectively, swollen in phosphate buffer, pH 5. At the same time, the latter shows a sharp change in water content when equilibrated with buffer of pH 6. It should be noted that the change of water content in sample-44 already achieves its maximum at pH 6 and further does not change. However, this sorbent demonstrates the highest change of water content on ionization of carboxylic groups (Figure 2). The phenomena observed are associated with the structure of polymer matrix and their swelling features under ionization. High water contents in Polygran sorbents ensure high sorption capacity of proteins differing in both isoelectric points and molecular weights. Moreover, Polygran sorbents with optimum chromatographic properties for separation of a wide range of bioactive substances can be synthesized by controlling the polymerization conditions.

Despite high values of the specific volume, a number of Polygran sorbents investigated in this study possess high structure stability within the pH range used (5-8). The relative swelling, i.e. the ratio of specific volumes of the sorbent at a given pH and pH 5 varies between 1.25 (sample-44) and 1.85 (sample-27) (Figure 2). Preliminary experiments involving sorption of human serum albumin from acetate buffer at pH 5 showed that the sorption capacity for protein increases in the following order: sample-27, 44, 26, 43, 36, 37.

Some typical chromatograms of separation of desalinated ascitic fluid components on carboxylic Polygran sorbents can be seen in Figures 3 and 4. At pH 5-5.1 virtually all IgG$_1$ is sorbed on the sample-27. Under such conditions only a part of carboxylic groups was ionized. The observed high sorption capacity for the protein mixture on sample-27 cannot be interpreted merely by the ion exchange mechanism. As demonstrated earlier, polyfunctional interactions (ion-dipole, dipole-dipole, hydrogen bonds) play an important role in protein retention on a similar type of carboxylic sorbents. Desorption of IgG$_1$ takes place in the pH 5.5-6 range (yield up to 84%, purity 66%).

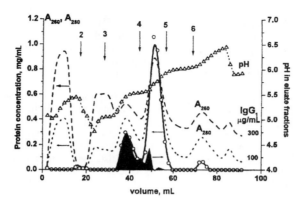

**Figure 3** Ion exchange chromatography of the desalted ascites fluid on Polygran sorbent-27. Conditions: loading – 3.15 mL of the fluid diluted with the same volume of buffer. Gradient desorption with 0.1 M citrate buffer: 2 pH 5; 3 pH 5.5; 4 pH 6; 5 pH 6.5; 6 pH 7.7 (containing 0.2 M NaCl). The shaded area indicates the peak of specific immunoglobulin against horse radish peroxidase

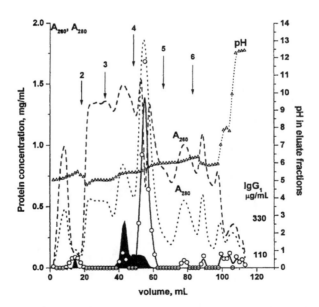

**Figure 4** Ion exchange chromatography of the desalted ascites fluid on Polygran sorbent-44. Conditions: loading 3.3 mL of the fluid diluted with the same volume of buffer and adjusted to pH 8.17. Gradient desorption with 0.1 M citrate buffer: 2 pH 5; 3 pH 5.5; 4 pH 6; 5 pH 6.5; 6 pH 7.7 (containing 0.2 M NaCl). The shaded area indicates the peak of specific immunoglobulin against horse radish peroxidase

At the same time, the IgG$_1$ zone can be successfully separated from the mouse serum albumin zone, as the latter is desorbed at higher pH. It should be pointed out that the sorption of IgG$_1$ proceeds with 94% recovery also at alkaline pH (8) on Polygran sorbent-44. In such case desorption of IgG$_1$ reaches 88% but its purity drops to 40%. It may be assumed that, at alkaline pH, the protein sorption is accompanied by a simultaneous ionization of carboxylic groups of the sorbent at the expense of ion exchange with buffer components. The buffer properties of the sorbent itself (due to carboxylic groups) may become extremely desirable in the separation of specific Igs tending to association and coalescence at acid pH. However, such a variant of chromatographic separation is possible only on the sorbents with a sufficiently high buffer capacity, thus being a great advantage of weakly acid ion exchangers. Polygran sorbent-44 is characterized with low water content in the ionized and unionized form. More hydrophobic properties of the sorbent influencing on ionization of carboxylic groups assists IgG$_1$ sorption at the higher pH. It was also found that maximum sorption of bovine serum albumin on Polygran sorbent-44 takes place under the acid pH (4). This fact indicates that the mechanism of protein sorption can be changed in dependence on pH. Polyfunctional interactions assist high protein sorption on H-form of the sorbent.

In the separation process of IgG$_1$ from the crude ascitic fluid (Figure 5) on Polygran sorbent-27, approx. 30% of accompanying proteins are already separated in the sorption at pH 5. The recovery of IgG$_1$ from crude ascitic fluid reaches 100%. Desorption of IgG$_1$ from the sorbent begins after 0.1 M citrate buffer (pH 5.5) has been introduced into the column.

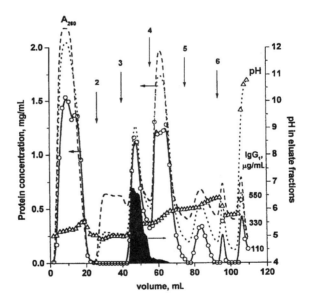

**Figure 5** Ion exchange chromatography on Polygran sorbent-27 of the initial ascites fluid containing immunoglobulin specific against horse radish peroxidase. Conditions: loading 3 mL of the fluid diluted with 6 mL of distilled water and adjusted to pH 5. Gradient desorption with 0.1 M citrate buffer: 2 pH 5; 3 pH 5.5; 4 pH 6; 5 pH 6.5; 6 pH 7.7 containing 1 M NaCl; 7 0.5 M sodium carbonate. The shaded area indicates the peak of specific immunoglobulin

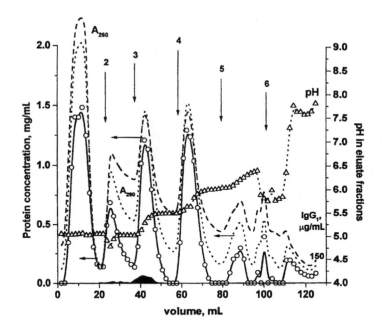

**Figure 6** Ion exchange chromatography on Polygran sorbent-44 of the initial ascites fluid containing immunoglobulin specific against horse radish peroxidase. Conditions: loading – 3 mL of the fluid diluted with 7 mL of distilled water and adjusted to pH 5. Gradient desorption with 0.1 M citrate buffer: 2 pH 5; 3 pH 5.5; 4 pH 6; 5 pH 6.5; 6 pH 7.7 containing 0.2 M NaCl; 7 0.01 M NaOH. The shaded area indicates the peak of specific immunoglobulin

The rear front of its zone overlaps the front of the mouse serum albumin zone. In this case the $IgG_1$ yield reaches 66% and its purity amounts to 53%. The sorption of $IgG_1$ on Polygran sorbent-44 proceeds similarly to its sorption from desalinated ascitic fluid (Figure 6). At pH 5, not more than 6% of $IgG_1$ is sorbed while with increasing pH, the sorption increases up to 9-16%. At the same time, a high sorption capacity of the accompanying components can be observed. These special features of sorption on Polygran sorbent-44 may be employed for prepurification of crude ascitic fluid. Then specific $IgG_1$ can be isolated on Polygran sorbent-27.

## 4 CONCLUSIONS

Summarizing the results of the investigation it should be pointed out that chromatographic properties of heterogeneous carboxylic Polygran sorbents are determined by both their structural and physicochemical properties.

The experiments have shown that, for some Polygran sorbents which tend to lower swellings with increasing salt concentration, the complete reversibility of protein sorption can be successfully achieved with alkaline buffers (pH 8-9).

Of the structures under study, the best chromatographic properties were found with Polygran sorbent-44 and Polygran sorbent-27. High sorption capacity of the former towards the accompanying low-molecular-weight proteins and of the latter towards the specific $IgG_1$ makes it possible to combine them for the use in the two-stage separation of target protein from crude ascitic fluid. High thermal and chemical stability of Polygran sorbents is their great advantage in comparison with cellulose chromatographic supports. Moreover, considerably higher concentrations of carboxylic groups in Polygran sorbents make it possible to control separation of protein zones easier in elution. The principal possibility of direct separation of $IgG_1$ from the crude ascitic fluid containing high concentrations of competing substances is also a considerable advantage of Polygran sorbent-27, since it rules out the necessity of additional prepurification stages and increases the $IgG_1$ yield. However, it may be a problem to achieve the high purity of $IgG_1$ suitable for medical applications in a one-stage process. At present, a combination of immobilized-metal-ion affinity chromatography followed by ion exchange chromatography on carboxylic sorbents is recommended for separation of highly purified specific Igs.[1] This variant of specific Ig purification by using Polygran sorbents will be studied in the near future.

**Acknowledgement**

This work was suppported by the Grant of Academy of Sciences of the Czech Republic (2050602-12).

**References**

1. P. Gagnon. 'Purification Tools For Monoclonal Antibodies'. Validated Biosystems, Tucson (AZ) 1996, 253 pp.
2. S. Oscarsson, J.H. Porath. Alkali-resistant protein adsorbent, suitable for medical procedures, PCT Int. Appl. WO 95.33.557.
3. L.K. Shataeva, A.A. Selezneva, O.V. Orlievskaya, D.I. Ostrovskii, V.V. Korshunov, R.B. Ponomareva, N.P. Kuznetsova, K.P. Papukova, G.V. Samsonov. `Synthesis, Structure and Properties of Polymers'. Leningrad, 1970, pp.225-228 (in Russian).
4. G.V. Samsonov, S.S. Koltsova, L.K. Shataeva, N.N. Kuznetsova, K.M. Rozhetsky, I.K. Paradeeva, Z.D. Fedorova. Patent Appl. SSSR 584034.
5. L.K. Shataeva, I.A. Chernova, G.V. Samsonov, G.D. Kobrinskii, V.D. Solovev, I.V.Domaradskii. *Vopr .Med. Khim.* 1978, **4**, 569-572.
6. B.V. Moskvichev, N.A Margolina, O.A. Mirgorodskaya, T.S. Velikanova, L.I. Oreshchenko, K.A. Kalunyants. *Mikrobiol. Prom.* 1974, **7**, 10-12.
7. G.V. Samsonov. *Pure Appl. Chem.* 1974, **38**, 151-172.
8. E.V. Vorobeva, L.K. Shataeva, E.N. Datunashvili, G.V. Samsonov. *Prikl. Biokhim. Mikrobiol.* 1976, **12**, 715-718.
9. M.S. Loitsanskii, V.N. Elisashvili, L.K. Shataeva. *Prikl. Biokhim. Mikrobiol.* 1974, **10**, 406-408.
10. K.E. Pappel, Kh.Ya. Kalyukha, Yu.I. Sakalauskaite, E.V. Letunov, A.S. Tikhomirova, L.K. Shataeva, G.V. Samsonov. *Prikl. Biokhim. Mikrobiol.* 1975, **11**, 598-601.
11. L.K. Shataeva, G.V. Samsonov. *Khim.-Fharm. Zh.* 1977, **4**, 78-90.
12. V.G. Morozov, V.Kh. Khavinson, O.A. Pisarev. *Dokl. Akad. Nauk.SSSR* 1977, **233**, 491-494.
13. V.V. Kozlov. *Med. Radiol.* 1975, **20**, 46-49.
14. L.K. Shataeva, N.N. Kuznetsova, G.E. Elkin. 'Carboxylic Sorbents in Biology'. Leningrad, 1979, p.286 (in Russian).

# Monitoring the Selectivity of Supercritical Fluid Extractions of Metal Ions using Ion Chromatography.

Mark O'Connell, Tom O'Mahony, Margaret O'Sullivan, Stephen J. Harris, W. Brian Jennings[¶] and Jeremy D. Glennon,

Analytical Chemistry, [¶]Organic Chemistry, Chemistry Department, University College Cork, Ireland.

## ABSTRACT

Ion chromatography is used to monitor the efficiency and selectivity of extraction of the metal ions $Fe^{3+}$, $Cu^{2+}$, $Zn^{2+}$, $Co^{2+}$ and $Mn^{2+}$, from spiked cellulose paper, using perfluoro-octanohydroxamic acid and an upper rim fluorinated calix[4]arene tetrahydroxamate reagent in unmodified supercritical $CO_2$. The supercritical fluid extraction (SFE) experiments were carried out at 300atm and 70°C and metal ions were separated and analysed on a Dionex CS5a column. The ion chromatograms obtained following supercritical fluid extraction (SFE) on injection of acidified aqueous samples, back extracted from MIBK collection solutions, show a single peak, consistent with extraction selectivity for $Fe^{3+}$.

## INTRODUCTION

The solvent extraction of metal ions as neutral metal chelates is one of the most widely used techniques for the preconcentration and separation of metal ions from aqueous solution. For example the extraction of trace metal ions from aqueous environmental samples into the solvent MIBK as dithiocarbamate complexes is a standard approach. Extraction with supercritical $CO_2$ is an attractive alternative to conventional solvent extraction techniques, where the solvating power of the fluid can be varied by varying the pressure and where the use of organic solvents for the extraction process is avoided [1]. The first demonstration of the solvating power of supercritical fluids actually involved metal species i.e. cobalt and iron chlorides in supercritical ethanol in 1879. However, the more recent revival in SFE has focused mainly on the extraction of organic compounds.

Extraction of free metal ions by supercritical $CO_2$ ( $SF-CO_2$ ) is known to be highly inefficient due to the polar nature of the metal ion and the non-polar nature of the supercritical fluid. As with solvent extraction, charge neutralistaion of the metal ion is required before it can be extracted by $SF-CO_2$. This can be achieved by binding the metal ions to designed organic ligands, resulting in neutral complexes and hence improved solubilities in $SF-CO_2$ [2]. In this manner, $Fe^{3+}$ can be solubilised in supercritical $CO_2$ through complexation with fluorinated hydroxamic acid ligands [3], while $Cu^{2+}$ and $Hg^{2+}$ can be extracted from solid matrices by SFE using lithium bis(trifluoroethyl)dithiocarbamate. The choice of the fluorinated ligands is based on the fact that the solubilities of the fluorinated metal complexes are several orders of magnitude higher than the non-fluorinated analogues.

This *in situ* chelation -SFE technique has a wide range of applications including trace metal analysis, the treatment of metal contaminated waste materials and mineral processing. The selective extraction of a contaminant or interferent metal ion from a matrix of desirable constituent metals can also be obtained by SFE. This selectivity can be achieved by careful reagent design or selection. Chelating reagents can be linear, macrocyclic or polymeric in nature. One potentially powerful approach is to use reagents based on molecular recognition technology. Molecular recognition of a metal ion occurs through host-guest complexation, where a host macrocyclic reagent provides a cavity size and donor atoms that complement the size and coordination chemistry of the guest metal ion. Recently, such selective extractions using macrocyclic reagents in SFE have been demonstrated [4,5]. The selective extraction of mercury from sand, and cellulose filter paper using ionisable dibenzobistriazolo crown ether in methanol modified supercritical $CO_2$ has been reported [5]. The synthesis and use of fluorinated molecular baskets for metal extraction in unmodified supercritical $CO_2$ has been described. The ability of one such molecular basket, a fluorinated calix[4]arene tetrahydroxamate, to selectively extract $Pb^{2+}$ and $Fe^{3+}$ from metal mixtures on cellulose paper has been monitored using atomic absorption analysis.

This paper examines the use of ion chromatography to monitor the SFE of a targeted metal ion, $Fe^{3+}$ from metal mixtures spiked on cellulose paper using fluorinated linear and macrocyclic complexing reagents.

## EXPERIMENTAL

### Reagents

All metal nitrate salts were supplied by Merck (Darmstadt, Germany). Perfluoro-octanohydroxamic acid and the fluorinated calix[4]arene tetrahydroxamate were synthesised as previously reported [3,4]. L-Ascorbic acid, $K_2SO_4$ and formic acid were obtained from BDH Chemicals Ltd. (Poole, Dorset, UK) with methyl isobutylketone (MIBK) from Riedel-de-Haen AG (D-30926 Seelze, Germany). 4-(2-Pyridylazo)resorcinol (PAR) was obtained from Sigma-Aldrich Chemical Co. (Steinheim, Germany). Pyridine-2,6-dicarboxylic acid (PDCA) was supplied by Dionex (Sunnyvale, California, USA). KOH was obtained from Prolabo, (Paris, France).

### Ion Chromatography

Metal ion analysis was carried out on a Dionex 4500i series ion chromatograph, which uses a six way air solenoid operated microinjection valve and a 25µL injection loop. Detection of the metal ions was accomplished using post column derivatisation with $5 \times 10^{-4}$M PAR, with visible detection at 520nm using a Dionex variable wavelength detector. The column employed was a Dionex CS5a separation column fitted with a CG5a guard column. The eluent consisted of 0.007 M PDCA, 0.066 M KOH, 0.074 M formic acid and 0.0056 $K_2SO_4$. The eluent flow rate was 1.2 ml/min with the PAR flow rate at 0.6 ml/min.

### Supercritical Fluid Extraction

All extractions were performed using an Isco SFX supercritical fluid extraction system (Isco Inc., supplied by Jones Chromatography, Hengoed, Mid. Glamorgan, UK). The SFE system was controlled by a 260D series pump controller. It consisted of a syringe pump, heated extractor block and adjustable restrictor. Reagent and filter paper samples were loaded into an open-ended glass tube (3.0 x 0.5 cm i.d. ), packed with glass wool at both ends and mounted inside a stainless steel cartridge (5.5 x 0.76cm i.d., volume 2.5ml). Extracts were collected in a liquid trap containing MIBK.

### Supercritical Fluid Extraction of Metal Ion Mixture from Cellulose Paper

For the SFE study of metal ion mixtures from cellulose paper, 26μL aliquots of $Fe^{3+}$, $Cu^{2+}$, $Zn^{2+}$ and $Co^{2+}$ and 52μL of $Mn^{2+}$ from 1000ppm metal nitrate standard solutions in 0.5M $HNO_3$ were spiked onto filter paper (4cm x 1cm). When air dry, an additional 40μL $H_2O$ was spiked on to aid the extraction [3]. The wetted filter paper along with 60mg of perfluoro-octanohydroxamic acid or fluorinated calix[4]arene tetra-hydroxamate was loaded into a glass tube, placed in the SFE system and statically extracted for 20 min. at 300atm and 70°C (optimum conditions). The extraction cell was vented into a collecting vessel containing 7mL of MIBK for 15 min.

### Back Extraction

In order to liberate the extracted metal ions into an aqueous environment suitable for direct injection into the ion chromatograph, it was necessary to decomplex the metal ions from any extracted complexes. The MIBK collecting solution was placed in a separatory funnel (50mL) with 10mL of water acidified to pH 0.85 with nitric acid. The funnel was gently shaken by hand for approximately 5 min. and the aqueous layer drawn off, made up to 20 mL with acidified water, 30mg L-ascorbic acid added and subsequently analysed using ion chromatography. The L-ascorbic acid was added to reduce any extracted ferric ion to ferrous, thus avoiding any chromatographic peak tailing that would arise with ferric ion present.

## RESULTS AND DISCUSSION

Hydroxamate reagents, while known to complex many transition metal ions, form very stable complexes with $Fe^{3+}$ and play an important role in microbial systems, where they act as metal sequestering agents. In general, acidification of aqueous metal-hydroxamate solutions decomplexes the metal ions from the ligand and thus acid back extraction of complexes collected in MIBK is expected to release any metal ions extracted by SFE into the acidified aqueous layer. Such acid extracts can then be analysed by a variety of techniques available in trace metal analysis, such as atomic absorption analysis and ICP spectroscopy. Ion chromatographic analysis is a useful approach, as it can provide a rapid multi-metal profile of ions extracted by ligands in supercritical $CO_2$. It should be particularly useful in demonstrating and optimising the selectivity of extraction. It is with this objective in mind, alongside that of reagent recovery, that ion chromatography of back extracted samples was investigated. Direct aspiration of MIBK collected extracts into a flame atomic absorption spectrometer was also carried out to confirm the extraction results.

Ion chromatography is thus used here to monitor the selectivity of supercritical fluid extraction of metal ions from cellulose-based filter paper, following decomplexation of the extracted metal from the reagent. This laboratory has designed two ligands capable of selectively extracting $Fe^{3+}$. These are the linear perfluoro-octanohydroxamic acid and the fluorinated calixarene tetrahydroxamic acid, both shown in Figure 1. Fluorinated calixarenes are the macrocyclic templates on which carefully selected chelating groups can be incorporated around the cavity to yield selective extractants for targeted metals.

(a) $CF_3(CF_2)_6CONHOH$

(b) Calix[4]arene Tetrahydroxamate
$R = CF_3(CF_2)_7(CH_2)_2S(CH_2)_3$-
$R' = -CH_2CONHOH$
$n = 1$

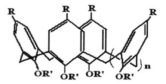

**Figure 1** **Fluorinated (a) linear monohydroxamate and (b) macrocyclic tetrahydroxamate reagents for SFE of metal ions.**

Figure 2 shows ion chromatograms obtained for a five metal standard mixture spiked onto cellulose filter paper prior to SFE. The iron is actually spiked as $Fe^{3+}$ but analysed following reduction to $Fe^{2+}$. The high selectivity of extraction of $Fe^{2+}$ from among the other metal ions is evident from the chromatogram obtained after SFE with perfluoro-octanohydroxamate. The percentage recovery of $Fe^{3+}$ is calculated as 84%, which could be improved with careful washing steps. The percentage recovery and the selectivity obtained were confirmed in separate analyses carried out using flame atomic absorption analysis. Similar selectivity was shown in the chromatograms obtained following SFE with the macrocyclic ligand, although recoveries were lower. This may be partially as a result of poorer efficiency of back extraction of iron from the calixarene reagent. The results show the utility of ion chromatography in determining the efficiency and selectivity of extraction of iron from a cellulose based matrix and the feasibility of using the technique to monitor the SFE of metal ions generally.

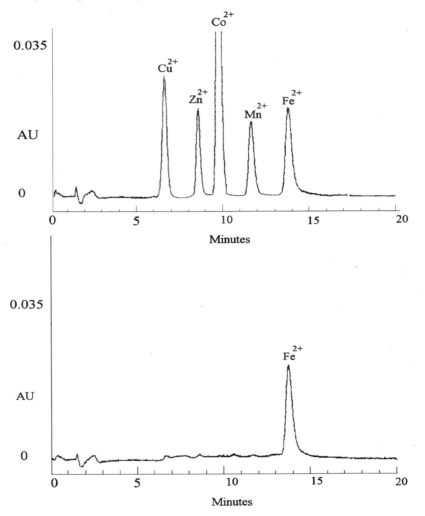

**Figure 2  Typical ion chromatograms obtained of standard metal mixture spiked onto cellulose paper (top), and of back-extracted MIBK collecting solution after SFE ( bottom).**

# REFERENCES

1. T. Clifford and K. Bartle, "Chemistry goes Supercritical" *Chemistry in Britain* (June) (1993) 499.

2. C.M. Wai and S. Wang, *J. Chromatography A*, **785** (1997) 369.

3. J.D. Glennon, S. Hutchinson, A. Walker, S.J. Harris and C.C. McSweeney, *J. Chromatography A*, **770** (1997) 85.

4. J.D. Glennon, S. Hutchinson, S.J. Harris, A. Walker, M.A. McKervey and C.C. McSweeney, *Anal. Chem.*, **69** (1997) 2207.

5. S. Wang, S. Elshani and C.M. Wai, *Anal. Chem.* **67** (1995) 919.

Email j.glennon@ucc.ie

**Novel Inorganic Ion Exchangers**

# THE STRUCTURES AND ION EXCHANGE BEHAVIOR OF SOME TUNNEL-TYPE INORGANIC ION EXCHANGERS

Abraham Clearfield

Department of Chemistry
Texas A&M University
P. O. Box 300012
College Station, TX  77842-3012
U.S.A.

## 1 INTRODUCTION

Inorganic ion exchange materials are widespread in nature. There are literally hundreds if not thousands of such compounds, including clays, zeolites, manganese oxides, hydrotalcites, certain phosphates, silicates and many more. They possess a number of advantages over organic resins which commend their use in special cases. They exhibit greater stability to high temperature and high radiation fields. However, it should be recognized that a number of these exchangers are susceptable to hydrolytic decomposition in acid or basic solutions at elevated temperatures. Of primary importance is the very high selectivities exhibited by many of the inorganics for specific ions. This specificity stems from the closeness of approach of the ions to the fixed groups along the lines proposed by Eisenman.[1,2]

In order to understand the behavior of inorganic ion exchangers we need to have a knowledge of their structure. They may be prepared as amorphous gels as crystalline phases and in many cases as semicrystalline materials. Structures can be layered as in clays and group 4 phosphates, three dimensional with cavities or tunnels and these can have a variety of shapes. Inorganic ion exchangers prepared as amorphous gels behave in much the same way as organic resins. That is they form solid solutions in which the ions are uniformly spread throughout the gel particle. These exchangers can usually be prepared as beads or macrosized particles for column use. Crystalline ion exchangers may undergo phase changes as the ion exchange process proceeds and these changes need to be taken into account when considering thermodynamic or kinetic aspects of the exchange reaction. A number of the major structure types and the influence of structure on their ion exchange properties has been described in some detail.[3] In this paper we will describe two types of tunnel structure ion exchangers that may find use in nuclear waste remediation programs. Their preparation and details of the structure determinations have been provided earlier.[4-7] The sodium titanium silicate (NaTS) was found to be highly selective for $Cs^+$ and $Sr^{2+}$ even in highly alkaline media.[8] A related family of compounds are the titanium silicate (germanates) with the pharmacosiderite structure.

## 2 SODIUM TITANIUM SILICATE (NaTS)

Sodium titanium silicate, $Na_2Ti_2O_3SiO_4 \bullet 2H_2O$, was prepared hydrothermally at 170°C from a titanium silicate gel in 6.3 M NaOH.[4] The potentiometric titration curves for $Na^+$ (0.05 M NaCl-NaOH) and $Cs^+$ (0.05 M CsCl-CsOH) are shown in Figure 1.

The exchanger was in the protonic form $H_2Ti_2O_3SiO_4 \cdot H_2O$ for which the theoretical ion exchange capacity (IEC) is 7.81 meq./g. Sodium ion uptake starts at a pH near 2 and at

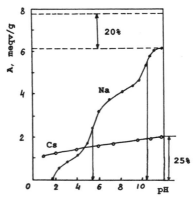

**Figure 1.** *Potentiometric titration curves for uptake of cations, A, as a function of pH. Titrants: 0.05M (NaCl + NaOH), O; 0.05M (CsCl + CsOH), O.*

pH12 attains a value of 6.2 meq/g. In contrast $Cs^+$ is exchanged at a pH below 1 and attains a maximum value of 25% of the theoretical capacity at pH12.

The reasons for these differences in ion exchange behavior became evident after the crystal structure was solved from its x-ray powder pattern. The compound is tetragonal a=7.8082 (2), c=11.974 (1)Å, with four formula units per unit cell. The titanium atoms occur in groups of four connected by oxygen atoms in a cubane-like structure (Figure 2).

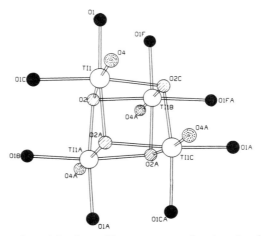

**Figure 2.** *A section of the titanosilicate structure showing the cluster of four titanium-oxygen octahedra. The 01 atoms are part of the $SiO_4$ tetrahedra linking the clusters along the a and b-axis directions, 02 oxygens are bonded to three titanium atoms to form the cubane-like structure and 04 oxygens connect the clusters in the c-axis direction.*

These cubane octahedra are linked together in the a and b-directions by silicate tetrahedra leaving an opening in the center as shown in Figure 3. In the c-direction the octahedra are

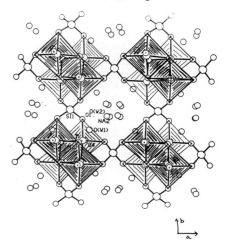

**Figure 3.** *A portion of the structure of sodium titanosilicate as viewed down the c-axis direction. The titania octahedra, two up and two down rotated 90° from each other, are lined to emphasize, with the connecting silicate groups, the formation of the tunnels. Half the Na⁺ and water molecules fill the tunnels.*

linked by oxo groups as shown in Figure 4. This arrangement creates tunnels parallel to

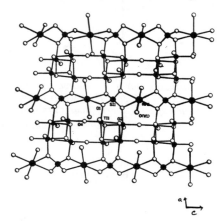

**Figure 4.** *Schematic representation of the ac or bc plane of the sodium titanosilicate that forms the walls of the tunnel. Note the octahedrally coordinated sodium ions in these framework sites.*

the c-axis direction. The sodium ions are located in two separate sites. Site I is located in the framework as seen in Figure 4. The sodium ions in this site are coordinated by four silicate oxygens and two water molecules. Each $Na^+$ is shared by two adjacent unit cells and there are eight sodiums in the four ac faces so these framework sodiums constitute half the total or four $Na^+$ per unit cell (Z=4). The remaining sodium ions are located in the tunnel as shown in Figure 3. This tunnel site was 82% occupied by $Na^+$ so the true formula was $Na_{1.64}H_{0.36}Ti_2O_3SiO_4 \bullet 2H_2O$.

X-ray studies of the half sodium ion exchanged phase[5] showed that all the sodium ions were in the framework sites. Thus, these are the preferred sodium ion sites and this fact was corroborated by the observed bond distances. The four bonds to silicate oxygens in the framework sites are 2.414 (5) Å which is almost exactly the sum of their ionic radii. The bonding of $Na^+$ in the tunnel sites is much weaker with Na-O silicate, 2.74(2) Å (2 bonds) and Na-O water, 2 bonds at 2.79 (2) Å.

Cesium ion is too large to fit into the framework sites and can only occupy tunnel sites. Two such sites were found, one at 1/4 C and 3/4 C related by symmetry and the other at 0.13 C and 0.87 C. In the former site the $Cs^+$ are perfectly coordinated by 8 silicate oxygens at a distance of 3.183 (5) Å (Figure 5).

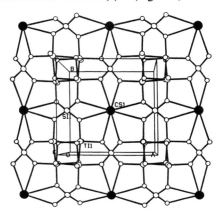

**Figure 5.** *Top view of $Cs^+$ titanosilicate showing the almost perfectly symmetrical coordination of $Cs^+$ in the tunnel sites at 1/4, 3/4. Four water molecules, two above and two below complete the coordination sphere. Steric factors limit the $Cs^+$ occupancy to* $CsH_3Ti_2O_3SiO_4 \bullet 2H_2O$ *in the absence of $Na^+$ and much lesser amount when sodium is present.*

In the second $Cs^+$ tunnel site there are only four bonds with silicate oxygens at 3.06 (1) Å and two to water molecules at 2.95 (2) Å. Both sitings are favorable from a bonding point of view but the 1/4, 3/4 site is preferred (80% occupancy) because the larger the number of strong bonds formed, the greater the negative free energy.

Titration of the proton phase of the titanium silicate with a mixture of 0.05 M NaCl - NaOH + CsCl + CsOH reduces the uptake of $Cs^+$ to about 0.35 meq/g.[9] A possible explanation for this reduction in $Cs^+$ uptake lies in the fact that the cesium ions are tightly held in the tunnel sites. Therefore they probably diffuse relatively slowly down the tunnels. In contrast $Na^+$ can diffuse rapidly down the tunnels but also in directions perpendicular to the tunnels through the framework sites. Thus, $Na^+$ can rapidly enter the tunnels from three directions, competing successfully for tunnel sites, which, because of space crowding, are then difficult to dislodge by the $Cs^+$ moving from both ends of the tunnels.

The explanation given above is kinetic but the composition at equilibrium is governed by thermodynamic considerations. For sodium ion exchange there are two separate phases formed and therefore two isotherms to determine.

(1) $H_2Ti_2O_3SiO_4 \bullet 2H_2O + Na^+$ (Aq) $\rightarrow Na_FHTi_2O_3SiO_4 \bullet 2H_2O + H^+$ (Aq)

(2) $Na_FHTi_2O_3(SiO_4) \bullet 2H_2O + Na^+$ (Aq) $\rightarrow NA_FNA_TTi_2O_3(SiO_4) \bullet H_2O + H^+$ (Aq)

where $Na_F$ refers to the framework sites and $Na_T$ refers to the tunnels sites.

In the case of $Cs^+$ only half the tunnel sites can be filled so the isotherm extends to the uptake of $Cs_{0.5}H_{1.5}$. However, there is a complication in evaluating the $Cs^+$ isotherm. The exchange is not reversible. That is, only a small fraction of the cesium ion can be removed by treatment even with very strong acid. Thus, the titanosilicate exchanger is not reusable as a $Cs^+$ extractant. Nevertheless, it can be used as an agent for removal of $^{137}Cs$ from nuclear waste solutions, even in highly basic tank wastes containing high levels of sodium ions. In such solutions the cesium ion capacity may be 0.05 - 0.1 meq/g but the level of $Cs^+$ present is very low so that it can be removed without resort to excessive amounts of the exchangers.

## 3 PHARMACOSIDERITES

There is another family of titanium silicate exchangers that have the pharmacosiderite structure. Pharmacosiderite is a mineral of composition $KFe_4(AsO_4)_3(OH)_4$ and a cubic structure.[10] The titanium silicate, $K_3H(TiO)_4(SiO_4)_3 \bullet 4H_2O$, is a 4-4 version of the 3-3 pharmacosiderite. Chapman and Roe[11] prepared a number of titanosilicates, among them were several pharmacosiderite phases. We duplicated their preparations of the potassium and cesium phases.[6] These phases are cubic with a=7.7644 (3) Å for the $K^+$ phase and a=7.8212(2) for the cesium phase. Their crystals consist of clusters of four titania octahedra that are very similar in structure to the cubane clusters in the sodium titanosilicate. Each corner of the cube contains one such cluster and they are linked to each other by silicate groups as shown in Figure 6.

Potentiometric titrations and Kd measurements yielded the following affinity sequences:

$$Cs^+ > Rb^+ > K^+ > Na^+ > Li^+$$
$$Ba^{2+} > Sr^{2+} > Ca^+ > Mg^{2+}$$

The fact that these pharmacosiderite phases are cubic rather than tetragonal has an important bearing on their ion exchange behavior. This cubic arrangement leads to similar intersecting channels parallel to the a, b, c unit cell directions. There is only one ion exchange site and that is located in the cubic face centers i.e. at 1/2 1/2 0, 0 1/2 1/2, 1/2 0 1/2. Since there are six faces each contributing half an ion, the unit cell contains three $M^+$ ions and a proton. All four protons have been exchanged by silver ion[12] with the fourth $Ag^+$ located in the center of the cube. Presumably the smaller silver ion is able to exchange all the protons but the electrostatic repulsions for the larger ($K^+$, $Rb^+$, $Cs^+$) ions prevents the exchange of the fourth proton. Both the potassium and cesium ion phases were prepared hydrothermally at 200°C in an excess of KOH and CsOH, respectively. Yet only the face centered sites were occupied. In the sodium titanosilicate the longer c-axis allows the $Na^+$ in the framework to be at O and 1/2 c with the $Cs^+$ ions halfway in between at 1/4, 3/4 c, thus minimizing electrostatic repulsions.

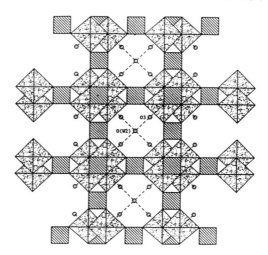

**Figure   6.**   *Schematic   polyhedral   representation   of   the   structure   of*
$H_4Ti_4O_4(SiO_4)_3 \bullet 8H_2O$. *The titania octahedra are designated by stippled octahedra
sharing edges, the $SiO_4$ tetrahedra are represented as striped squares and one level of
water molecules in the tunnels are shown as dotted circles. Dashed lines are probable H-
bonds.*

The $K^+$ ions are twelve coordinate forming 8 long bonds to silicate oxygens at
3.234 (2) Å and four water oxygens at 3.17(1) Å.  It is interesting to note that the $Cs^+$
ions do not reside exactly in the face centers but are moved to either side of the plane by
about ±0.46 Å. This location results in four strong Cs-O bonds of 3.14(1) Å and four
weaker ones at 3.41 Å.  there are also two strong bonds to water molecules at 2.82 Å.
Our study was carried out using X-ray powder data but concurrent with our work a
single crystal structure of $Cs_3H(TiO)_4SiO_4 \bullet 4H_2O$ published.[13]  The crystals were grown
at 750°C and 30,000 psi and confirmed our results obtained by the Rietveld-powder
method.
    Recently, we prepared two additional pharmacosiderite phases in which Ge
partially or completely substituted for Ti or Si.[14]  These phases had the compositions
(approximate), as determined from site occupancy based on their X-ray powder patterns,
$M_3H(TiO)_4(GeO_4)_3 \bullet 4H_2O$   and   $M_3H(TiO)_{3.5}(GeO)_{0.5}(SiO_4)_{0.5}(GeO_4)_{2.5} \bullet 4H_2O$, where
$M=K^+$ or $Cs^+$. Although there are some changes in space groups from phase to phase
the structures are essentially the same as described for the non-germanium containing
phase.  In Table 1 we have listed a number of Kd values for alkali metal and alkaline
earth ions for the several phases.  The $K_d$s in ml/g were determined for 1 x $10^{-3}$ M
chloride solutions with V:m=200.

**Table 1**  Distribution Coefficients, $K_d$s, and Equilibrium pHs (in parentheses), for the Alkali and Alkaline Earth Metals using $HK_3(TiO)_{3.5}(GeO)_{0.5}(GeO_4)_{2.5}(SiO_4)_{0.5} \bullet 4H_2O$ (KTiSiGe), $HCs_3(TiO)_{3.5}(GeO)_{0.5}(GeO_4)_{2.5}(SiO_4)_{0.5} \bullet 4H_2O$ (CsTiSiGe), $HK_3(TiO)_4(SiO_4)_3 \bullet 4H_2O$ (KTiSi), and $HK_3(TiO)_4(GeO_4)_3 \bullet 4H_2O$ (KTiGe),

| Sample | Li | Na | Cs | Mg | Ca | Sr | Ba |
|---|---|---|---|---|---|---|---|
| KTiSiGe | 120 | 280 | 32,000 | 4,900 | 5,480 | 79,00 | 72,700 |
|  | (8.4) | (9.3) | (9.3) | (8.1) | (8.2) | (7.8) | (7.9) |
| CsTiSiGe | 10 | 90 | n/a | 100 | 100 | 680 | 280 |
|  | (8.9) | (7.9) |  | (8.3) | (7.7) | (7.6) | (7.3) |
| KTiGe | 200 | 100 | 46,200 | 6,100 | >6,990 | >>100,000 | >100,000 |
|  | (8.5) | (8.2) | (8.5) | (8.2) | (8.3) | (7.2) | (7.3) |
| KTiSi | 750 | 220 | 5,800 | 31,000 | 12,000 | >52,000 | 3,100 |
|  | (10.1) | (10.7) | (9.7) | (10.6) | (10.7) | (9.6) | (9.4) |

It is seen that there is a dramatic increase in Kd values for $Cs^+$ and $Sr^{2+}$ for the KTiGe phase $[K_3H(TiO)_4(GeO_4)_3 \bullet 4H_2O]$. This phase was tetragonal with a=b=11.215 (1) Å and c=7.9705 (2) Å. These cell dimensions are the largest of the three types of phases prepared. It should be noted that the 11.2 Å distance is the diagonal of the original cubic cell which translates to a'=b'=7.931 Å. This larger tetragonal cell contains two formulas (Z=2) or six $K^+$ ions per unit cell. There are two $K^+$ positions, the first, designated K1, contains 4 ions and the second, K2 only two potassium ions. The K1 oxygen bonds are large, ranging from 3.231(6) Å to 3.397(7) Å. Because of this weak bonding it would be expected that 2/3 of the $K^+$ would readily be displaced by $Cs^+$. The cesium ions, on the other hand form strong bonds in the range of 2.92 Å 3.22 Å. Thus, the large increase in Kd results from the displacement of eight weak bonds per $K^+$ by six strong bonds and two moderately strong Cs-O bonds.

## 4 CONCLUSIONS

The structures and ion exchange behavior of two types of titanium silicates, possessing frameworks that create tunnels and cavities within the framework, have been described. The size of the tunnels and cavities are of the same order of magnitude as the cations for which they show a high selectivity. These ions form symmetrical coordination compounds of high coordination number with the framework oxygen atoms. The selectivity sequence for alkali cations was shown to depend upon cation size. The greatest affinity was shown for the cation whose cation-oxygen bonds are closest to the sum of their ionic radii. Smaller cations form weaker bonds with the framework oxygens and are less preferred and more easily displaced by the larger cation. This principle was demonstrated in the case of the germanium substituted pharmacosiderites. Increased substitution of Ge in the framework enlarged the tunnel and allowing for formation of Cs-O bond distances very close to the sum of their ionic radii. At the same time the larger tunnel size produced weaker K-O bonds as shown by the increased K-O bond distances. As a result the K$^+$ was more readily displaced by Cs$^+$ in the titanogermanate.

Further application of these principles could lead to methods of producing highly selective exchangers for ions of different size. For example, reducing the tunnel size to fit a smaller cation such as Na$^+$ would produce an ion sieving effect where ions larger than sodium would be rejected and smaller ions would be less preferred. However, we still need to take into effect the role of charge and hydration-dehydration phenomena. Continued studies along these lines are in progress.

Acknowledgment:  We acknowledge with thanks financial support for this study by DOE's Basic Energy Sciences Grant No. 434741-0001, with funds supplied by environmental management.

### References

1.     G. Eisenman, Biophys. J. Suppl. 1962, **2**, 259.

2.     G. Eisenman in Glass Electrodes for Hydrogen and Other Cations; G. Eisenman, (Ed.), M. Dekker Inc., New York, 1967. Ch. 3.

3.     A. Clearfield, I & EC Research 1995, **34**, 2865.

4.     D. M. Poojary, R. A. Cahill and A. Clearfield, Chem. Mater. 1994, **6**, 2364.

5.     D. M. Poojary, A. I. Bortun, L. N. Bortun and A. Clearfield, Inorg. Chem. 1996, **35**, 6131.

6.     E. A. Behrens, D. M. Poojary and A. Clearfield, Chem. Mater. 1996, **8**, 1236.

7.     E. A. Behrens, D. M. Poojary and A. Clearfield, Chem. Mater., 1998, **10**, 959.

8.     R. G. Anthony, C. V. Philip and R. G. Dosch, Waste Manage. 1993, **13**, 503.

9.     A. I. Bortun, L. N. Bortun and A. Clearfield, Solv. Extr. Ion Exch. 1996, **14**, 341.

10.    M. J. Buerger, W. A. Dollase and Garaycochea, Z. Kristallogr. 1967, **125**, 92.

11.    D. M. Chapman and A. L. Roe, Zeolites 1990, **10**, 730.

12.    M. A. Roberts and A. N. Fitch, J. Phys. Chem. Solids, 1991, **52**, 1209.

13.    W.T.A. Harrison, T. E. Gier and G. D. Stucky, Zeolites, 1995, **15**, 408.

14.    E. A. Behrens, D. M. Poojary and A. Clearfield, Chem. Mater. 1998, **10**, 959.

# COMPUTER SIMULATIONS OF LAYERED DOUBLE HYDROXIDES

S. P. Newman and W. Jones

Department of Chemistry, University of Cambridge, Lensfield Road, Cambridge, U.K., CB2 1EW

P. V. Coveney

Schlumberger Cambridge Research, High Cross, Madingley Road, Cambridge, U.K., CB3 0EL.

## ABSTRACT

Layered double hydroxides (LDHs) represent an interesting class of anion-exchange materials. The arrangement and dynamics of the exchange anions in LDHs, along with the interlayer water, have been studied using molecular dynamics computer simulations. Good agreement with experimentally observed values is obtained for the simulated interlayer spacing of a model LDH containing the inorganic anions carbonate, nitrate or chloride as well as the organic anions terephthalate or benzenesulphonate. The dynamics of the anions and water molecules in the interlayer region is also studied.

## 1 INTRODUCTION

Layered double hydroxides (LDHs), also known as anionic clays, are anion-exchange materials that have received much attention.[1,2]

The structure of LDHs is most clearly described by considering the brucite structure, $Mg(OH)_2$, which consists of $Mg^{2+}$ ions coordinated octahedrally by hydroxyl groups, with the octahedral units sharing edges to form infinite, charge neutral layers. In a LDH, isomorphous replacement of a fraction of the divalent cations with a trivalent cation such as $Al^{3+}$ occurs and generates a positive charge on the layers that necessitates the presence of interlayer, charge balancing, anions (Figure 1). A general formula for such LDHs is, therefore, $[M^{2+}_{1-x} M^{3+}_x (OH)_2][^x/_n A^{n-}.mH_2O]$, where $M^{2+}$ and $M^{3+}$ are divalent and trivalent cations, respectively; x is equal to the ratio $M^{3+} / (M^{2+} + M^{3+})$ and A is an anion of charge n. Observed $M^{2+}$ and $M^{3+}$ species include $Mg^{2+}$, $Fe^{2+}$, $Co^{2+}$, $Cu^{2+}$, $Ni^{2+}$, or $Zn^{2+}$ and $Al^{3+}$, $Cr^{3+}$, $Ga^{3+}$ or $Fe^{3+}$.

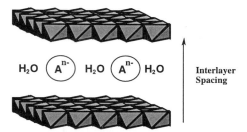

**Figure 1.** *Schematic representation of the structure of a LDH.*

The remaining free space of the interlayer may be occupied by varying numbers of water molecules. The charge density on the hydroxide layers of a LDH is dependent upon the $M^{2+}$ / $M^{3+}$ ratio (R) of the matrix cations in the layers. The anion-exchange capacity, and hence number and arrangement of the charge balancing anions in the LDH, may therefore be controlled by varying R.

There is no significant restriction to the nature of the charge balancing anion that can occupy the interlayer region; for example, inorganic anions such as halides,[3] oxo-anions[4] and polyoxometalate anions[5] as well as organic anions[6] may be incorporated. The interlayer separation (the perpendicular distance between the hydroxide layers) of LDHs, depends upon the size and orientation of the charge balancing anion incorporated between the layers. Furthermore, it has been found experimentally that the interlayer water content and the charge density on the hydroxide layers also have an effect upon the interlayer separation as well as on the orientation of the interlayer anion. In particular, the interlayer arrangement of LDHs containing organic anions has been found to be extremely sensitive to the value of R and the interlayer water content.[7-10]

We will demonstrate in this paper how computer simulations have allowed us to rationalize and predict the interlayer structure and dynamics of LDHs containing a variety of inorganic (carbonate, nitrate or chloride) and organic (terephthalate or benzenesulphonate) interlayer anions. An objective has been to determine the accuracy with which the arrangement and orientation of the exchange anions, as a function of the anion-exchange capacity and water content, may be simulated. The properties of the interlayer water, such as the O-O radial distribution function (RDF) and the water self-diffusion constant (D) have also been studied. Additionally, in the case of inorganic interlayer anions ($NO_3^-$, $Cl^-$ and $CO_3^{2-}$), diffusion coefficients for the anions between the layers have also been computed. In the following Sections the notation MgAl(A) R(X) will be used to represent the LDH, where A is the interlayer anion and X is the Mg/Al ratio.

## 2 SIMULATION METHODS

### 2.1 Model Construction

The hydroxide layers of the simulation cell were constructed using the atomic positions from the crystal structure of naturally occurring hydrotalcite (a $MgAl(CO_3)$ LDH).[11] Hydroxide layers of size 18.30 x 12.20 Å$^2$ (equivalent to 6 x 4 unit cells in the *ab* plane) were used and the proportions of Mg and Al in the hydroxide layers adjusted to give a Mg / Al ratio of 3 [R(3)]. The periodic simulation cell contained two interlayer units.

The initial interlayer separation employed was varied according to the identity of the interlayer anion. For carbonate, nitrate and chloride a value of 8.0 Å was used - a value which is close to that observed experimentally. A sufficient number of the anions to charge-balance the layers were placed in the simulation cell. For the carbonate, nitrate and chloride models, 24 water molecules were included in the simulation cell (12 per interlayer), which corresponds to $m = 0.5$. For the case of terephthalate, the initial interlayer separation was set at 14.4 Å, with the anions in an orientation approximately 45° to the hydroxide layers. For the MgAl(benzenesulphonate) R(3) model, the starting interlayer separation was 16.0 Å, with the anions placed in a bilayer arrangement between the hydroxide layers. A total of 48 water molecules were included in the simulation cell (24 per interlayer), corresponding to a water content of 15.5 weight percent which is close to experimentally observed values $(m = 1)$.[12]

To study the influence of layer charge on the interlayer arrangement of MgAl(terephthalate) LDHs an additional MgAl(terephthalate) model with a Mg/Al ratio of 2 [R(2)] was constructed with hydroxide layers of size 18.30 x 9.15 Å (equivalent to 6 x 3 unit cells in the *ab*-plane). By using simulation cells of different size for the R(2) and R(3) models, the same number of terephthalate anions per interlayer of each model are required for charge neutrality. In addition, the effect of varying the water content on the interlayer arrangement of a MgAl(terephthalate) LDH was investigated by systematically adjusting the number of water molecules included in the R(3) and R(2) simulation cells.

## 2.2 Simulation Details

All simulations were performed using the commercially available software package Cerius[2] on a single node of a Silicon Graphics Origin 2000 machine comprising twenty R10000, 195 MHz processors.

As a result of the large number of atoms in the simulation cell, quantum mechanical methods are not appropriate and empirical, classically-based, force field methods must be used. Full details of the simulation procedure have previously reported previously.[13-16] In the present work, a modified version of the Dreiding force field was used for the molecular dynamics (MD) simulations.[17] The modification, originally introduced by Aicken *et al*,[13] includes the addition of parameters for Mg which are not present in the Dreiding force field. The initial geometry and charges of the terephthalate anion were calculated using the MOPAC semi-empirical method, employing the PM3 Hamiltonian.[18] The initial water geometry and charges were taken from the dedicated TIP3P water force field. Recent MD simulations of bulk water have found that a combination of the Dreiding force field with the water geometry and charges from the TIP3P force field, as used in the present work, yields structural parameters (such as radial distribution function and the water self-diffusion constant) in good agreement with experimental values.[19] Charges on the atoms in the hydroxide layers were calculated using the Charge Equilibration (QEq) method.[20] The long range Coulombic interactions and the attractive van der Waals interactions were computed using the Ewald summation technique. Repulsive van der Waals interactions were treated with a direct cut-off radius of approximately 7 Å. Prior to the molecular dynamics simulations, energy minimization was performed to remove atom-atom close contacts created during the building of the model.

MD simulations, with a time step of 0.001 ps, were performed in the constant composition, isothermal-isobaric (NPT) ensemble at 300 K. The temperature was maintained using the Hoover thermostat and the equivalent hydrostatic pressure was set to

$10^5$ Pa (1 atmosphere).[21] Periodic boundary conditions were applied in three dimensions. In the NPT ensemble, the volume of the system is allowed to vary, which enables the interlayer spacing and arrangement to vary during the simulation for a particular anion and/or water content.

The total MD simulation time was generally 40 ps. Snapshots of the model were collected with a frequency of 0.1 ps, and the interlayer spacing averaged over the last 10 ps of the MD simulation. Variations in the calculated interlayer spacing were observed to be small during the final 10 ps of the MD simulation, thus indicating that equilibration had occurred.

The diffusion coefficient of the interlayer carbonate, nitrate or chloride anions were examined using data extracted from extended trajectory files of 500 ps MD simulations. The increased length of the simulation enables extraction of a reliable mean squared displacement (MSD) of the anions, owing to the improved statistics of longer MD runs. The MSDs of the interlayer anions were calculated by monitoring the displacement of all atoms in the anions. The self-diffusion constant of the anion may then be determined from a linear fit of the computed MSD.

The radial distribution function (RDF) (O-O distances) and MSD of the water molecules in the MgAl(terephthalate) R(2) model were also examined using an extended 500 ps simulation.

## 3 RESULTS

### 3.1 Inorganic Interlayer Anions

Figure 2 is a snapshot of the simulation cell of the $MgAl(CO_3)$ R(3) model, following 40 ps of MD simulation at 300 K. The simulated interlayer separation is 7.46 Å, in good agreement with experimentally observed values (7.65 Å).[4]

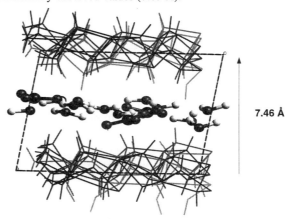

7.46 Å

**Figure 2.** *Snapshot of the simulation cell of the $MgAl(CO_3)$ R(3) model following 40 ps of MD simulation at 300 K. For clarity, only one interlayer of the simulation cell is displayed.*

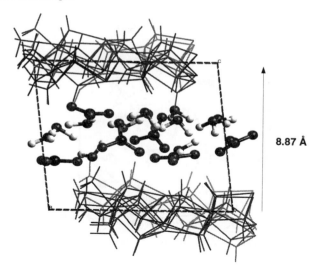

**Figure 3.** *Snapshot of the simulation cell of the MgAl(NO₃) R(3) model following 40 ps of MD simulation at 300 K. For clarity, only one interlayer of the simulation cell is displayed.*

The carbonate anions lie midway between, and coplanar with, the hydroxide layers. There is a strong hydrogen-bonding interaction between the OH groups on the layer surfaces and the anions in this orientation. For the case of nitrate (Figure 3), the simulated interlayer separation is 8.87 Å which is also in good agreement with experiment (8.79 Å).[4]

Charge balancing requires twice as many nitrate ($NO_3^-$) anions as carbonate ($CO_3^{2-}$), with the resulting increase in anion density creating greater steric repulsion. The simulations show that the nitrate anions therefore tilt away from coplanarity with the hydroxide layers to reduce the repulsion, with a concomitant increase in interlayer spacing compared with the coplanar carbonate case. For the case of chloride (Figure 4), the simulated layer spacing is 7.92 Å (7.86 Å) which is once again in good agreement with experiment.[4] For all three anions, the water is located in the mid-plane of the interlayer.

The calculated interlayer spacing and internal energy (U) of the carbonate, nitrate and chloride models from the initial 40 ps of MD simulation are unchanged following the extended (500 ps) simulations. The invariance of the interlayer spacing and internal energy confirm that equilibration of the system has been reached following the initial 40 ps of MD simulation. Figure 5 compares the simulated MSD of the carbonate, nitrate and chloride anions calculated from data extracted from the 500 ps MD simulations. It is clear that the mobility of the anions between the hydroxide layers increases in the order $NO_3^-$ > $Cl^-$ > $CO_3^{2-}$. The self-diffusion coefficients for the three anions, as determined from a linear fit to the MSD computed, are recorded in Table 1. To our knowledge there is no experimental data to compare with these values. It has been reported, however, that the ease of exchange of monovalent anions in LDHs increases in the order $NO_3^-$ > $Br^-$ > $Cl^-$ > $F^-$ > $OH^-$ and that divalent anions such as $SO_4^{2-}$ and $CO_3^{2-}$ have higher selectivity than monovalent anions.[4] If it is considered that the magnitude of the simulated anion diffusion coefficient is related to the strength of the binding of the anion to the hydroxide layers in the LDH, then the simulated trend agrees qualitatively with the experimentally determined relative selectivity.

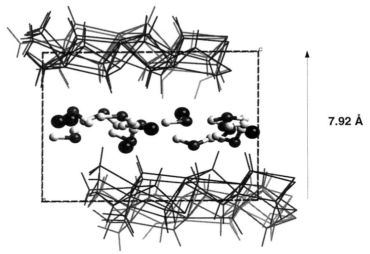

**7.92 Å**

**Figure 4.** *Snapshot of the simulation cell of the MgAl(Cl) R(3) model following 40 ps of MD simulation at 300 K. For clarity, only one interlayer of the simulation cell is displayed.*

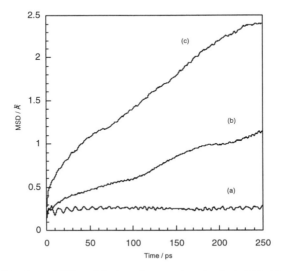

**Figure 5.** *Simulated MSDs of the interlayer anions in the (a) MgAl(CO₃) (b) MgAl(Cl) and (c) MgAl(NO₃) R(3) LDH models at 300 K.*

**Table 1.** Diffusion coefficients calculated for the interlayer anions in the model $MgAl(CO_3)$, $MgAl(NO_3)$ and $MgAl(Cl)$ R(3) LDHs.

| Interlayer anion | $D / cm^2s^{-1}$ per atom |
|---|---|
| carbonate | $0.79 \times 10^{-9}$ |
| nitrate | $0.13 \times 10^{-6}$ |
| chloride | $0.59 \times 10^{-7}$ |

## 3.2 Organic Interlayer Anions

*3.2.1 MgAl(terephthalate).* For a MgAl(terephthalate) LDH it has been found experimentally that the interlayer arrangement of the LDH is very sensitive to the layer charge density and water content of the LDH.[6,8,16] Two extreme orientations of the terephthalate anion with respect to the hydroxide layers of the LDH have been observed by powder X-ray diffraction: either vertical (corresponding to an interlayer spacing of approximately 14.0 Å) or horizontal (corresponding to an interlayer spacing of approximately 8.3 Å). In general, it has been observed that a vertical orientation is preferred for high layer charge (low R) and high interlayer water content, i.e., when the density of interlayer species is high. A horizontal orientation is preferred for low layer charge density (high R) and low water content.

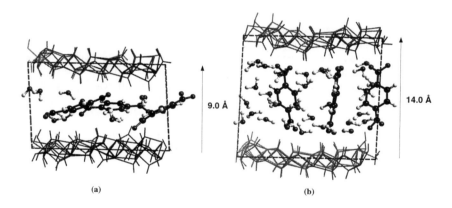

9.0 Å

14.0 Å

(a)

(b)

**Figure 6.** *Snapshot of the simulation cell of the MgAl(terephthalate) R(3) LDH following 40 ps of MD simulation at 300 K with (a) low interlayer water content (8 water molecules per interlayer) and (b) high water content (32 water molecules per interlayer).*

Although the models have the same initial interlayer spacing (14.4 Å) and terephthalate orientation (45°), the MD simulations predict substantially different interlayer arrangements depending upon the number of water molecules included in the simulation cell. Figure 6a is a snapshot of the simulation cell of the MgAl(terephthalate) R(3) model containing 8 water molecules per interlayer ($m = 0.33$), following 40 ps of MD simulation at 300 K. The simulated layer spacing is 9.0 Å which is in reasonable agreement with experimental values (8.4 Å) observed for a MgAl(terephthalate) R(3) LDH with low water content.[6,8] The computer simulation confirms, therefore, that for low interlayer water contents the terephthalate anions adopt an approximately horizontal arrangement with respect to the hydroxide layers. Figure 6b is a snapshot of the simulation cell of the MgAl(terephthalate) R(3) model containing 32 water molecules per interlayer ($m$ = 1.33), following 40 ps of MD simulation at 300 K. The terephthalate anion adopts a vertical orientation with respect to the hydroxide layers and the simulated layer spacing is 14.0 Å, which is in good agreement with experimental observations (14.0 Å).[6,8] It is interesting to notice that the majority of water molecules form a monolayer on the hydroxide surfaces, in contrast to the carbonate, nitrate and chloride models in which the water molecules lie in the mid-plane of the interlayer

For the MgAl(terephthalate) R(2) model, the MD simulations suggest that fewer water molecules (22 per interlayer) are required to ensure a vertical orientation of the terephthalate anion, presumably due to the higher layer charge and interlayer anion packing density compared with the R(3) model. There is, therefore, good agreement between the computer simulations and the experimental observations.

The properties of the interlayer water in the MgAl(terephthalate) R(2) LDH model containing 22 water molecules per interlayer were examined using data extracted from a 500 ps MD simulation. The simulated water O-O RDF for the model is very similar to that obtained by Boek *et al.*[19] from simulations of bulk water (Figure 7). A large, sharp peak at 2.9 Å and a broad, weak peak at approximately 5.6 Å in the O-O RDF represent the first and second coordination shells of water, respectively - probably within a single surface water monolayer. An additional broad and weak peak at approximately 7.5 Å, which corresponds to O-O distances between water monolayers, is also observed. The water self-diffusion constant, calculated for all the water molecules in the model ($1.14 \times 10^{-7}$ $cm^2 s^{-1}$ per atom) is the same order of magnitude as experimental data ($2.47 \times 10^{-7}$ per atom at 50 °C) obtained using quasi elastic neutron scattering.[22] As would be expected for water constrained between the layers of a LDH, the calculated water self-diffusion constant is significantly lower than the value obtained from simulations of bulk water (D = $1.88 \times 10^{-5}$ $cm^2 s^{-1}$ per atom).[19]

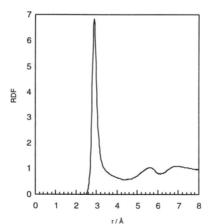

**Figure 7.** *Simulated water O-O RDF for the model MgAl(terephthalate) R(2) LDH containing 44 water molecules.*

*3.2.2 MgAl(benzenesulphonate).* Following 40 ps of MD simulation of the MgAl(benzenesulphonate) R(3) LDH model, there is good agreement between the calculated interlayer spacing (15.3 Å) and experimentally measured values (~ 15.5 Å).[12] The simulation shows that the benzenesulphonate anions remain in a bilayer arrangement, with the sulphonate group hydrogen-bonded to the hydroxide layers and the phenyl ring in the interlayer mid-plane (Figure 8). The anions are inclined at an angle of approximately 30° to the normal direction from the layers, similar to that suggested for a MgAl(benzoate) LDH.[23]

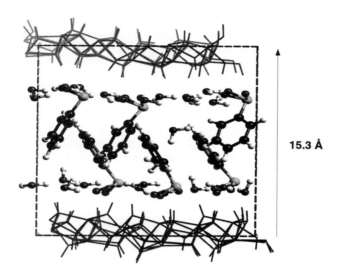

15.3 Å

**Figure 8.** *Snapshot of the simulation cell of the model MgAl(benzenesulphonate) LDH following 40 ps of MD simulation at 300 K. For clarity, only one interlayer of the simulation cell is displayed.*

A particularly interesting observation is that the simulation also shows that the geometry of the sulphur atom of the interlayer anion is significantly distorted from the tetrahedral geometry expected for the free anion. The observed distortion is an unexpected and interesting result, although it is not clear whether the distortion actually occurs in the real system, or is an artifact of the simulation. An additional observation is that when the distorted anion was extracted from the model and energy-minimized, the sulphur atom geometry remained distorted - indicating that it is in a local energy minimum. The geometry of the distorted molecule was also optimized using the MOPAC semi-empirical method; again the sulphur atom geometry remained distorted.

## 4 CONCLUSIONS

Molecular dynamics simulations, using a modified Dreiding force field, have been successful in predicting the interlayer spacing and anion orientation in a family of anion-exchange materials known as layered double hydroxides. Good agreement with experimentally observed values has been obtained for the simulated interlayer spacing of a model LDH containing the inorganic anions carbonate, nitrate or chloride as well as the organic anions terephthalate or benzenesulphonate.

In addition, diffusion coefficients for the exchange anions have been obtained for the models containing inorganic interlayer anions. The calculated mobility of the anions between the hydroxide layers increases in the order $NO_3^- > Cl^- > CO_3^{2-}$ - a trend which agrees with the experimentally observed relative selectivity of these anions.

For the model LDH containing interlayer terepthalate anions, the simulations successfully reproduce experimental observations that the interlayer arrangement of the LDH is very sensitive to the charge density on the layers (Mg / Al ratio) and the interlayer water content. The MD simulations have also provided insight to the structure and dynamics of the interlayer water molecules in these materials.

## ACKNOWLEDGEMENTS

The authors would like to thank Molecular Simulations Inc. for facilitating access to some of the simulation software, and Schlumberger Cambridge Research Ltd. and the EPSRC for financial support (CASE Award to SPN).

## REFERENCES

1. F. Cavani, F. Trifirò and A. Vaccari, *Catal. Today*, 1991, **11**, 173.
2. A. Vaccari, *Appl. Clay Sci.*, 1995, **10**, 1.
3. I. C. Chisem and W. Jones, *J. Mater. Chem.*, 1994, **4**, 1737.
4. S. Miyata, *Clays Clay Miner.*, 1983, **31**, 305.
5. M. A. Drezdzon, *Inorg. Chem.*, 1988, **27**, 4628.
6. S. P. Newman and W. Jones, *New J. Chem.*, 1998, **22**, 105.
7. S. Bonnet, C. Forano, A. de Roy and J. P. Besse, *Chem. Mater.*, 1996, **8**, 1962.
8. F. Kooli, I. C. Chisem, M. Vucelic and W. Jones, *Chem. Mater.*, 1996, **8**, 1969.
9. E. Kanezaki, S. Sugiyama and Y. Ishikawa, *J. Mater. Chem.*, 1995, **5**, 1969.
10. E. Kanezaki, K. Kinugawa and Y. Ishikawa, *Chem. Phys. Lett.*, 1994, **226**, 325.
11. R. Allmann and H. P. Jepsen, *Neues Jahrb. Miner. Monatsch*, 1969, **12**, 544.
12. S. P. Newman, *PhD Thesis, University of Cambridge*, 1998.
13. M. Aicken, I. S. Bell, P. V. Coveney and W. Jones, *Adv. Mater.*, 1997, **9**, 496.
14. I. S. Bell, F. Kooli, W. Jones and P. V. Coveney, *Materials Research Society Proceedings*, Spring 1996.
15. S. J. Williams, P. V. Coveney and W. Jones, *Mol. Sim.*, accepted, 1998.
16. S. P. Newman, S. J. Williams, P. V. Coveney and W. Jones, *J. Phys. Chem.*, accepted, 1998.
17. S. L. Mayo, B. D. Olafson and W. A. Goddard III, *J. Phys. Chem.*, 1990, **94**, 8897.
18. J. J. P. Stewart, MOPAC version 6.0, QCPE No. 455, 1990, Department of Chemistry, Indiana University.
19. E. S. Boek, P. V. Coveney, S. J. Williams and A. Bains, *Mol. Sim.*, 1996, **18**, 145.
20. A. K. Rappé and W. A. Goddard III, *J. Phys. Chem.*, 1991, **95**, 3358.

21. W. H. Hoover, *Phys. Rev. A*, 1985, 31.
22. W. Kagunya, P. K. Dutta and Z. Lei, *Physica B*, 1997, **234-236**, 910.
23. G. D. Moggridge, P. Parent and G. Tourillon, *Clays Clay Miner.*, 1994, **42**, 462.

# ORGANIC ANION INTERCALATION IN LAYERED DOUBLE HYDROXIDES

Sylvie Bonnet and William Jones

Department of Chemistry, University of Cambridge
Lensfield Road, Cambridge CB2 1EW UK

ABSTRACT

The synthesis of several types of organic intercalates of Mg-Al layered double hydroxides are described. The materials are made by co-precipitation. Anions considered are benzoate, benzenesulphonate and phenylphosphonate. The crystallinity of the products depends on the Mg/Al ratio within the hydroxide layer. At high Mg/Al values competition with nitrate anions during synthesis inhibits incorporation of the organic anion. In competitive uptake experiments the doubly charged phosphonate anion dominates.

## 1 INTRODUCTION

Layered double hydroxides (LDHs, also known as anionic clays and hydrotalcite-like materials) are lamellar compounds and are one class of anion exchanger.[1,2] They can be represented by the formula: $[M^{2+}_{1-x}M^{3+}_x(OH)_2]A^{n-}_{x/n} \cdot nH_2O$. Possible $M^{2+}$ species are Mg, Fe, Co, Cu, Ni, Cd or Zn and $M^{3+}$ are Al, Cr, Ga or Fe. An extensively studied example is the LDH resulting from combination of $Mg^{2+}$ and $Al^{3+}$. This LDH may be considered as derived from brucite $[Mg(OH)_2]$ by replacement of a fraction of the $Mg^{2+}$ by $Al^{3+}$.

The intercalation of organic anions into LDHs is of interest because of potential applications in catalysis, photochemistry, adsorption and electrochemistry.[3-5] Organic anion intercalated LDHs, prepared either directly or by anion exchange, have also been used as intermediate compounds in the preparation of pillared materials resulting from exchange of the organic with heptamolybdate and decavanadate polyoxoanions.[6] In this paper we compare the properties of LDHs containing the following types of anions: carboxylates,[7-9] sulfonates[10-12] and phosphonates[13-15] and in particular a Mg-Al LDH containing phenylphosphonate (PP), benzenesulfonate (BS) and benzoate (BZ) as interlayer anions.

We investigate how the Mg/Al ratio (R) of the metal salt solution used for co-precipitation (and hence charge density on the layers and anion exchange capacity (aec)) as well as the synthesis temperature influence uptake of the anion and the crystallinity of the product. We also consider the competitive intercalation between the three anions.

## 2 EXPERIMENTAL

The sample nomenclature used is as follows: $[Mg(R)-Al-X]_{cop}$ - where R is the $Mg^{2+}/Al^{3+}$ ratio of the metallic salt reagent solution used during co-precipitation, and X identifies the intercalated anion (PP, BS and BZ). For all preparations metal nitrates and sodium hydroxide (Aldrich, analytical grade reagents) were used along with deionised water.

## 2.1 Co-precipitation at constant pH

The constant pH co-precipitation method permits the direct synthesis of LDHs containing organic anions and is performed by adding an aqueous solution of the metal salt to an aqueous solution of the organic anions. Constant pH is maintained by the simultaneous addition of NaOH solution. In this study the pH was 10 - at this pH the Mg/Al ratio in the LDH which is formed is known to be close to the Mg/Al ratio in the starting salt solution.[2]

An example of the synthesis procedure is as follows. For an Mg/Al equal to R, V mL of a 0.5 M aluminium nitrate solution (x mol) and V' mL of a 0.5 M magnesium nitrate solution (y mol) were added to 50 mL of a PP solution (m g, z mol). Values of V, V', x, y, m and z are listed in Table 1. Addition of the metallic salt solution was over a 30 minute interval. In these preparations the amount of PP within the solution was ten times the theoretical value required to balance the positive charge of the layer. The LDH was recovered by filtration, washed with 1 L of deionised water and air-dried.

Using the same general synthesis procedure different parameters were systematically changed: temperature of synthesis (0, 25 and 65 °C ) or after formation of the gel at room temperature, the gel was hydrothermally aged (using a CEM microwave system) for one hour at approximately 200 °C. Time of aging was varied between one and two days for atmospheric preparations or one hour for hydrothermal.

Similar procedures were used for BS and BZ (Table 1).

## 2.2 Co-intercalation of mixed anions

Co-intercalates were prepared as follows: 10 mL of a 0.5 M aluminium nitrate solution (5.0 $10^{-3}$ mol) and 20 mL of a 0.5 M magnesium nitrate solution (1.0 $10^{-2}$ mol) were added to 50 mL of a mixed anion solution containing 7.21 g of BZ (5.0 $10^{-2}$ mol sodium salt) and 9.0 g of BS (5.0 $10^{-2}$ mol) or 7.21 g of BZ (5.0 $10^{-2}$ mol sodium salt) and 3.95 g of PP (2.5 $10^{-2}$ mol) or BS (5.0 $10^{-2}$ mol) and 3.95 g of PP (2.5 $10^{-2}$ mol). Aging was at room temperature for two days. The LDH product was recovered by filtration, washed with 1 L of deionised water and air-dried.

Table 2 presents the unit cell parameters of the compounds obtained. The PXRD patterns were indexed using an hexagonal lattice with a 3 R stacking sequence.

## 3 RESULTS

### 3.1 Co-precipitation

<u>BS</u>

Figure 1 shows PXRD patterns for different values of R for the [Mg(R)-Al-BS]$_{cop}$ series prepared at room temperature. The PXRD pattern of [Mg(1)-Al-BS]$_{cop}$ (Figure 1(a)) confirms the formation of an intercalated LDH (with an interlayer repeat of 15.67 Å). This value is similar to that reported by Meyn et al[11] for [Mg(2)-Al-BS] obtained by an anionic exchange process (15.9 Å). Certain reflections present in the pattern may be attributed to bayerite and gibbsite - the simultaneous precipitation of these Al(OH)$_3$ phases increases the Mg/Al ratio in the LDH as shown by the value of the d$_{110}$ reflection (1.512 Å). For R = 2 and 3 (Figures 1(b) and (c)) a well ordered and expanded intercalate with an interlayer repeat of 16.11 Å is produced.

**Table 1.** Preparative conditions for $[Mg_R\text{-Al-PP}]_{cop}$, $[Mg_R\text{-Al-BS}]_{cop}$ and $[Mg_R\text{-Al-BZ}]_{cop}$ : for Mg/Al = R, V mL of a 0.5 M aluminium nitrate solution (x mol) and V' mL of a 0.5 M magnesium nitrate solution (y mol) are added to 50 mL of a anion (PP, BS or BZ) solution (m mg, z mol).

| anion | R | V | x | V' | y | m | z |
|-------|------|------|------|------|------|------|------|
| | Mg/Al | (mL) | (mol) | (mL) | (mol) | (g) | (mol) |
| | 1 | 10 | $5.0 \times 10^{-3}$ | 10 | $5.0 \times 10^{-3}$ | 3.95 | $2.50 \times 10^{-2}$ |
| | 2 | 10 | $5.0 \times 10^{-3}$ | 20 | $1.0 \times 10^{-2}$ | 3.95 | $2.50 \times 10^{-2}$ |
| **PP** | 3 | 10 | $5.0 \times 10^{-3}$ | 30 | $1.5 \times 10^{-2}$ | 3.95 | $2.50 \times 10^{-2}$ |
| | 4 | 5.0 | $2.5 \times 10^{-3}$ | 20 | $1.0 \times 10^{-2}$ | 1.98 | $1.25 \times 10^{-2}$ |
| | 5 | 5.0 | $2.5 \times 10^{-3}$ | 25 | $1.25 \times 10^{-2}$ | 1.98 | $1.25 \times 10^{-2}$ |
| | 1 | 10 | $5.0 \times 10^{-3}$ | 10 | $5.0 \times 10^{-3}$ | 7.21 | $5.0 \times 10^{-2}$ |
| | 2 | 10 | $5.0 \times 10^{-3}$ | 20 | $1.0 \times 10^{-2}$ | 7.21 | $5.0 \times 10^{-2}$ |
| **BS** | 3 | 10 | $5.0 \times 10^{-3}$ | 30 | $1.5 \times 10^{-2}$ | 7.21 | $5.0 \times 10^{-2}$ |
| | 4 | 10 | $5.0 \times 10^{-3}$ | 40 | $2.0 \times 10^{-2}$ | 7.21 | $5.0 \times 10^{-2}$ |
| | 5 | 10 | $5.0 \times 10^{-3}$ | 50 | $2.5 \times 10^{-2}$ | 7.21 | $5.0 \times 10^{-2}$ |
| | 1 | 10 | $5.0 \times 10^{-3}$ | 10 | $5.0 \times 10^{-3}$ | 9.0 | $5.0 \times 10^{-2}$ |
| | 2 | 10 | $5.0 \times 10^{-3}$ | 20 | $1.0 \times 10^{-2}$ | 9.0 | $5.0 \times 10^{-2}$ |
| **BZ** | 3 | 5 | $2.5 \times 10^{-3}$ | 15 | $7.5 \times 10^{-3}$ | 4.5 | $2.5 \times 10^{-2}$ |
| | 4 | 5 | $2.5 \times 10^{-3}$ | 20 | $1.0 \times 10^{-2}$ | 4.5 | $2.5 \times 10^{-2}$ |
| | 5 | 5 | $2.5 \times 10^{-3}$ | 25 | $1.25 \times 10^{-2}$ | 4.5 | $2.5 \times 10^{-2}$ |

For the case of R = 3 (Figure 1(c)) the (003) reflection is broader than for R = 1 or 2 and the (006) reflection has a higher relative intensity compared with the patterns for R = 1 and 2. This enhanced relative intensity of the (006) reflection may be attributed to the presence of a nitrate-LDH phase - with the (003) of the nitrate phase overlapping with the (006) of the sulphonate LDH.

Support for this interpretation comes from the fact that in the case of R = 4 and 5 (Figure 1(d) and (e)) the LDHs obtained are not intercalated by BS anions but by nitrate anions - the basal spacings observed being 7.97 Å and 8.08 Å, respectively. The fact that nitrate rather than BS is intercalated may be related to the relative amounts of nitrate and BS within the synthesis mixture. All the syntheses are made with the same BS/Al ratio (equal to 10) in the solution. As R increases the amount of nitrate relative to BS also increases - such that at high values of R nitrate dominates.

We conclude, therefore, that for BS under these conditions a well crystallized LDH phase is produced when the Mg/Al ratio of the starting metal salt solution is close to 2. The interlayer repeat of ca. 16 Å suggests furthermore that a bilayer-like structure has been formed, with the sulphonate group linked by hydrogen bonding to the hydroxide sheets and the phenyl ring located within the centre of the interlayer.

## BZ

PXRD patterns of [MgR-Al-BZ]$_{cop}$ are shown in Figure 2. The intercalation process for BZ generally agrees with that for BS. One difference, however, concerns the synthesis with a Mg/Al ratio equal to 1 (Figure 2(a)). For BZ, intercalation is not complete and a LDH phase intercalated by nitrate anions is also obtained. The most crystalline LDH appears to be formed at an Mg/Al ratio equal to 2 (Figure 2(b)) with the anions again present as a bilayer. For R = 3 (Figure 2(c)) the relative intensity of the second diffraction line ((003) of [Mg-Al-NO$_3$] and (006) of [Mg-Al-BZ]) is higher compared to the one observed for [Mg3-Al-BS]$_{cop}$ (Figure 1(c)) and again suggests that nitrate competition becomes important. (Figures 2 (d) and (e)).

## PP

The PXRD patterns of the PP products are shown in Figure 3. For R = 1 (Figure 3(a)) an intercalate with d$_{003}$ equal to 17.05 Å is observed. This intercalate is less crystalline than for R = 2 (Figure 3(b)) For R = 3 (Figure 3(c)) the diffraction lines are generally very broad, although some (for example the first reflection) are asymmetric suggesting the superposition of a broad and a pronounced sharp diffraction line. For R = 4 and 5, the PXRD patterns (Figure 3(d) and (e)) suggest a biphasic compound. The first phase is typified by the sharp and intense reflection present at low 2θ (spacing of 14.20 Å). Other reflections which are present and which are of lower intensity and also broader correspond to an LDH intercalated by nitrate anions with a basal spacing equal to 8.24 and 7.90 Å for R = 4 and 5, respectively. Increasing the value of R from 3 to 5, leads to an increase in the intensity of the sharp diffraction lines.

The phase characterised by the sharp reflection at 14.2 Å is identified as a magnesium phosphonate salt $Mg(O_3PC_6H_5).H_2O$. In the crystal lattice of this salt the phosphonate anion is covalently bonded via the oxygen atoms to a layer of Mg octahedrally coordinated to five phosphonate oxygen atoms and one water molecule.[16]

**Table 2.** PXRD data of LDH.

| Compound | T (°C) | aging | $d_{003}$ (Å) | c ‡ (Å) | $d_{110}$ (Å) | a ‡ (Å) |
|---|---|---|---|---|---|---|
| [Mg1-Al-PP]cop | 25 | 2 days | 17.05 | 51.15 | 1.51 | 3.02 |
| [Mg2-Al-PP]cop | 25 | 2 days | 16.11 | 48.33 | 1.514 | 3.03 |
| [Mg3-Al-PP]cop | 25 | 2 days | 14.20 | 42.60 | 1.518 | 3.04 |
| [Mg4-Al-PP]cop | 25 | 2 days | 14.15 | 24.72 | 1.531 | 3.06 |
| [Mg5-Al-PP]cop | 25 | 2 days | 14.28 | 23.70 | 1.538 | 3.08 |
| [Mg1-Al-BS]cop | 25 | 2 days | 15.67 | 47.01 | 1.512 | 3.02 |
| [Mg2-Al-BS]cop | 25 | 2 days | 16.11 | 48.33 | 1.518 | 3.03 |
| [Mg3-Al-BS]cop | 25 | 2 days | 16.11 | 48.33 | 1.530 | 3.06 |
| [Mg4-Al-BS]cop | 25 | 2 days | 7.97 | 23.91 | 1.536 | 3.07 |
| [Mg5-Al-BS]cop | 25 | 2 days | 8.08 | 24.24 | 1.541 | 3.08 |
| [Mg1-Al-BZ]cop | 25 | 2 days | 15.10 | 45.30 | 1.518 | 3.04 |
| [Mg2-Al-BZ]cop | 25 | 2 days | 16.11 | 48.33 | 1.510 | 3.02 |
| [Mg3-Al-BZ]cop | 25 | 2 days | 16.11 | 48.33 | 1.520 | 3.04 |
| [Mg4-Al-BZ]cop | 25 | 2 days | 7.97 | 23.91 | 1.530 | 3.07 |
| [Mg5-Al-BZ]cop | 25 | 2 days | 7.97 | 23.91 | 1.540 | 3.08 |
| [Mg3-Al-PP]cop | MW * | 1 hour | 11.12 | 33.36 | 1.532 | 3.06 |
| [Mg3-Al-PP]cop | 65 | 1 day | 14.15 | 42.42 | 1.527 | 3.05 |
| [Mg3-Al-PP]cop | 25 | 1 day | 15.67 | 47.01 | 1.521 | 3.05 |
| [Mg3-Al-PP]cop | 0 | 1 day | 16.57 | 49.71 | 1.510 | 3.02 |
| [Mg3-Al-PP]cop | 0 | 2 day | 16.57 | 49.71 | 1.510 | 3.02 |
| [Mg2-Al-PP]cop | 0 | 1 day | / | / | / | / |
| [Mg2-Al-PP]cop | 0 | 2 days | 17.05 | 51.15 | 1.500 | 3.00 |
| [Mg4-Al-PP]cop | 0 | 1 day | 17.05 | 51.15 | 1.51 | 3.02 |
| *Mixed anion intercalation* | | | | | | |
| [Mg2-Al-(PP+BZ)]cop | 25 | 2 days | 15.67 | 47.01 | 1.507 | 3.01 |
| [Mg2-Al-(PP+BS)]cop | 25 | 2 days | 15.67 | 47.01 | 1.507 | 3.01 |
| [Mg2-Al-(BS+BZ)]cop | 25 | 2 days | 16.11 | 48.33 | 1.510 | 3.02 |

* MW stands for microwave, ‡ a and c parameter are calculated assuming an hexagonal symmetry
of the LDH.

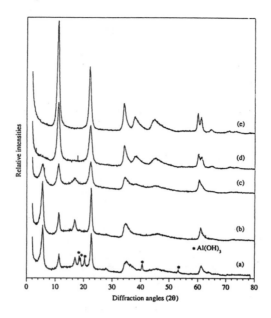

Figure 1. PXRD patterns for [Mg(R)-Al-BS]$_{cop}$ for R values of (a) 1, (b) 2, (c) 3, (d) 4 and (e) 5.

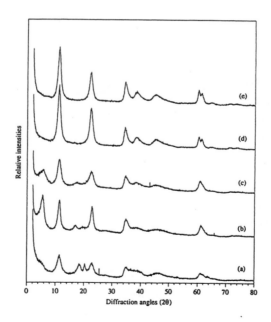

Figure 2. PXRD patterns for [Mg(R)-Al-BZ]$_{cop}$ for R values of (a) 1, (b) 2, (c) 3, (d) 4 and (e) 5.

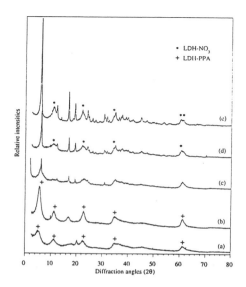

Figure 3. PXRD patterns for [Mg(R)-Al-PP]$_{cop}$ for R values of (a) 1, (b) 2, (c) 3, (d) 4 and (e) 5.

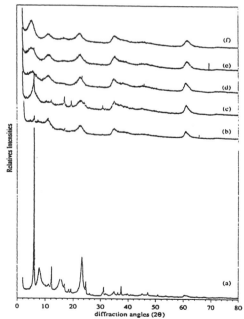

Figure 4. PXRD patterns for [Mg(3)-Al-PP]$_{cop}$ for (a) hydrothermal, (b) 65 °C 1 day, (c) RT 2 days, (d) RT 1 day and (e) 0 °C 1 day.

### 3.2 Influence of synthesis temperature on PP incorporation

Since R = 3 is close to the point where salt formation appears to become significant in PP incorporation, this ratio was examined in more detail and the synthesis of [Mg(3)-Al-PP]$_{cop}$ repeated at a range of temperatures between 0 and 65 °C as well as hydrothermally.

The PXRD patterns for a sample prepared by microwave-hydrothermal aging (200 °C autogeneous pressure) is shown in Figure 4(a). The product is clearly biphasic. The LDH phase has a basal spacing equal to 11.12 Å. Such a spacing leads to a gallery height of 6.32 Å which is not compatible with a bilayer arrangement of the PP or even to a single monolayer of PP perpendicular to the hydroxylated sheets. With a bifunctional phenyl phosphonate anion, an LDH phase with a spacing equal to 9.8 Å has been reported, and it was suggested that a grafting of the anion to the hydroxylated layer had occurred.[14] In the case of PP, however, grafting is unlikely since the PP anions are readily exchanged by carbonate - with a 7.77 Å carbonate phase being produced. It is likely that the hydrothermal product may be best described as a tilted PP monolayer.

Figure 4(b-f) shows PXRD patterns of [Mg(3)-Al-PP]$_{cop}$ compounds prepared at different temperatures of co-precipitation. A pure LDH phase is obtained when prepared at 0 °C with aging of either one or two days. If the synthesis gel is allowed to age for one day at room temperature an LDH with a basal spacing of 15.67 Å is obtained although the crystallinity of the LDH is poor. Increasing the temperature to 65 °C, with one day aging, gives rise to a very poorly crystallized LDH with the emergence of the magnesium phosphonate salt.

### 3.3 Pair-wise competitive anion intercalation

Room temperature competitive anion intercalation experiments with BZ, BS and PP were undertaken at pH 10. A value of R = 2 was chosen since fully intercalated and well crystallized LDHs were obtained for PP, BZ or BS at this ratio. The anions were mixed in a 50/50 ratio, each anion being in 10 fold charge excess. PXRD patterns of the products obtained are shown on Figure 5. All the diffractograms are characteristic of an intercalated LDH, exhibiting basal spacings between 15.67 and 16.11 Å and a-parameter between 3.01 - 3.02 Å.

For the mixing of PP with either BS or BZ the infrared spectra do not contain the characteristic bands of -CO$_2^-$ or -SO$_3^-$. The incorporation of PP dominates that of BZ and BS - it being likely that the higher charge of PP facilitates the intercalation of this anion.

When the mixed anions are BZ and BS, the C/S ratio of the LDH obtained is equal to 12.2 which is close to the theoretical value of 13 assuming equal uptake. The infrared spectrum of this compound exhibits bands for both anions: $v_s$COO$^-$ at 1371 cm$^{-1}$, $v_{as}$COO$^-$ at 1545 cm$^{-1}$, $v_s$S=O at 1040 cm$^{-1}$ and $v_{as}$S=O at 1182 cm$^{-1}$. At this stage it is not possible to distinguish between the mixing of the anions within the same or separate interlayers or indeed separate crystallites.

### 4 DISCUSSION

Organic LDH intercalates for PP, BZ and BS were prepared using a wide range of Mg/Al ratios. For LDHs with a low (about 1) or high value of R (R > 3) it was difficult to obtain pure phases by co-precipitation at room temperature:

LDHs with high values of R have a lower anionic exchange capacity (aec) than those with low R,[3] and typically a [Mg(5)-Al-NO$_3$] LDH has an aec of 240 meq/100 g of

dried LDH - with the value increasing to 550 meq/100 g of dried LDH for [Mg(1)-Al-NO$_3$].
The composition of the LDH layers greatly influences organic anion intercalation process.
In the case of BZ or BS, if a high ratio is used (typically 4 or 5) there is no intercalation of
the anion because of the high excess of nitrate anions present in the solution. Intercalation
of BX and BS is possible for [Mg(3)-Al] and [Mg(2)-Al] LDH. For R = 1 the presence of
aluminium hydroxide is usually observed. A similar influence of R is see for PP
intercalation. For R equal to 1, 2 or 3, LDH-PP compounds are formed though for R = 1
the presence of a small amount of aluminium hydroxide is detected. At high Mg/Al ratios,
as for BZ and BS, it was impossible to form a PP intercalated LDH. PP intercalation is
further complicated by the competitive formation of a M(II) phenylphosphonate salt. In the
synthesis mixture used here [(Mg(NO$_3$)$_2$, Al(NO$_3$)$_3$, PP and NaOH)] there is a competition
between LDH formation (at low ratio Mg/Al) and Mg phenylphosphonate salt formation
(dominant for high R).

Despite the possible formation of a phosphonate salt, the intercalation of PP into the
LDH with high Mg/Al ratio (typically 3 or 4) can be obtained at 0 °C. while intercalation of
BZ and BS at R = 3 or 4 does not occur even at higher temperatures. The three anions have
very similar dimensions, and we conclude that the charge on the anions is crucial (PP, -2:
BZ and BS -1). Competitive intercalation experiments also show the preference for PP over
BZ and BS.

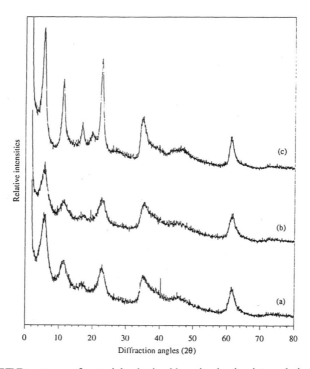

Figure 5. PXRD patterns of materials obtained by mixed anion intercalation (a) PP + BZ,
(b) PP + BS and (c) BZ + BS.

## Acknowledgment

We are grateful to Region Auvergne for support for SB.

## References

1.    F. Cavani, F. Trifiro and A Vaccari, *Catal. Today,* **1991**, *11, 173*
2.    A. de Roy, C. Forano, K.E. Malki and J. P. Besse, in *Anionic Clays: Trends in Pillaring Chemistry, Synthesis of Microporous Materials*; ed. M. I. Occelli and H. E. Robson, Van Nostrand Reinhold: New York, 1992; Vol. 2, p 108.
3.    S. Newman and W. Jones, *New J. Chem.,* **1998**, *22,* 105.
4.    S. Bonnet, C. Forano, A. de Roy and J. P. Besse, *Chem. Mater.,* **1996**, *8,* 1962.
5.    S. Carlino, *Solid State Ionics* **1997**, *98,* 73.
6.    M. A. Drezdzon, *Inorg. Chem.,* **1988**, *27,* 4628.
7.    F. Kooli, F and W. Jones, in *Synthesis of Porous Materials. Zeolites, Clays and Nanocomposites,* ed. M. Occelli, H. Kessler; Marcel Dekker, New York, 1996, pp 641.
8.    M. Chibwe, J. B. Valim and W. Jones, in *Multifunctional Mesoporous Inorganic Solids,* ed. C. A. C. Sequeira and M. J. Hudson; Kluwer Academic Publishers: Dordrecht, 1992; Vol. 400, p 191.
9.    M. Vucelic and W. Jones, in *Multifunctional Mesoporous Inorganic Solids,* ed. C. A. C. Sequeira and M. J. Hudson, Kluwer Academic, Dordrecht, 1992, vol 400, p 373.
10.   T. Kuwahara, O. Onitsuka; H. Tagaya, J. Kadokawa, and K. Chiba, *J. Incl. Phenom., Mol. Recog. Chem.,* **1994**, *18,* 59.
11.   M. Meyn, K. Beneke and G. Lagaly, *Inorg. Chem.,* **1990**, *29,* 5201.
12.   M. Meyn, K. Beneke and G. Lagaly, *Inorg. Chem.,* **1993**, *32,* 1209.
13.   F. M. Vichi, O. L. Alves, *J. Mat. Chem.,* **1997**, *7,* 1631.
14.   J. D. Wang, G. Serrette, Y. Tian and A. Clearfield, *Appl. Clay Sci.,* **1995**, *10,* 103.
15.   S. Carlino, M. J. Hudson, S. W. Husain and J. A. Knowles, *Solid State Ionics* **1996**, *84,* 117.
16.   G. Cao, L. Haiwon, V. M. Lynch and T. E. Mallouk, *Inorg. Chem.,* **1988**, *27,* 2781.

E-mail: wj10@cam.ac.uk

# CERIUM(IV) PHOSPHATE: CHARACTERISATION AND INTERCALATION OF AMINES

Masami Kaneyoshi and William Jones

Department of Chemistry, University of Cambridge
Lensfield Road, Cambridge, CB2 1EW, U.K.

## ABSTRACT

Using previously reported methods several cerium(IV) phosphates (CePs) were prepared. On the basis of elemental analysis, PXRD, TGA, IR and amine uptake, we conclude that for those phosphates for which P/Ce ~ 2 there are at least two layered structures, neither of which is isostructural with the well-known $\alpha$- or $\gamma$- M(IV) phosphate structures. A structure with P/Ce~1.5 also appears to be layered. Powder X-ray measurements monitoring amine uptake by the phosphates are also reported.

## 1 INTRODUCTION

Cerium(IV) phosphates (CePs) have been studied since the late 1960's.[1-5] They have received attention because of their cation exchange capacity and potential application for trapping of radioactive cations.[6,7] In addition, proton conductivity studies have been reported.[8,9] Compared with other metal(IV) phosphates, however, and especially zirconium(IV) phosphates, CePs have not been as fully characterised. Although crystal structures have not been reported, it was suggested that some CePs have layered structures.[10] Support for this layered nature was provided by the fact that changes in basal spacing occur as a function of relative humidity.[2,11] More recently a new hydrothermal synthetic route to CeP formation was described and a new CeP phase reported.[11]

The purpose of this work is to further characterise CeP and to study the intercalation of amine into the CePs in a manner similar to that reported for ZrP and SnP.[12,13]

## 2 EXPERIMENTAL

### 2.1 Synthesis of Materials

$CeO_2$ and $H_3PO_4$, were hydrothermally reacted following the method of Tsuhako et al.[11] 9.45 g (0.055 mole) of $CeO_2$ and 25.27 g of 85% phosphoric acid (0.22 mole $H_3PO_4$)

were mixed in a 25 ml beaker and stirred. The beaker was placed in an autoclave and heated at 180°C for seven hours. The pressure inside the reactor was about 500 kPa. The reaction product was cooled and filtered. The precipitate was collected and dispersed in distilled water, filtered, washed repeatedly with water and air dried. The yield based on Ce was 98 %. A part of this resulting phosphate was stored at room temperature in a desiccator with a saturated aqueous solution of $K_2SO_4$. The relative humidity in the desiccator was 97 %. This sample was designated CePH. Another part of the phosphate was stored at r.t. in a desiccator together with solid $P_2O_5$ - relative humidity close to 0 % (CePD). A third part was calcined in an oven at 170°C in air, and then kept in a desiccator under conditions similar to those for CePD - sample designated CePB.

Fibrous cerium(IV) phosphate (CePF) was synthesised by precipitation from solution, using the same method as Alberti et al.[1]

Another crystalline CeP (CePL) was synthesised hydrothermally as follows: 0.04 mole of freshly precipitated cerium(IV) hydroxide was dissolved with a stoichiometric amount of 70% nitric acid. This solution and 85% phosphoric acid (0.06 mole $H_3PO_4$) were placed in a Teflon tube and mixed. The resulting yellow viscous solution was heated in a sealed reactor (using microwaves) for one hour, with the pressure inside the tube being kept at 1.5 MPa - temperature estimated to be 200°C. The product was filtered, washed with 10 ml of 7% nitric acid, then repeatedly with distilled water and finally dried and stored in a desiccator with $P_2O_5$. The yield based on Ce was 65%.

Amine intercalated compounds were prepared by solid-vapour reaction, following literature methods.[13]

## 2.2 Characterisation Techniques

The Ce and P content was analysed by ICP-AES, using a SEIKO SPS-4000 spectrometer. C, H, N content for the amine intercalated compounds were analysed by combustion in oxygen, using an Exeter Analytical CE-400 Elemental Analyser.

Thermogravimetric analysis (TGA) was carried out with a Polymer Laboratories TGA1500 analyser. Data were recorded from r.t. to 450°C under 20 ml/min $N_2$ flow with heating rate of 5°C/min.

Powder X-ray diffraction (PXRD) was carried out using a Phillips PW1710 diffractometer with $CuK\alpha$ radiation (Ni filtered). Samples were usually ground and packed onto a glass sample holder. The patterns were collected with a step scan of 0.04° $2\theta$, count time of 1 sec at each step. For the determination of unit cell parameters, patterns were recorded with a step scan of 0.01° with a 8 sec count at each step. The relative humidity in the sample chamber was controlled at appropriate values.

Powder XRD patterns for some CePs were also measured in a sedimented form. The CeP was dispersed with ultrasonic radiation in water or ethanol, and the suspension deposited onto a glass plate and air dried. If necessary, the procedure was repeated. The sample was stored at an appropriate relative humidity.

Fourier transform infrared(FT-IR) spectra were recorded with a Perkin-Elmer Paragon 1000 spectrometer using the KBr pellet technique. Resolution was 1 cm$^{-1}$ and 64 scans were averaged.

## 3 RESULTS

### 3.1 Synthesised Materials

Table 1 gives details of P/Ce and H/Ce ratios as determined by chemical analysis. The TGA mass loss curves are shown in Figure 1. For CePH and CePD, the curves consist of two mass losses with a plateau centered around 200°C. The first loss is associated with desorption of interlayer water (see later). In CePB, this loss is not present, indicating that interlayer water is absent. The second loss corresponds to condensation of $HPO_4^{2-}$ to $P_2O_7^{4-}$. The curve of CePL basically consists of one mass loss assigned to condensation between $HPO_4^{2-}$ and $OH^-$. CePF shows a large amount of mass loss broadly throughout the measured temperature range. The powder X-ray diffraction patterns are shown in Figure 2. We conclude that CePH corresponds to the materials made by Tsuhako et al.[11]; CePB, CePF and CePL correspond to those reported by Alberti.[1,2] CePD is likely to be a intermediate between CePH and CePB.

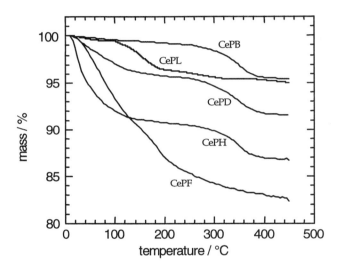

**Figure 1**  *TGA mass loss curves of CePs*

**Table 1** *The chemical compositions of CePs*

| Species | P/Ce atomic ratio | H/Ce atomic ratio | chemical formula |
|---------|-------------------|-------------------|------------------|
| CePH | 1.97 | 5.70 | $Ce(HPO_4)_2 \cdot 1.85H_2O$ |
| CePD | 1.97 | 3.34 | $Ce(HPO_4)_2 \cdot 0.60H_2O$ |
| CePB | 1.97 | 2.04 | $Ce(HPO_4)_2$ |
| CePF | 2.01 | 6.98 | $CeO(H_2PO_4)_2 \cdot 1.50H_2O^a$ |
| CePL | 1.43 | 1.57 | $Ce(HPO_4)(PO_4)_{0.5}(OH)_{0.5}{}^a$ |

a based on previously reported study (ref. 4 and 7)

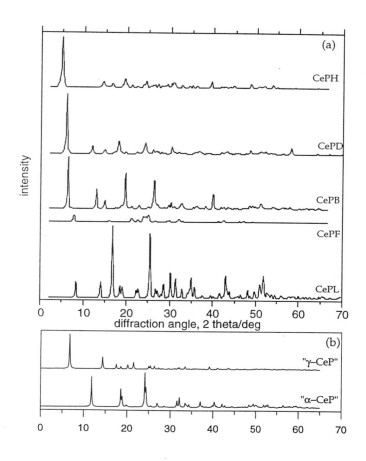

**Figure 2** *Powder XRD patterns of CePs: (a) observed (b) calculated for hypothetical structures. The low intensity of the CePF pattern results from the poor crystallinity of the material.*

## 3.2 Unit Cells of CePs

The PXRD patterns were indexed using the auto indexing program TREOR90[14] and the resulting cells were refined by least-squares methods using the program LAPOD.[15] Proposed unit cells are shown in Table 2, along with reliability indices.[16,17] Paticularly noteworthy is the fact that of a total of 86 reflections observed for CePL only one could not be indexed [F(85) = 36.0 (0.0107, 220)]. High resolution data for this phase is currently being obtained for Rietveld analysis.

**Table 2** *Proposed unit cells for CePs and reliability of indexing*

|  | CePH | CePD | CePB | CePL |
|---|---|---|---|---|
| system | triclinic | triclinic | triclinic | monoclinic |
| space group | P1 or P $\bar{1}$ | P1 or P $\bar{1}$ | P1 or P $\bar{1}$ | Cc or C2/c |
| a / Å[a] | 18.361(0.0046) | 14.667(0.0049) | 13.594(0.0023) | 21.043(0.0023) |
| b / Å | 7.566(0.0026) | 7.247(0.0027) | 8.466(0.0029) | 6.5579(0.00080) |
| c / Å | 9.690(0.0031) | 9.648(0.0025) | 8.971(0.0022) | 6.9491(0.00089) |
| $\alpha$ / ° | 99.17(0.032) | 109.52(0.035) | 97.42(0.031) | 90.00 |
| $\beta$ / ° | 96.98(0.032) | 88.27(0.029) | 94.75(0.025) | 91.95(0.013) |
| $\gamma$ / ° | 88.99(0.024) | 91.53(0.041) | 94.30(0.035) | 90.00 |
| cell volume/Å³ | 1319.17 | 947.95 | 1016.52 | 958.38 |
| basal spacing/Å | 18.20 | 14.66 | 13.50 | 10.52 |
| M(20)[b] | 5.9 | 5.7 | 7.1 | 44.1 |
| X(20)[c] | 2{7.0,3.0} | 3{3.9,1.6,1.5} | 1{1.6} | 0 |
| F(30)[d] | 8.5[0.0189,186] | 4.9[0.0137,448] | 10.4[0.0135,212] | 76.3[0.0096,41] |

a The figures in parentheses are the errors in the cell parameters

b The value which indicates the reliability of indexing. The larger this value the more reliable the indexing. (ref. 16)

c The number of peaks which could not be indexed, the diffraction angle of which is smaller than 20th indexed peak. The figure in { } is the relative intensity of the unindexed peaks (strongest = 100).

d The value which indicates the reliability of indexing. The larger this value the more reliable the indexing. The meaning of the figure in [ ] is also defined in reference 17.

### 3.3 Fourier Transform Infrared (FT-IR) Spectra

FT-IR spectra are shown in Figure 3. A band around 3180 cm$^{-1}$, due to hydration water, is well resolved in CePH, and as a shoulder in the spectrum of CePD. It is absent in the spectrum of CePB. The strong absorption at 1230-1240 cm$^{-1}$ which was observed in all CePs other than CePF is characteristic of HPO$_4^{2-}$.[4] The absence of this band in CePF confirms that HPO$_4^{2-}$ is absent. All spectra other than that for CePF have several bands in the 1150-900 cm$^{-1}$ region. In the case of CePF a band at 1064 cm$^{-1}$ has been attributed by Hayashi et al. to H$_2$PO$_4^-$.[7]

### 3.4 Sedimentation PXRD

In order to determine whether CePL is layered, sedimented samples were investigated by PXRD. Since layered materials generally consist of thin plates such materials will deposit as oriented crystals on sedimentation (Figure 4) and consequently, in the PXRD pattern, (h00) reflections will be emphasised and reflections such as (0kl) will be weaker or absent.

The PXRD patterns of packed and sedimented samples of CePD and CePL are shown in Figure 4. There is a clear difference in peak intensities between two sampling methods. That CePD has a layered structure is supported by the result of amine intercalation (see later), and the fact that CePL shows a similar trend suggests that it also has a layered structure. That CePL does not show ion-exchange and amine uptake[7] has led to the suggestion that it does not have a layered structure. The inertness of CePL to amine uptake, however, may be due to it not having sufficiently acidic protons in the interlayer region to drive amine intercalation.

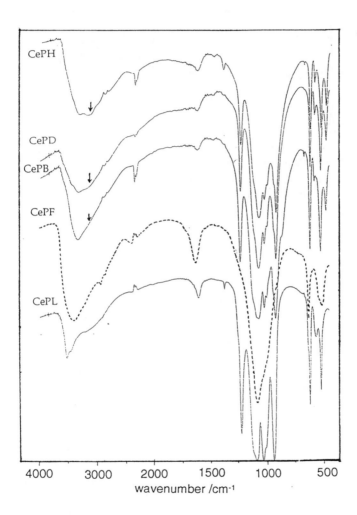

**Figure 3** *FT-IR spectra of CePs. The spectra of CePF clearly differs from those of the other samples.*

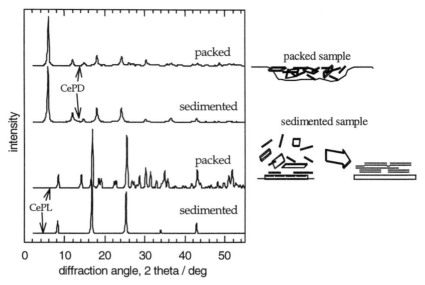

**Figure 4**    *XRD patterns of packed and sedimented samples (left)*
*the sampling methods (right)*

### 3.5 Intercalation of Amines into CePs

The chemical compositions and basal spacings of products of reaction of amines and CePs are shown in Table 3. All CePs other than CePL absorbed n-butylamine and resulted in an increased basal spacing - i.e. intercalation has occurred providing further evidence for layered structures. Mata-Arjona et al.[18] studied the retention of n-butylamine by CeP from cyclohexane solution. Our result for CePD is consistent with theirs. For CePB, pyridine and piperidine uptake was compared. Pyridine was only slightly adsorbed, in contrast to piperidine, which has a similar vapour pressure and molecular size to pyridine. This difference probably results from the lower basicity of pyridine ($pK_b$ = 8.8) compared to piperidine ($pK_b$ = 2.8). We conclude that the acidity of CePB is weak.  Water was co-adsorbed with ammonia, ethylamine, and piperidine. For the former two amines, aqueous solutions were used to expose CePB to vapour and water could have adsorbed at the same time as the amine. In the case of piperidine water must have adsorbed after amine intercalation and as a result of expose of the sample to air. Hayashi et al. have reported the intercalation of n-octylamine into CePF.[7] However in the

**Table 3** *Results and experimental conditions of reaction of amines and CePs*

| | temp. | time/day | composition amine/Ce | $H_2O^a$ /Ce | basal spacing/Å | increase of bas.sp./Å |
|---|---|---|---|---|---|---|
| CePH n-butylamine | r. t. | 2 | 1.35 | 0.56 | 25.2 | 7.0 |
| CePD — | r. t. | 2 | 0.68 | 0.19 | 24.5 | 9.8 |
| CePB — | r. t. | 2 | 0.95 | 0.00 | 25.2 | 11.7 |
| CePF — | r. t. | 2 | (not analysed) | | (no peak observed) | |
| CePL — | r. t. | 2 | 0.00 | 0.00 | 10.5 | 0.0 |
| CePB ammonia | r. t. | 1 | 1.21 | 0.50 | 13.6 | 0.1 |
| — ethylamine | r. t. | 1 | 1.61 | 1.01 | 18.3, 21.2[b] | 4.8, 7.7[b] |
| — n-butylamine | r. t. | 2 | 0.95 | 0.00 | 25.2 | 11.7 |
| — n-hexylamine | 60°C | 1 | 1.11 | 0.00 | 30.0 | 16.5 |
| — n-octylamine | 60°C | 2 | 0.75 | 0.00 | 33.2 | 19.7 |
| — pyridine | 60°C | 1 | 0.05 | 0.38 | 16.0 | 2.5 |
| — piperidine | 60°C | 1 | 1.54 | 0.58 | 21.5 | 8.0 |

a interlayer water        b two phases observed

present work, the reaction of n-butylamine and CePF does not result in clear evidence for intercalation although a reaction between the amine and CePF occurs (the PXRD pattern becoming amorphous).

Figure 5 shows the powder XRD patterns of various amine intercalated CePBs. The relationship between the length of the amine alkyl-chain and the basal spacing of the intercalation compound is shown in Figure 6. A linear relation is observed for which the slope of line, $\Delta d/\Delta Nc$, is 2.46 Å. This suggests, following similar arguments for uptake by $\alpha$-Sn(IV) phosphate,[13] that alkylamines are intercalated as a bimolecular film, Figure 7.

TGA mass loss curves of n-butylamine intercalated into the CePs are shown in Figure 8. A large mass loss occurs in the region 150 to 200°C, associated with the desorption of the amine. This is a higher temperature than the boiling point (78 °C) of the amine, and results from the interaction between amine and phosphate layers. The difference between the curve of CePH-Bu and the others indicates that water is desorbed prior to the amine. The 1700 to 1500 cm$^{-1}$ region of the FT-IR spectra of various alkylamine intercalated into CePB is shown in Figure 9. For all except ethylamine a band at 1553 to 1560 cm$^{-1}$ was observed. This band is characteristic of the -NH$_3^+$ group. There was also a band at 1607-1611 cm$^{-1}$. This band is assigned to -NH$_2$ deformation of the neutral form of the amine. Thus the amines exist in the interlayer both as protonated and neutral forms. Ethylamine seems to exist entirely in the neutral form. This may be related to the existence of co-intercalated water.

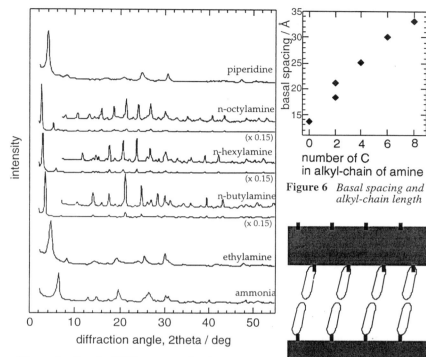

**Figure 5**  *Powder XRD patterns of amine-intercalated CePBs*
*For some phases, (hoo) reflections are very clear - well-ordered layered structures.*

**Figure 6**  *Basal spacing and alkyl-chain length*

**Figure 7**  *Model of amine-intercalated CePs*

## 4 DISCUSSION

It is important to consider the structure of CePs compared to those of other M(IV) phosphates. The structure of zirconium phosphate(ZrP) of $\alpha$-type[19] and $\gamma$-type[20] are known. Isostructural families of $\alpha$-type (for TiP, SnP, PbP)[21] and $\gamma$-type (TiP) phosphate[22] are also known. Unit cell parameters of hypothetical "$\alpha$-CeP" and "$\gamma$-CeP" may be predicted through extrapolation from the reported parameters of TiP and ZrP, using appropriate ionic radii of Ce(IV). Such hypothetical unit cells are shown in Table 4. Powder XRD patterns of possible $\alpha$-and $\gamma$- CeP may be calculated[23] using the crystal data of $\alpha$-ZrP[19] and $\gamma$-TiP.[22] The calculated patterns are shown in Figure 2 and compared with the observed patterns. It is concluded that none of the CePs of the present work are isostructural with either $\alpha$- or $\gamma$-ZrP. It was reported that two moles of amines were intercalated into one mole of $\alpha$-ZrP.[12] This is in contrast with this work (see Table 3) and provides further negative evidence for an $\alpha$-type CeP.

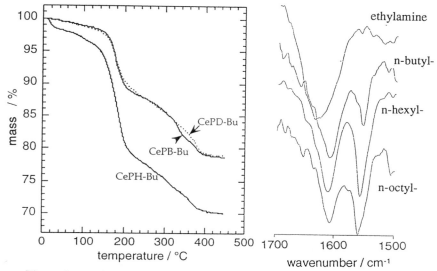

**Figure 8** *TGA mass loss curves of n-butylamine-intercalated CePs*

**Figure 9** *FT-IR spectra of amine-intercalated CePBs*

## 5 CONCLUSION

Cerium(IV) phosphate was synthesised and characterised by means of chemical analysis, TGA, PXRD and FT-IR. We conclude that: CePH is the same material as that synthesised by Tsuhako et al.[11] CePD and CePB are equivalent to materials reported by Alberti et al.[2] These three have layered structures (amine-intercalation) and contain $HPO_4^{2-}$ (FT-IR). They have similar layered structure and differ only in the amount of interlayer water (and hence basal spacing). Each was derived from the same raw phosphate by conditioning in different humidities. CePF is the fibrous phosphate first reported by Alberti et al.[1] CePL has been prepared directly through a new route, and is essentially the same as that first prepared by Alberti et al.[1] Amines did not intercalate into CePL although it is believed to

**Table 4** *Hypothetical unit cells of CePs of α- and γ- structure extraporated from data of TiP and ZrP[19-22]*

| | | α-type :monoclinic P2₁/c | | | | γ-type :monoclinic P2₁ | | | |
|---|---|---|---|---|---|---|---|---|---|
| | ionic rad./Å[a] | a / Å | b / Å | c / Å | β / ° | a / Å | b / Å | c / Å | β / ° |
| Ti | 0.745 | 8.640 | 5.009 | 16.204 | 110.20 | 5.181 | 6.347 | 11.881 | 102.59 |
| Zr | 0.86 | 9.060 | 5.297 | 16.217 | 111.45 | 5.386 | 6.636 | 12.403 | 98.70 |
| Ce | 1.01 | 9.608 | 5.671 | 16.247 | 113.09 | 5.651 | 7.010 | 13.085 | 93.65 |

a The Shannon's ionic radius for 6-coordinated M⁴⁺.

have a layered structure (sedimentation PXRD). Unit cell parameters are proposed for the first time for CePH, CePD, CePB and CePL. The intercalation of various amines was studied. The amine is present as a bimolecular film. In general, the amines are present both in the protonated and neutral forms. Pyridine is poorly intercalated.

In summary, cerium(IV) phosphate may be classified to three groups according to the nature of their intralayer structure: a. CePH, CePD, CePB; b. CePF; c. CePL  It is proposed that none of these is isostructural to the $\alpha$- or $\gamma$- M(IV) phosphates.

### Acknowledgement

The authors would like to thank Shin-Etsu Chemical Co. Ltd. for granting study leave to M.K. Thanks are also given to Messrs. H. Yamaguchi, K. Maeda and Ms. N. Takaoka of Shin-Etsu for chemical analysis of Ce and P.

### References

1. G. Alberti, U. Costantino, F. di Gregorio, P. Galli and E. Torracca,
   *J. Inorg. Nucl. Chem.*, **1968**, *30,* 295.
2. G. Alberti, U. Costantino and L. Zsinka, *J. Inorg. Nucl. Chem.*, **1972**, *34,* 3549.
3. K. H. König and E. Meyn, *J. Inorg. Nucl. Chem.*, **1967**, *29,* 1153.
4. R. G. Herman and A. Clearfield, *J. Inorg. Nucl. Chem.*, **1975**, *37,* 1697.
5. R. G. Herman and A. Clearfield, *J. Inorg. Nucl. Chem.*, **1976**, *38,* 853.
6. H. Hayashi, T. Ebina and K. Torii, *Chem. Lett.*, **1995**, 951.
7. H. Hayashi, K. Torii and S. Nakata, *J. Mater. Chem.*, **1997**, *7,* 557.
8. M. Casciola, U. Costantino and S. d'Amico, *Solid State Ionics*, **1988**, *28-30,* 617.
9. Y. Xu, S. Feng, W. Pang and G. Pang, *Chem. Commun.*, **1996**, 1305.
10. L. V. So and L. Szirtes, *Radiochem. Radioanal. Lett.*, **1981**, *47,* 383.
11. M. Tsuhako, M. Danjo, Y. Baba, M. Murakami, H. Nariai and I. Motooka,
    *Bull. Chem. Soc. Jpn.*, **1997**, *70,* 143.
12. A. Clearfield and R. M. Tindwa, *J. Inorg. Nucl. Chem.*, **1979**, *41,* 871.
13. E. Rodriguez-Castellon, S. Bruque and A. Rodriguez-Garcia,
    *J. Chem. Soc. Dalton Trans.*, **1985**, 213.
14. P.-E. Werner, L. Eriksson and M. Westdahl, *J. Appl. Crystallogr.*, **1985**, *18,* 367.
15. J. I. Langford, *J. Appl. Crystallogr.*, **1971**, *4,* 259, **1973**, *6,* 190.
16. P. M. de Wolff, *J. Appl. Crystallogr.*, **1968**, *1,* 108.
17. G. S. Smith and R. L. Snyder, *J. Appl. Crystallogr.*, **1979**, *12,* 60.
18. M. Mata-Arjona, A. Garcia-Rodriguez, F. del Rey-Bueno, E. Villafranca-Sanchez
    and G. Lopez-Conejo, *J. Chem. Tech. Biotech,.*, **1988**, *42,* 83.
19. J. M. Troup and A. Clearfield, *Inorg. Chem.*, **1977**, *16,* 3311.
20. D. M. Poojary, B. Zhang, Y. Dong, G. Peng and A. Clearfield,
    *J. Phys. Chem.*, **1994**, *98,* 13616.
21. S. Bruque, M. A. G. Aranda, E. R. Losilla, P. Olivera-Pastor and
    P. Maireles-Torres, *Inorg. Chem.*, **1995**, *34,* 893.
22. A. N. Christensen, E. K. Andersen, I. G. K. Andersen, G. Alberti, M. Nielsen
    and M. S. Lehman, *Acta Chem. Scand.*, **1990**, *44,* 865.
23. Cerius² v.3.5., Molecular Simulations Inc., **1996**

E-mail: wj10@cam.ac.uk  Fax.: +44 (0) 1223 336362

# Cation Exchange Characteristics of Modified Synthetic Hydroxyapatites and Calcium Carbonates

T. Hatsushika, H. Sakane and T. Suzuki

Department of Applied Chemistry & Biotechnology
Yamanashi University
Takeda 4-3-11, Kofu, Yamanashi, 400-0016, Japan

## Introduction

The crystals of biominerals resemble their synthetic counterparts in structure and nature, are extremely minute, therefore, these materials have a large specific surface area. The advantage of this large surface area has been attractive as with other systems possessing large specific surfaces, such as activated carbons and synthetic zeolites. It has been demonstrated that the exchangeable ions of bone mineral reside in these surfaces by using isotopic $Ca^{45}$, i.e. the old bone mineral is replaced by the new one under "metabolism"[1]. During investigation of the surface characteristics of various synthetic hydroxyapatites(HAp) and calcium carbonates, we have shown that cations in solution are easily incorporated into the crystals by ion-exchange reaction at normal temperature and pressure[2]. We have also examined the feasibility of employing the apatites and calcium carbonates as new inorganic ion-exchangers for the treatment of various toxic cations in aqueous solutions[3,5,6,7]. The purpose of this paper was to report the cation exchange characteristics of modified synthetic apatites and calcium carbonates.

## Experimental

The starting materials were all analytical reagent grade chemicals from Wako Chem. Co. Ltd., and used without further purification. The samples were prepared by the methods described elsewhere[2,7]. An active carbon(Takeda Chem Co.Ltd.) and a synthetic faujasiteY(Toso Co.Ltd.) were used for comparison. The cation exchange characteristics of samples for heavy metal ions in solutions were examined by using normal batch and column methods at 25°C.   Quantitative analyses of the various ions were made by the EDTA titration method, atomic absorption spectroscopy and ICP. The structural changes occurring in the samples before and after the uptake of the ions were investigated using XRD and FT-IR.

## Results and Discussion

### 1. Cation exchange equilibria in parent materials

In order to define the time necessary for the attainment of ion-exchange equilibrium, a study of the uptake rate of the various cations investigated by the samples at 25°C, was performed.   Fig.1 shows that equilibrium was almost attained in 1-3 hr for $Cd^{2+}$ions, but 5 hr after the reaction, the decrease in the concentration of $Cd^{2+}$ions by adding the crystalline hydroxyapatite was abruptly increased and reached the lowest concentration in those by adding active carbon or synthetic faujasite Y.   X-ray powder diffraction patters of the noncrystalline hydroxyapatite before and after the reaction showed changes in peak height and sharpness from the noncrystalline pattern to that of a new crystalline phase $[Cd_5H_2(PO_4)_4]$ through the hydroxyapatite $(Ca_{10}(PO_4)_6(OH)_2)$ structure with increasing the reaction time[2,6].   The molar ratios of $Ca^{2+}$ions(released) / $M^{2+}$ions(uptake) during these reactions were found by chemical analyses to be ca. 1.0,

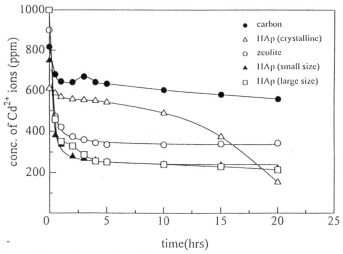

**Fig.1**   Rates of ion-exchange for $Cd^{2+}$ions on HAps in aqueous solution

i.e. the cation uptake phenomenon appears to be a cation exchange reaction between the lattice $Ca^{2+}$ions in hydroxyapatite and the $M^{2+}$ions in solution as concluded from the stoichiometry.    From these results, it was concluded that $Ca^{2+}$ions of hydroxyapatite (crystalline or non-crystalline(S<200, L $\leqq$ 100 mesh) are quickly exchanged by $Cd^{2+}$ions, and after the uptake, $Cd^{2+}$ions are incorporated into the crystal lattice of hydroxyapatite under the following reaction mechanism described as "remodeling" in bone [4].

(1) first step        adsorption and ion-exchange on the crystal surface

(2) second step      reconstruction in the crystal lattice

In first step, the release of $Ca^{2+}$ions from the surface is influenced by the hydration of $Ca^{2+}$ions which ultimately determined the rate of the reaction, and the ion selectivity of hydroxyapatites.    The exchangeable $Ca^{2+}$ions are released to hydrate by the thermal movement from the lattice of the surface, and are intensively influenced by temperature, pH, concentration of $Ca^{2+}$ions and ionic strength of the solution.    Eventually, the first step approaches the second step if the cation is similar in ionic radius to that of $Ca^{2+}$ions.

In the case of calcium carbonates, $Pb^{2+}$ions were found to be very reactive, and the same behavior with $PbCl_2$ as that of the hydroxyapatite described above.    The equilibria of the ion-exchange reactions on the calcium carbonate were attained at longer time than that of hydroxyapatites, and were dependent on the kind of counter-anions such as $CH_3COO^-$, $NO_3^-$ and $Cl^-$ion[7].

## 2. Cation exchange characteristics of modified calcium compounds

The samples prepared, HAp and silicate containing apatite(SiAp) have smaller surface areas than that of carbonate containing apatite(CAp).    These were characterized by XRD, FT-IR, B.E.T. and chemical analysis.    The specific surface areas and Ca/P molar ratios of the samples are summarized in Table 1.    The differences are twice as large as in specific surface areas between CAp and HAp, SiAp, and correspond to their preparation methods.    CAp was produced from aqueous solution by the homogeneous precipitation technique using magnetic stirrer in air.    On the other hand, HAp and SiAp were prepared by hydrothermal methods in a glass autoclave under a nitrogen atmosphere and at low temperatures

**Table 1**. Specific surface areas and Ca/P molar ratios of HAp, CAp, SiAp

|  | HAp | CAp | SiAp |
| --- | --- | --- | --- |
| specific surface($m^2/g$) | 31.2 | 87.2 | 32.2 |
| Ca / P | 1.66 | 2.32 | 2.27 |

below 200°C.    As small amount of $CO_3^{2-}$ ion was contained in SiAp in spite of using nitrogen, and affected the Ca/P ratios. These subtle matters changed the cation exchange characteristics of modified apatites by isomorphic substitutions.

The orders of cations according to the amounts of exchange by the original hydroxyapatite and the modified apatites respectively, were as follows,

HAp : $Fe^{3+} \gg Pb^{2+} > Cr^{3+} > Y^{3+} \geqq La^{3+} > Al^{3+} > Cu^{2+} > Zn^{2+} > Mn^{2+} > Fe^{2+} > Ni^{2+} > Cd^{2+} = Co^{2+}$

CAp : $Cu^{2+} > Pb^{2+} \gg Cr^{3+} > La^{3+} > Fe^{3+} > Al^{3+} > Y^{3+} > Zn^{2+} > Mn^{2+} = Fe^{2+} > Cd^{2+} > Co^{2+} > Ni^{2+}$

SiAp : $Cr^{3+} > Al^{3+} > Fe^{3+} > Fe^{2+} > Cu^{2+} > Co^{2+} = Zn^{2+} = Pb^{2+} > Mn^{2+} = Ni^{2+} = Cd^{2+} > La^{3+} > Y^{3+}$

In addition, cation exchange characteristics of hydroxyapatites were enhanced by modification of the crystal lattice, and surpassed the original ones for some cations.

It was found that the cation exchange characteristics of calcium carbonates strikingly differed from each other for $Cd^{2+}$ ions owing to their isomorphous structures. Aragonite and vaterite were more reactive for other cations($Pb^{2+}$, $Mn^{2+}$) than calcite.

## Conclusion

The cation exchange characteristics between $Ca^{2+}$ lattice ions and cations in solution were discovered to be improved with increasing contents of $CO_3^{2-}$ or $SiO_4^{4-}$ ions in the modified apatites. The results obtained from the substitution of $CO_3^{2-}$ and $SiO_4^{4-}$ ions for $PO_4^{3-}$ ions in hydroxyapatites showed that bones should intrinsically change cation exchange characteristics by substituting their component cations by taking into the body if necessary under "metabolism". Calcium carbonates were also changed in structure by isomorphous substitution to absorb $Ca^{2+}$ ions and $M^{n+}$ ions as in the growth of shell. Calcium compounds such as apatite and aragonite, are the synthetic counterpart of biominerals (bone and shell) so that they satisfactorily inherit the structure and the nature of the parent materials.

## References

1.J.H.Weikel,Jr.,W.F.Neuman and I. Feldman, J.Phys.Chem.,**76**,5202(1954)

2.T.Suzuki,T.Hatsushika andY.Hayakawa,J.Chem.Soc.,Faraday Trans.1,**77**,1059 (1981)

3. T.Suzuki and T.Hatsushika, Gypsum & Lime Japan, No.**224** 635(1988)

4.S.Mamn, J.Webb, and R.J.P.Williams, "Biomineralization", V.C.H., New York(1989)

5.Y.Tanizawa,K.Sawamura and T. Suzuki, J.Chem.Soc., Faraday Trans.1,**86**, 4025 (1990)

6.M.Miyake,Y.Nagayama,A.Goto and T.Suzuki, J.Chem.Soc. Jpn., No.**4**, 230 (1988)

7.M.Miyake,K.Ishigaki and T.Suzuki, J. Solid State Chem., **61**, 230 (1986)

# THE REMOVAL OF IODIDE AND CHLORIDE IONS FROM SOLUTION USING BISMUTH OXIDE NITRATE

H. Kodama

National Institute for Research in Inorganic Materials
Namiki 1-1, Tsukuba, Ibaraki 305-0044 Japan

## 1 INTRODUCTION

Various radioactive elements are produced in a nuclear reactor, and their removal and solidification are very important problems. The present paper discusses a method for removing radioactive iodide and chloride ions by fixing them onto an inorganic compound using anion exchange.

Various inorganic compounds such as $Bi_2O_3$[1-3], Ag compounds[4], $Bi_5O_7(NO_3)$[5-7], and $BiPbO_2(NO_3)$[8-10] have been studied for immobilizing radioactive iodide or chloride.

The present paper reports an inorganic compound for removing and solidifying radioactive iodide and chloride from solution. The synthesis of the inorganic compound, $BiO(NO_3) \cdot 0.5H_2O$ and its anion exchange properties with iodide and chloride ions are studied in detail. Since the compounds which contain $(NO_3)$ have a high possibility for removing and solidifying halogenide ions in solution, we can expect $BiO(NO_3) \cdot 0.5H_2O$ to be a useful material for removing and solidifying radioactive iodide or chloride ions.

## 2 SYNTHESIS OF $BiO(NO_3) \cdot 0.5H_2O$

$BiO(NO_3) \cdot 0.5H_2O$ was prepared by solid-state reaction. Powdered $Bi_2O_3$ and $Bi(NO_3)_3 \cdot 5H_2O$ were mixed in a mortar. The reaction can be written as follows:

$$Bi_2O_3 + Bi(NO_3)_3 \cdot 5H_2O \rightarrow 3BiO(NO_3) \cdot XH_2O \ (5/3 \geq X \geq 0) + (5/3 - X)H_2O \ (1)$$

This reaction proceeds at room temperature. The reaction product was examined using a powder x-ray diffraction (XRD) method and thermogravimetry. The XRD patterns of starting materials and a reaction product are shown in Figures 1A, B, and C. Pattern C is clearly different from patterns A and B. Specifically, pattern C is not a mixed pattern of starting materials but a pattern of a reaction product. The XRD pattern of the reaction product is also shown in Figures 2A, B, and C as a function of time. The XRD patterns of starting materials change within 2 hours, and the reaction almost ends within 6 hours then completely ends after 20 days.

We can also confirm by thermogravimetric (TG) analysis that the compound which has the XRD pattern shown in Figure 1C is not the mixture of starting materials but a reacted product . The experimental result of thermal analysis is shown in Figure 3, where curves A and B correspond to TG curves observed for $Bi(NO_3)_3 \cdot 5H_2O$ and $BiO(NO_3) \cdot 0.5H_2O$. The experiments were carried out under the same experimental conditions (atmosphere, Ar flow; heating rate, 10 K/min.). Since bismuth oxide shows no mass change under these experimental conditions, the two curves should exhibit fundamentally

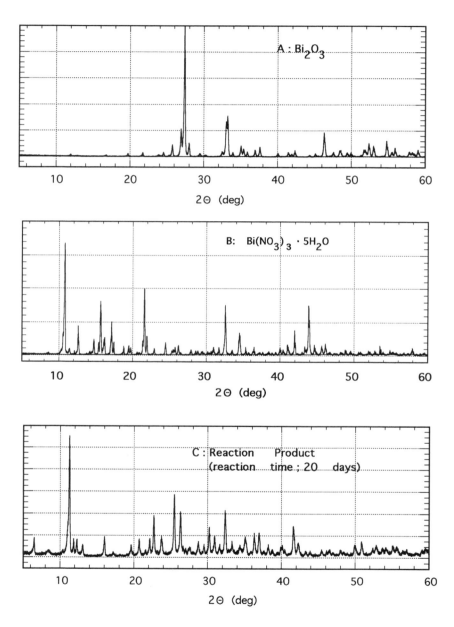

**Figure 1** *XRD patterns of $Bi_2O_3$, $Bi(NO_3)_35H_2O$ and a reaction product.*

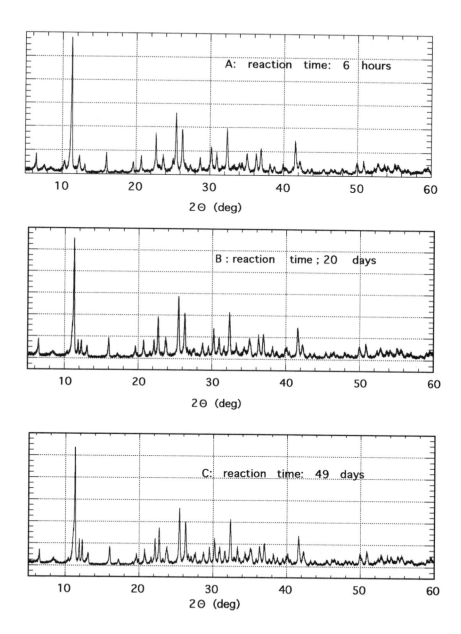

**Figure 2** *XRD patterns of a reaction product as a function of time.*

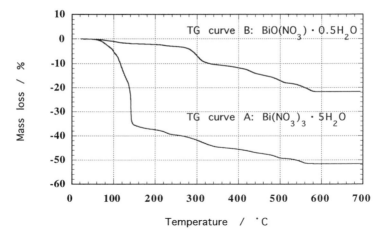

**Figure 3** *TG curves of Bi(NO₃)₃ · 5H₂O and BiO(NO₃) · 0.5H₂O.*

similar changes to similar temperature changes, if $BiO(NO_3) \cdot 0.5H_2O$ is a mixture of $Bi_2O_3$ and $Bi(NO_3)_3 \cdot 5H_2O$. The two observed curves are not similar, so they are not the mixture.

In the Gmelins Handbook[11] four compounds, $BiNO_4 \cdot 2H_2O$, $BiNO_4 \cdot H_2O$, $BiNO_4 \cdot 0.5H_2O$ and $BiO(NO_3)$ are reported as a single phase belonging to the $Bi_2O_3$-$N_2O_5$-$H_2O$ system. Two structures, $BiNO_4 \cdot H_2O$ and $BiNO_4 \cdot 0.5H_2O$, are reported as determined. They were prepared differently from the present one by reaction of bismuth nitrate with nitric acid. Later, the structure of $BiNO_4 \cdot 0.5H_2O$ was also refined by other authors[12].

The structure of the present product is not the same as the structure of one of the compounds or their mixture. Moreover, it was not possible to determine whether the reaction product is a single phase or mixed phases.

We determined only the average composition of the compound prepared by reaction (1). The chemical analysis of the product prepared by the reaction was carried out using a Perkin Elmer PE 2400 CHNS Analyzer.

The results were as follows:

N, 4.32 wt%; H, 0.34 wt%; C and S, negligible.

Based on the above chemical analysis, the chemical composition of the product was proposed to be $BiO(NO_3) \cdot 0.5H_2O$. This composition of the sample was used in the all experiments in the present paper.

### 3 REACTION WITH IODIDE AND CHLORIDE IONS

Accurately weighed quantities of samples were taken in different test tubes and equilibrated with NaI or NaCl solution for definite periods. The reaction was carried out with shaking in a plastic test tube stopped tightly with a lid and placed in a thermostatic container. After the reaction, the solids were separated from solution and identified by their XRD patterns. The iodide and chloride ion concentrations were determined by ion chromatography using a DIONEX 4500 i instrument.

#### 3.1 Rate of Exchange

The rate of ion exchange reaction of $BiO(NO_3) \cdot 0.5H_2O$ with aqueous iodide and chloride ions was examined as a function of time at 25 ℃. The experiments were performed in solutions previously adjusted to pH=1, 7, and 13. The experimental conditions are as follows: mass of $BiO(NO_3) \cdot 0.5H_2O$, 200 mg; concentration of NaX (X=I, Cl) solution, 0.2 mol dm$^{-3}$; volume of NaX solution, 4 ml.

The results are shown in Figures 4 and 5. Figure 4 shows the reaction with NaI solution with pH= 1.02, 6.63, and 12.94; Figure 5 shows the reaction with NaCl solution with pH=1.07, 6.72, and 13.03. The solution pH was adjusted by addition of either NaOH or $HNO_3$ and was determined before equilibrium. The three curves in each figure represent the results measured in the solutions of pH=1, 7, and 13.

Figure 4 gives the following results. (1) The rate of reaction in the initial stage proceeds very fast in every solution. (2) The reaction in the solution with pH= 6.63 needs more time to come to an end than in the other two solutions. (3) The reaction in the

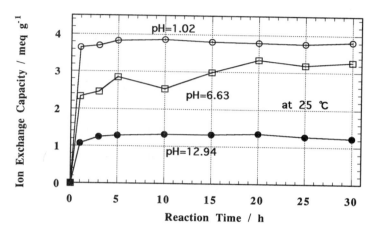

**Figure 4** *Ion exchange capacity of iodide as a function of time.*

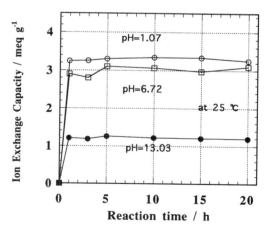

**Figure 5** *Ion exchange capacity of chloride as a function of time.*

solution with pH=12.94 has smaller ion exchange capacity in comparison with the reactions in other two solutions.

Figure 5 yields almost the same results as described on Figure 4. Results (1) in Figure 4 (the rate of reaction in the initial stage proceeds very fast in every solution) and (3) (the reaction in the solution with pH=12.94 has smaller ion exchange capacity in comparison with the reactions in other two solutions) are also shown in Figure 5. However, result (2) in Figure 4 (the reaction in the solution with pH= 6.63 needs more time to come to an end than in the other two solutions)  does not appear in Figure 5, and the reaction in the solution of pH=6.67 comes to an end in almost the same time as in the other two solutions.

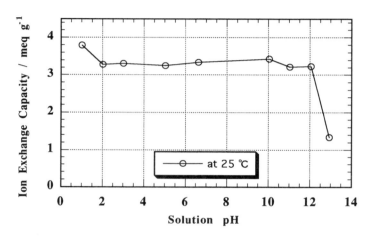

**Figure 6** *Ion exchange capacity of iodide as a function of solution pH.*

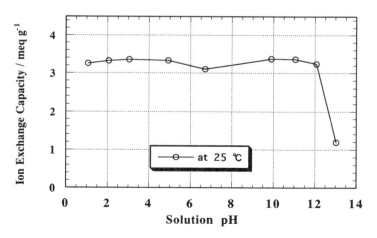

**Figure 7** *Ion exchange capacity of chloride as a function of solution pH.*

## 3.2 Ion Exchange Capacity

Accurately weighed samples (200 mg) were equilibrated with 4 ml and 0.2 mol dm$^{-3}$ solutions of NaI or NaCl for 20 hours at 25°C. The ion exchange capacity was measured as a function of solution pH. The results are shown in Figures 6 and 7. Figure 6 is the result for the NaI solution and shows that the ion exchange capacity of iodide has a maximum value at pH=1.07. The values between pH=2.01 and 12.05 are a bit smaller, and almost constant in the pH range. The smallest value was observed at pH=12.94.

Figure 7 is the result for the NaCl solution and shows that the ion exchange capacity of chloride has a large and almost constant value between pH=1.07 and 12.07, and the value decreases abruptly at pH=13.03.

This abrupt decrease of the ion exchange capacity at pH=13 in Figures 6 and 7 suggests that the ion exchange reaction proceeds by different mechanisms. This will be discussed in the next paragraph in more detail.

## 3.3 Reaction Mechanism

*3.3.1 The Reaction in NaI Solution.* The solids after reaction were identified by their XRD patterns. Their patterns showed that the compound produced in the solutions adjusted to pH values from 1 to 12 was a single phase BiOI and the starting materials were not left. The reaction in the solution with pH=13 gave a different result. The compounds produced in solution pH=12.94 were mixed phases of $Bi_4O_5I_2$ and $Bi_5O_7I$.

We can then propose the following equations for the ion exchange reactions. One is for the reaction in the solution with pH values from 1 to 12:

$$BiO(NO_3) \cdot 0.5H_2O + I^- \rightarrow BiOI + 0.5H_2O + NO_3^- \qquad (2)$$

The calculated maximum value of the ion exchange capacity for the reaction (2) is 3.38 meq/g. The observed average values between pH=2 and 12 is 3.29 meq/g, and this is very close to the calculated value. This result supports the above equation.

The following equations are proposed for the reaction in solution with pH=13:

$$4BiO(NO_3) \cdot 0.5H_2O + 2I^- + 2(OH)^- \rightarrow Bi_4O_5I_2 + 3H_2O + 4(NO_3)^- \qquad (3)$$
$$5BiO(NO_3) \cdot 0.5H_2O + I^- + 4(OH)^- \rightarrow Bi_5O_7I + 5(NO_3)^- + 4.5H_2O \qquad (4)$$

In this case, the calculated maximum value of the ion exchange capacity should be less than half of 3.38 meq/g.

*3.3.2 The Reaction in NaCl Solution.* In this case, the observed XRD patterns showed that the compound produced in the solution with pH =1 - 12 was also a single phase of BiOCl and the starting material was not left. The compound produced in the solution with pH = 13 was a mixed phase of $Bi_3O_4Cl$ and $Bi_2O_2CO_3$.

For the reaction with chloride in the solution with pH=1 - 12, we can also propose a similar exchange reaction.

$$BiO(NO_3) \cdot 0.5H_2O + Cl^- \rightarrow BiOCl + 0.5H_2O + NO_3^- \qquad (5)$$

The calculated maximum value of the ion exchange capacity for the reaction (5) is also 3.38 meq/g. The observed value at pH values between 1 and 12 is 3.30, and this is very close to the value.

For the reaction in the solution with pH=13, we can propose the following equation:

$$3BiO(NO_3) \cdot 0.5H_2O + Cl^- + 2(OH)^- \rightarrow Bi_3O_4Cl + 3(NO_3)^- + 2.5H_2O \qquad (6)$$

This reaction and the formation of $Bi_2O_2(CO_3)$ significantly decrease the value of the ion exchange capacity, 3.30 meq/g.

## 3.4 Examples of Ion Exchange Reaction

The ion exchange reaction of $BiO(NO_3) \cdot 0.5H_2O$ with iodide or chloride ions was

**Table 1** *Results of Ion Exchange Reaction.*

Halide ions remained in solution after reaction / %

|        | pH=1.02 | pH=6.63 | pH=12.94 |
|--------|---------|---------|----------|
| I⁻     | 0.4     | 0.1     | 0.3      |

|        | pH=1.07 | pH=6.72 | pH=13.02 |
|--------|---------|---------|----------|
| Cl⁻    | 0.8     | 0.5     | 0.1      |

examined to determine the minimum concentration of halide ion remaining in solution after reaction. About 0.76 g of $BiO(NO_3) \cdot 0.5H_2O$ was placed at 25 ℃ for 25 hours in 5 ml of NaI or NaCl solution previously adjusted to pH=1, 7 and 13. The results are shown in Table 1 and indicate that the present ion exchanger is capable of removing more than 99 % of the iodide or chloride ions from solution with pH=1 - 13.

### References

1. P. Taylor, AECL-1990, AECL-10163
2. P. Taylor, D. D. Wood and V. J .Lopata, AECL-1988, AECL-9554
3. H. Kodama, *Bull. Chem. Soc. Jpn.,* 1992, **65**, 3011
4. K. Funabashi, T. Fukasawa, M. Kikuchi, F. Kawamura and Y.Kondo, Proceeding of 23rd DOE/NRC Nuclear Air Cleaning and Treatment Conference, 10-3 (the number of paper)
5. H. Kodama, "Ion Exchange Progress: Advances and Applications" (Eds, A. Dyer, M. J. Hudson and P. A .Williams) Royal Soc. Chem., Wrexham UK, 1993, p-55
6. H. Kodama, *J. Solid State Chem.,* 1994, **112**, 27
7. H. Kodama, *Bull. Chem. Soc. Jpn.,* 1994, **67**, 1788
8. H. Kodama: "Progress in Ion Exchange: Advances and Applications" (Eds, A. Dyer, M. J. Hudson and P. A .Williams) Royal Soc. Chem. Wrexham UK, 1997, p-39
9. H. Kodama: in Proceedings of the Ion ICIE'95 Conference, Jpn Association of Ion Exchange, Takamatsu Japan, 1995, p-285
10. H. Kodama: the 13 th Radiochemical Conference Proceedings in the Czechoslovak Journal of Physics, Marianske Lazne Chech, 1998, to be published
11. Gmelins Handbuch der Anorganishen Chemie, Wismut, System-Nummer 19, p-655, 1964, Verlag Chemie GMBH Weinheim/Bergster
12. F. Lazarini, *Cryst. Struct. Comm.,* 1997, **8**, 69

# ION EXCHANGE PROPERTIES OF OXIDES AND HYDROUS OXIDES OF PENTAVALENT METALS

Mitsuo Abe[*1,*2], Takashi Kotani[*2] and Sachio Awano [*2]

[*1] Department of Chemistry, Faculty of Science, Tokyo Institute of Technology,
2-12-1, Ookayama, Meguro-ku, Tokyo 152-0033, JAPAN

[*2] Department of Material Industry, Tsuruoka National College Technology,
104 Sawada, Inooka, Tsuruoka, Yamagata Pref. 997-8511, JAPAN.

## 1. INTRODUCTION

Until now, many investigators have continued to seek new inorganic ion exchange materials because of their specialized properties such as resistance to high temperature and radiation fields. The most important groups of inorganic ion exchangers are clay mineral, zeolites, hydrous oxides, insoluble acid salts, heteropolyacids, and hexacyanoferrates[1-5]. Some inorganic ion exchangers exhibit excellent high selectivities with respect to certain elements or groups of elements. Inorganic ion exchangers usually have a rigid structure and do not undergo any appreciable dimensional changes during the ion exchange reaction. The rigid structure causes specific and unusual selectivities[6]. The effective ionic radii of the central metal in the oxides are plotted against the valency of the metal in Fig. 1. The metals with low valency and large ionic radius give soluble hydroxides(group I), while metals with high valency and small ionic radius give soluble oxyacids(group III). Metals in the intermediate range give insoluble hydroxide or hydroxides. The oxides listed in the region of group II are best suited to use as ion exchange materials. Among many hydrous oxides of various metals, hydrous oxides of pentavalent metals show cation exchange properties with relatively high acidity and cation exchange capacity[7,8].

This paper describes ion exchange properties of oxides and hydrous oxides of pentavalent metals such as niobium, tantalum and antimony.

### 1.1 Acidities of Hydrous Oxides

Abe and Ito have reported the uptakes of $K^+$ and $Cl^-$ as a function of pH on various hydrous oxides[8,9]. If we consider the polarizing power of a cation $M^{n+}$ to be proportional to $Z/r^2$ where Z is ionic charge and r is the effective ionic radii, various hydrous oxides of polyvalent metals can be classified as anion exchangers (I), amphoteric exchangers (II) and cation exchangers(III) in Figure 2. The soluble hydroxides are also included in this Figure.

The presence of additional oxygen attached to M increases the acid strength of the compound. Thus, series; $MO < M_2O_3 < MO_2 < M_2O_5 < MO_3$ can be represented with an approximate order of increasing acidity and increasing cation exchange character for the insoluble hydrous oxides[9,10].

---

[*1] To whom correspondence should be addressed . Mailing address: Tsuruoka National College Technology, 104 Sawada, Inooka, Tsuruoka, Yamagata Pref. 997-8511, JAPAN.

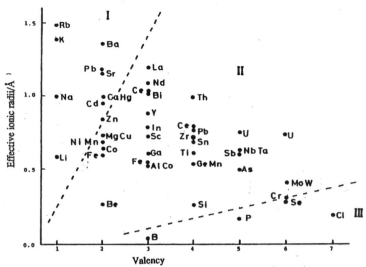

**Figure 1** Relation between valency and effective ionic radii of elements
(From M.Abe, in "Oxides and hydrous oxide of multivalent metals as inorganic ion
exchangers,A. Clearfield ed., Inorganic Ion Exchange Materials, CRC Press,
Boca Raton ,1982,. p.163. With permission)

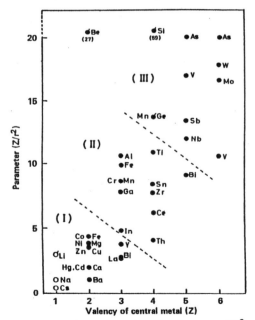

**Figure 2** Relation between valency of central metals and parameter($Z/r^2$)
(From M.Abe, in "Oxides and hydrous oxide of multivalent metals as inorganic ion
exchangers,A. Clearfield ed., Inorganic Ion Exchange Materials, CRC Press,
Boca Raton ,1982,. p.193. With permission)

## 1.2  Preparation of Amorphous Hydrous Oxide of Pentavalent metals

**Amorphous Hydrous Oxide**   Generally, the hydrous oxides of Sb, Ta and Nb are obtained by adding an excess of base to solutions of their salts[9,10]. Amorphous products are the most often obtained from these precipitations at ordinary temperature. The hydrous oxides of Sb, Ta and Nb show cation exchange behavior[9,10].

**Crystalline Antimonic Acids**   Two crystal systems of the antimonic acid have also been known, as cubic[11,12] and monoclinic[13].   It is possible to obtained a measure of the degree of crystallinity of particular exchanger by carrying out an alkali metal ion titration with it.

**Cubic antimonic Acid(C-SbA)**   It has been known that amorphous antimonic (V) acid(A-SbA) is gradually transformed into crystalline (cubic) material (C-SbA) with space group of Fd3m by aging in the acidic solution. The lattice parameter and general intensity variation of powder pattern is typical of the pyrochlore structure[14].

**Monoclinic Antimonic Acid(M-SbA)**   The monoclinic antimonic acid(M-SbA) was prepared by $Li^+/H^+$ ion exchange reaction with concentrated nitric acid solution from $LiSbO_3$[15]. The $LiSbO_3$ was obtained by heating $LiSb(OH)_6$ was prepared by the addition of LiOH solution to Sb(V) chloride solution at 60°C[15,16].   The M-SbA can be synthesized by $Li^+/H^+$ ion exchange reaction with a 6 M nitric acid solution from $LiSbO_3$. The $LiSbO_3$ reactant was obtained by heating a mixture of $LiCO_3$ and $Sb_2O_3$ with molar ratio of 1: 1 at 900°C[14,15].

**Preparation of Cubic Tantalic acid and Cubic Niobic Acid**   Crystal ionic radius of antimony ion with hexagonal coordination bonding is 0.74Å, which is the almost same radius of 0.78Å for both of niobium ion and tantalum ions, respectively[17].   $HTaO_3$ (C-TaA) and $HNbO_3$ (C-NbA) may be very hopeful to show an extremely high selectivity for lithium ions.

**Cubic Tantalic acid(C-TaA)**   $LiTaO_3$ used as a starting material has been known to a lithium ion conducting material[18], and its $H^+$-exchanged material is a proton conducting material. A cubic tantalic acid (C-TaA) was synthesized by $Li^+/H^+$ ion exchange reaction with a 3 M (mol dm$^{-3}$) sulfuric acid solution from $LiTaO_3$(hexagonal system)[19,20]. The chemical analysis shows that the empirical formula of C-TaA is $HTaO_3 \cdot 0.03H_2O$ and X-ray diffraction pattern of C-NbA was indexed to a cubic structure (space group I3m).

**Cubic Niobic Acid(C-NbA)**   was obtained as the same manner as C-TaA[20]. The chemical analysis shows that the empirical formula of C-TaA is $HNbO_3 \cdot 0.5H_2O$.

## 2  CHARACTERIZATIONS

### 2.1  XRD Paterns  of Hyrous Oxides of Pentavalent Metals

**A-SbA and C-SbA**   X-ray diffraction patterns(XRD) of amorphous and cubic antimonic acid are shown in Figure 3. The amorphous antimonic (V) acid(A-SbA) is gradually transformed into crystalline (cubic) material (C-SbA) with space group of Fd3m by aging in the acidic solution.  The crystallization is proceeded by increased tenperature of aging and increased concentration of acid solution.   The empirical formula of C-SbA shows $Sb_2O_5 \cdot 4H_2O$.   The Sb cations occupy centers of oxygen octahedera and overall structure is of the pyrochlore type.  From ion exchange studies of organic ions, the size of the window of the cavity of C-SbA can be estimated as 6Å[6].

**X-ray diffraction patterns of M-SbA**   The pattern of M-SbA showed a slightly different modification to which the orthorhombic structure with a space group Pncn. X-ray diffraction patterns of M-SbA were indexed to a monoclinic cell (space group $P2_1/m$ or $P2_1$ with a = 8.676Å(Å=100pm), b = 4.752Å, c = 5.263Å and b = 90.75°.   The exchange of $Li^+$ in $LiSbO_3$ with $H^+$ resulted in a slightly expansion of unit cell volume, i.e., from 215.28Å$^3$ to 216.98Å$^3$. The structure of M-SbA is essentially  the same as that of $LiSbO_3$

(Figure 4). Chemical analysis shows that the empirical formula of M-SbA is $HSbO_3 \cdot 0.12H_2O$. M-SbA has tunnel structure in the monoclinic cell.

**X-ray diffraction pattern of C-TaA** The XRD analysis of the $LiTaO_3$ obtained showed very sharp patterns and good agreement with those reported earlier[18]. The crystal system of the $LiTaO_3$ reactant belongs to a hexagonal lattice. Figure 5 shows the results of XRD patterns of the samples during the reaction; $LiTaO_3 \rightarrow HTaO_3$ by treatment of a sulfuric acid solution. The $X_{Li}$ showed the equivalent fraction of $Li^+$ in the samples. An appreciable change was observed in the XRD pattern of the $H^+$ exchange samples up to 40%. The XRD patterns belonging to cubic system (space group I3m) were increased by

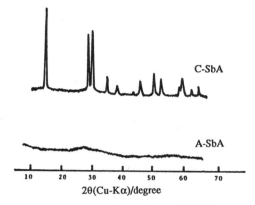

**Figure 3** XRD Patterns of amorphous and C-SbA
A-SbA: amorphous, C-SbA( cubic antimonic acid)

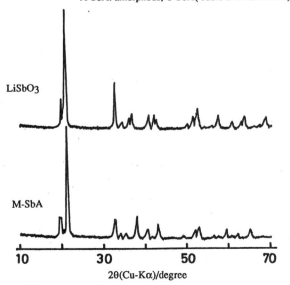

**Figure 4** XRD Patterns of $LiSbO_3$ and M-SbA
Top: $LiSbO_3$, M-SbA( monoclinic antimonic acid)
(From R. Chitrakar and M. Abe, Mat. Res. Bull., 1989, 23:1231. With permission).

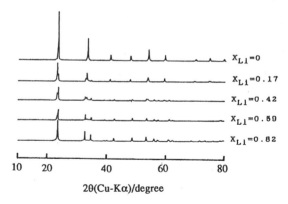

Figure 5  XRD Patterns of LiTaO$_3$ and ion exchanged with H$^+$.
X$_{Li}$ : equivalent fraction of Li$^+$ in the LiTaO$_3$ ,
X$_{Li}$ = 0: C-TaA( cubic tantanic acid)
(From Y. Inoue and M. Abe, Mat. Res. Bull., 1988, 23, 1231. With permission)

pyrochlore  Structure          Monoclinic  Structure

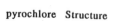● :antimony,  ○ : Li$^+$(2.7 meq/g  exchanged)
Figure 6  Schematic representation of C-SbA(left) and M-SbA(right)
(left: from W. A. Engrand et al., Solid State Ionics, 1980, 1, 231.Right: Y. Kanzaki,
R. Chitrakar, and M. Abe, J. Phys. Chem., 1990, 94: 2206.With permission)

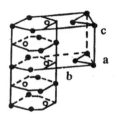

Trigonal structure          Perovskite Structure

Figure 7  Schematic representation of LiNbO$_3$(left) and HNbO$_3$(right).
(From  M. Weller and P. Dickens, J. Solid State Chem., 1985, 58, 164.With permission)

decreasing equivalent fraction of $Li^+$ ($X_{Li}$) in the $LiTaO_3$ sample. A single phase was observed at $X_{Li}$ <0.1. The X-ray diffraction patterns and lattice parameters were in a good agreement with results of Rice and Jackel[19] (Table 1 ). The lattice constant of $HTaO_3$ sample from $LiTaO_3$ sample heated at 900°C was in good agreement with that reported by Weller and Dickens[21].

Table 1   *The Lattice Constants of $LiTaO_3$, $HTaO_3$, $LiNbO_3$ and $HNbO_3$*

|  |  | crystal systrem | space group | $a_0$(Å) | $c_0$(Å) |
|---|---|---|---|---|---|
| $LiTaO_3$ | observed | Trigonal | R3c | 5.24* | 14.02* |
|  |  | Trigonal | R3c | 5.1530*[1] | 13.755* |
| $HTaO_3$ | observed | Cubic | Im3 | 7.62 |  |
|  | $Ta_2O_5 \cdot 0.97H_2O$ | Cubic | Im3 | $7.622_5$*[2] |  |
| $LiNbO_3$ | observed | Trigonal | R3c | $5.147_3$* | $13.849_7$* |
|  |  | Trigonal | R3c | $5.149_4$* | $13.862_0$* |
| $HNbO_3$ | observed | Cubic | Im3 | $7.644_6$ |  |
|  | $Nb_2O_5 \cdot 0.96H_2O$ | Cubic | Im3 | $7.645_2$*[2] |  |

*: as hexagonal lattice, *[1] : Ref.22.   *[2]: Ref. 21.

**X-ray diffraction pattern of C-NbA**  The XRD analysis of the $LiNbO_3$ and $HNbO_3$ showed essentially the same as those of the $LiTaO_3$ and $HTaO_3$.

## 2.2  Crystal Structure of C-SbA, M-SbA, C-TaA and C-NbA

**C-SbA**  The lattice parameters, and intensity of the powder pattern, of C-SbA are typical of pyrochlore structure, which corresponds to the compositional formula $A_2M_2O_6$[23]. This structure contains as $M_2O_6$ three-dimensional framework constructed from infinite chains of vertex-linked octahedra, as illustrated in Figure 6(left). A pyrochlore structure of $Sb_2O_5 .4H_2O$ would require two $H^+$ with two hydrated water molecules and two free water molecules distributed probably in the cage in the structure.

**M-SbA**  According to Estrand and Ingri, the $Li^+$ position in orthorhombic structure was estimated to be (0, 0.16, 0.5), etc.   These positions have an octahedral coordination in narrow tunnel estimated about 0.6Å. This tunnel can be entered only by $Li^+$, so the $H^+$ in M-SbA can be only exchanged with $Li^+$. The crystal structure of M-SbA transformed from monoclinic to orthorhombic structure by proceeding with $Li^+$ exchange. Figure 6(right)[23] illustrates the tunnel structure of M-SbA exchanged with 2.7 meq/g of $Li^+$.

**C-TaA and C-NbA**  $HTaO_3$ and $HNbO_3$ can be considered to have the perovskite structure, depending on whether they are written $HMO_3$ or $MO_2OH$.  Figure 7 illustrates the $LiNbO_3$ and perovskite structure showing the $BO_6$ octahedral unit and A ions (shaded circle)[19].

## 2.3  The pH titration curves of the hydrous oxides

The pH titration curves are important to characterize inorganic ion exchangers. However, the pH titration curves are strongly influenced by their crystallinities, crystalline system, supporting electrolyte, ratio of the solution and exchanger phases, ionic strength and temperature. The pH titration curves showed that the acidities of these amorphous oxides increased in the order; Sb >Ta >Nb (Figure 8-left).   The uptake of potassium ions increased with increasing pH of solution. Apparent capacities of potassium ions increased in the order Sb >Ta >Nb, lower than pH 7, and then Sb >Nb >Ta at higher than pH 7(Figure 8-right).

**Amorphous hydrous oxides of pentavalent metals**
The pH titration curves of amorphous hydrous oxides for $K^+$ ions were those of apparently monobasic acids.  The acidities show increase  in the order; $Nb_2O_5$ <$Ta_2O_5$ <

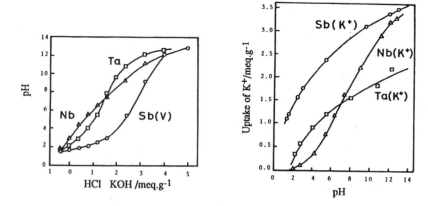

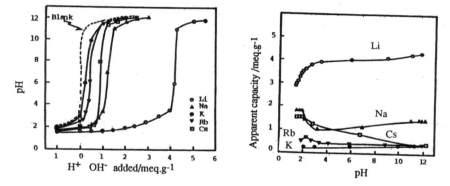

**Figure 8** The pH titration curves and apparent capacity of amorphous hydroxides of
antimony, tantalum and Niobium.
(From M. Abe and T. Ito, Nippon Kagaku Zasshi, 1965, **86**, 1259. With permission)

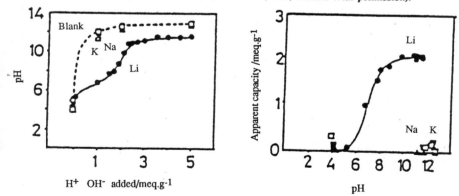

**Figure 9** The pH titration curves and apparent capacity of M-SbA
(From R. Chitrakar and M. Abe, Mat. Res. Bull., 1989, 23:1231. With permission).

**Figure 10** The pH titration curves and apparent capacity of C-TaA
(From Y. Inoue and M. Abe, Mat. Res. Bull., 1988, **23**, 1231. With permission)

$Sb_2O_5$[24]. The apparent capacities of $K^+$ determined agreed well with the amount estimated form titration curves, indicating 1: 1 ion exchange reaction.

## Crystalline acids of pentavalent metals

**C-SbA**  The pH titration curves of C-SbA for alkali metal ions were those of apparently monobasic acids.  The shapes of the titration curves much depended  on the nature of the alkali metal ions. The apparent capacities of C-SbA  increased in the order Li <K <Na[25].

In the absence of space  for ingoing large ions within the cavities in the exchangers, the ion exchange would become increasing difficult. This is called the "steric effect" and its extent increases generally with increasing difference in the crystalline ionic radii between two exchanging cations. Incomplete exchange can also arise with large ions from lack of interstitial space in the window of cavity, even when no ion sieve effect occurs. Alkyl ammonium ions are usually available in a wide range of different diameters. Uptakes of $(CH_3)_4N^+$ and $(C_2H_5)_4N^+$ are very small, indicating that adsorption may occur only on the surface of the exchanger because of the ion sieve effect.  A similar sequence was found for organic cation/$H^+$ systems on clinoptilolite[26].

**M-SbA**  The pH titration curves of M-SbA were those of an apparently strong monobasic acid for the systems of alkali metal ions/$H^+$ (Figure 9)[13]. The  M-SbA product shows an exchange reaction equivalent to one proton per one $Li^+$ in a mixed solution of lithium nitrate  and lithium hydroxide. M-SbA had an extremely high selectivity for lithium ions due to the ion memory effect. The uptake order of the metal ions was $K^+$ <$Rb^+$ <$Cs^+$ <$Na^+$ <<$Li^+$ throughout the pH range studied.     A very low capacity was observed for $K^+$ and $Rb^+$ over the entire pH range studied, while the capacities for $Na^+$ and $Cs^+$ at pH 1.6 were higher than those at pH 3 and then decreased with increase in the pH which was reverse phenomena of those observed on $Li^+$ exchange. The XRD patterns of the products exchanged with $Na^+$ and $Cs^+$ at lower pH showed a mixed pattern of C-SbA  in M-SbA. The high capacities for $Na^+$ and $Cs^+$ at low pH can be explained by partial transformation from M-SbA to C-SbA. No phase transformation was observed at pH >5.

**C-TaA and C-NbA**     The  C-TaA product showed an exchange reaction equivalent  to one proton  per  one $Li^+$ in a mixed solution of lithium  nitrate  and  lithium hydroxide. The pH titration curves of C-TaA on the $Li^+$/$H^+$ exchange  system indicated an apparently  weak monobasic acid( Figure 10).  The C-TaA had a high uptake for lithium ions, whereas no appreciable uptake  was observed for other alkali metal ions in the range pH 4-12. Almost the same result was observed on the C-NbA.

## 2.4   Selectivities of A-SbA, C-SbA and M-SbA

When small cations in C-SbA are exchanged with large cations, the ion exchange becomes progressively more difficult as the proportion of one of the large cations in the solid phase is increased (i.e., steric effect).  This may arise in a rigid structure which undergoes relatively  little swelling if there is a large difference in the size of two cations. If there is no site available for the ingoing large cations within the inter crystalline volume, incomplete exchange results.   When the size of the cations is much larger than that of the window of the cage structure, an ion-sieve effect will occur even if there is some lattice vibration. The affinity series of the inorganic exchangers for alkali metal ions vary depending on the crystalline form of exchanger and the nature of the sorption media. At low loading the A-SbA exhibits the usual selectivity sequence $Li^+$ <$Na^+$ <$K^+$ <$Rb^+$ <$Cs^+$. The C-SbA showed an unusual selectivity sequence; $Li^+$ <<$K^+$ <$Cs^+$ <$Rb^+$ <<$Na^+$ for micro amounts in acid media[27]. The values obtained on Bio-Rad AG50W-X8 are included for comparison. The separation factor between  $Na^+$ and other alkali  metal ions, $\alpha_M^{Na}$, showed a value higher than 10 in $0.1MHNO_3$. A separation factor ($\alpha$) higher than $10^4$ was observed for  $Na^+$ to $Li^+$ ions.

Table 2  *Distribution Coefficients ($K_d$) and Separation Factor ($\alpha$) For Alkali Metal Ions on A-SbA, C-SbA, M-SbA and Bio-Rad AG50W-X8*

| Ion Exch | Solution | Parameter | | | | | | | | |
|---|---|---|---|---|---|---|---|---|---|---|
| | | | Li | Na | | K | | Rb | | Cs |
| A-SbA | 0.1MHNO$_3$ | $K_d$ | 6.4 | 27.8 | | 123 | | 196 | | 226 |
| | | $\alpha$ | | 4.3 | 4.4 | | 1.6 | | 1.2 | |
| | | | Li | K | | Cs | | Rb | | Na |
| C-SbA | 0.1MHNO$_3$ | $K_d$ | 0.9 | 4.5x10$^2$ | | 1.4x10$^3$ | | 8.1x10$^3$ | | 8.3x10$^4$ |
| | | $\alpha$ | | 500 | 3.1 | | 5.8 | | 10.2 | |
| | | | | | | | | Na | | Li |
| M-SbA | 0.1MHNO$_3$ | $K_d$ | | | | | | 2.0 | | 1.0x10$^4$ |
| | | $\alpha$ | | | | | | 5x10$^3$ | | |
| | | | Li | Na | | K | | Rb | | Cs |
| Bio-Rad AG50W-X8 | 0.1MHNO$_3$ | $K_d$ | 33 | 54 | | 99 | | 118 | | 148 |
| | | $\alpha$ | | 1.6 | 1.8 | | 1.1 | | 1.2 | |

## 2.5 Selectivities of C-SbA for various metal ions

For the ion exchange reaction for polyvalent metal ions

$$n \overline{H^+} + M^{n+} \rightleftarrows \overline{M^{n+}} + n H^+ \tag{3}$$

$K_H^M$ is an equilibrium constant and $(\ln K_H^M)\overline{X}_{M^{n+} \to 0}$ values can be plotted against effective ionic radii(EIR) of exchanging metal ions(Figure 11 ).

The $(\ln K_H^M)\overline{X}_{M^{n+} \to 0}$ is the value of $\ln K_H^M$ obtained by extrapolation of $\overline{X}_{M^{n+}}$ to zero.

Free energy of ion exchange reaction can be represented by;

$$\Delta G = ( \Delta G_e^B - \Delta G_e^A ) - (\Delta G_h^B - \Delta G_h^A) \tag{4}$$

The first expression on the right hand side is attributable to the difference in $\Delta G_e$ resulting from the electrostatic interaction of two cations to anionic field of the site, and in the second one, $\Delta G_h$ represents the difference in free energies of hydration of two cations. Small cations having a high charge density are accompanied with the large hydration shell. These hydrated cations undergo large steric hindrance, if the hydration energy of the ions is larger than the energy for the removals of part, or all, water molecules from the ions by electrostatic interaction. For example, hydrated water is bonded more strongly to the Li$^+$ than to Na$^+$ on C-SbA. These hydrated Li$^+$ undergo more steric effect than hydrated Na$^+$. A similar tendency is observed for Mg$^{2+}$ and Al$^{3+}$/H$^+$ exchange systems. Thus, minimum steric effect and maximum selectivity coefficient are observed for the exchanging ions having about 0.12 nm of the effective ionic radii (Fig. 11)[6].

An extremely high selectivity was observed for Na$^+$ in alkali metal ions, Ca$^{2+}$ and Sr$^{2+}$ in alkaline earth metal ions, Pb$^{2+}$ and Cd$^{2+}$ in transition metal ions. These exchanging ions have EIR of about 0.12nm.

**Selectivity of Alkali metal ions on M-SbA**    The pH titration curves of M-SbA for alkali metal ions showed a single large jump, with MOH added, indicating that M-SbA behaves as a strong monobasic acid type exchanger[14]. The apparent capacity was found to be 4.3 meq/g for lithium ions, but less than 0.5 meq/g for potassium ion. The lithium ions can enter the tunnel of the M-SbA crystal. Even through K$^+$ ion cannot, because of the ion sieve effect, when a potassium chloride solution is added to M-SbA the decreased pH can be ascribed to the displacement of surface protons by the K$^+$.

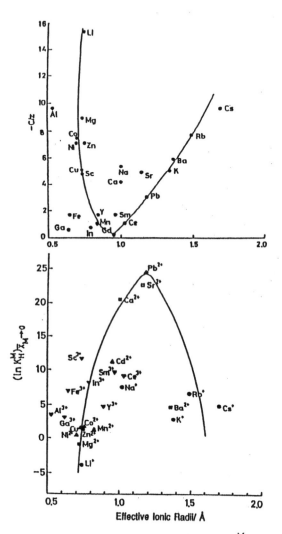

**Figure 11** Plot of equivalent of Kieland coefficients(top) and $(\ln K_H^M)\overline{X}_{M^{n+}\to 0}$ (bottom)

values versus effective ionic radii(EIR) for C-SbA(From M. Abe, M. Abe, Ion Exchange
Selectivities of Inorganic Ion Exchangers in Ion Exchange and Solvent Extraction, Vol. 12,
J. A. Marinsky and Y. Marcus Eds., M. Dekker, NY, 1995, p. 381, With permission)

## 2.6 Ion Exchange Selectivities on Hydrous Oxides of Niobium and Tantalum

**Amorphous Hydrous Oxides of Niobium and Tantalum**    Hydrous oxides of Ta and Nb are obtained by adding an excess of base to solutions of their salts.   Amorphous products are most often obtained from these precipitations.   The affinity series of the inorganic exchangers for alkali metal ions vary, depending on the nature of the sorption media.     At low loading the amorphous oxides exhibit the usual selectivity sequence $Li^+$ $<Na^+ <K^+ <Rb^+ <Cs^+$ in acidic or neutral solution. This selectivity is observed with the strong acid organic ion exchange resins.

**Niobic Acid and Tantalic acid as Lithium ion memory exchangers**
The retention of high selectivity for a particular cation after exchanging with other cation can be sought by endowing the crystal structure of inorganic ion exchanger with a fixed and well-defined structure. The crystal structure was observed to remain essentially unchanged after exchanging a particular cation in these oxides with $H^+$.

### Selectivities of C-TaA and C-NbA

**Distribution coefficient ($K_d$)**   The plot of $K_d$ vs. pH (Figure 12) showed a linear relationship with a slope of -1 for alkali metals except $Na^+$, indicating the "ideal" 1:1 ion exchange reaction . Thus, ion exchange mechanism on C-TaA was established for micro- and macro- amounts of $Li^+$. An extremely  high $K_d$ value was observed for $Li^+$ on all  the samples studied. A similarly high selectivity for $Li^+$ was observed  on M-SbA[14] and C-NbA[28]. Crystal structure   and $^1H$ NMR studies on M-SbA and C-TaA indicated the presence of $Li^+$ in the small  cavity of the tunnel arrangement in the  structure. Alkali metal

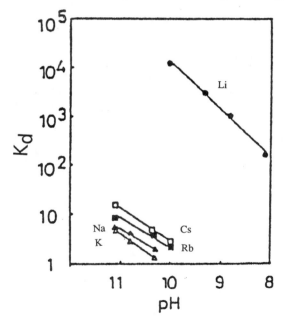

**Figure 12**  Distribution coefficients($K_d$) of alkali metal ions on C-TaA vs. pH of
   the solution
     Alkali metal ion conc.: 1 x 10⁻¹M, C-TaA: 1.0g, Total Vol.:10.0mL
   (From Y. Inoue and M. Abe, Mat. Res. Bull., 1988, **23**, 1231. With permission)

ions having an ionic crystal radii larger than $Li^+$ cannot enter the cavity. Thus, ion sieve properties arise on the C-NbA. The high selectivity for $Li^+$ may be due to the same situation as that on M-SbA[14] and C-TaA[25]. Very low $K_d$ values were obtained for the micro-amounts of other alkali ions at lower than pH 6. The uptake of the alkali metal ions may be due to an ion exchange reaction on the surface of the C-NbA.

Table 3.   *Distribution Coefficients ($K_d$) and Separation Factor ($\alpha$) for Alkali Metal Ions on C-TaA and C-NbA*

| Ion Exch. | Solution | | Parameter | | | | |
|---|---|---|---|---|---|---|---|
| | $(NH_3 + NH_4NO_3)$ | | Rb | K | Cs | Na | Li |
| C-TaA | (pH = 10) | $K_d$ | 1.0 | 3.56 | 5.72 | 10.1 | $2.31 \times 10^4$ |
| | | $\alpha$ | | 1.3 | 2.3 | 1.11 | $5.8 \times 10^3$ |
| | $(NH_3 + NH_4NO_3)$ | | K | Rb | Na | Cs | Li |
| C-NbA | ( pH = 10) | $K_d$ | 1.21 | 1.60 | 3.61 | 4.00 | $2.31 \times 10^4$ |
| | | $\alpha$ | | 1.3 | 2.3 | 1.11 | $5.8 \times 10^3$ |

The selectivity sequences, the $K_d$ value and separation factors $K_d(M_1{}^+) / K_d(M_2{}^+)$ for alkali metal ions are summarized in Table 3. The values obtained on Bio-Rad AG50W-X8 are included for comparison. The separation factor between $Li^+$ and other alkali metal ions, $\alpha_M{}^{Li}$, was larger than $10^3$ at pH=10. This indicated that selective adsorption $Li^+$ is feasible for micro amounts other alkali metal ions.

An extremely high selectivity was found for lithium ions and separation factor $(\alpha)^{20}$ of lithium ion to other alkali metal ions was higher than $10^3$.

## 3. CONCLUDING REMARKS

The major objective of this paper was to demonstrate the selectivities of hydrous oxides of pentavalent as inorganic ion exchange materials. The characteristics of hydrous oxides greatly depend on the preparative conditions; e. g. , starting substances, temperature of preparation, concentration, aging, drying temperature etc.
1. The acidities of these amorphous oxides are increased in the order; Sb >Ta >Nb.
2. The selectivity depends on the crystallinity of the hydrous oxides.
3. The rigid structure of C-SbA causes specific and unusual selectivities, because of limited space of the cavity in the pyrochlore structure. Thus, there is an ion size preference on the C-SbA by a balance between electrostatic and dehydration effect.
4. The lithium ion memory preference can be seen on the M-SbA, C-TaA and C-NbA, which have all tunnel structures. The $Li^+$ is exchanged as dehydrated ions in these acids.

### References

1. C. B. Amphlett, *Inorganic Ion Exchangers, Elsevier,* New York, 1964.
2. H. S. Sherry, *The Ion Exchange Properties of Zeolites,* Chapter 3 in *Ion Exchange* Vol. 2. (J. A. Marinsky ed. ), Marcel Dekker, Inc., (1969).
3. V. Vesely and V. Pekarek, *Talanta,* **19**, 219 (1972).
4. A. Clearfield, G. H. Nancollas and R. H. Blessing, *New Inorganic Ion Exchangers,* Chapter1, *Ion Exchange and Solvent Extraction* Vol. 5. (J. A. Marinsky ed. ) Marcel Dekker, Inc., 1973.
5. A. Clearfield, *Inorganic Ion Exchange Materials,* CRC Press, Boca Raton (1982).
6. M. Abe, *Ion Exchange Selectivities of Inorganic Ion Exchangers* in *Ion Exchange*

*and Solvent Extraction,* Vol. 12, J. A. Marinsky and Y. Marcus Eds., M. Dekker, NY, 1995, p. 381.

7.  M. Abe and T. Ito, *Kogyo Kagaku Zasshi,* 1967, **79,** 440.
8.  M. Abe, 'Oxides and Hydrous Oxides of Multivalent Metals as Inorganic Ion Exchangers' in "Inorganic Ion Exchange Materials", A. Clearfield, Ed., CRC Press, Fl., 1982. Chap. 6.
9.  M. Abe and T. Ito, *Nippon Kagaku Zasshi,* 1965, **86,** 1259.
10. M. Abe and T. Ito, *Nippon Kagaku Zasshi,*1966, **87,** 417.
11. M. Abe and T. Ito, *Bull. Chem. Soc. Jpn.,* 1968, **41,** 333.
12. L. W. Baetsle and D. Huys, *J, Inorg. Nucl. Chem.,* 1968, **30,** 639.
13. R. Chitrakar and M. Abe, *Mat. Res. Bull.,* 1989, **23,**1231.
14. W. A. Engrand, M. G. Cross, A. Hamnett, P. J. Wiseman and J. B. Goodenough, *Solid State Ionics,* 1980, **1,** 231
15. M. Edstrand and N. Ingri, *Acta Chem. Scand.,* 1954, **8,** 1021.
16. R. Franck, *Thermochimica Acta* 1970, **1,** 261.
17. R. D. Shannon and C. Prewitt, *Acta Crystallogr.,* 1965, **B25,** 925.
18. B. Elouadi,M. Zrioul,J. Ravez and P. Haenmuller, *Mat. Res.Bull.,* 1981, **16,** 1099.
19. C. E. Rice and L. Jackel, *J. Solid State Chem.,* 1982, **41,** 308
20. Y. Inoue and M. Abe, *Mat. Res. Bull.,* 1988, **23,** 1231.
21. M. Weller and P. Dickens, *J. Solid State Chem.,* 1985, **58,** 164.
22. Nat. Bur. Stand.(US) Monog. 1977, **25,** Sect. 14.
23. Y. Kanzaki, R. Chitrakar, and M. Abe, J. Phys. Chem., 1990, **94**: 2206.
24. M. Abe, Kogyo Kagaku Zasshi, 1967, **79,** 2226.
25. B. G. Novikov, F. A. Belinskaya, and E. A. Mateova, *Vestn. Leningrad Univ. Fiz. Khim,* 1971, **4,** 29.
26. R. M. Barrer, R. Papadopoulos and L. V. C. Rees, J. Inorg. Nucl. Chem., 1967, **29,** 2047.
27. M. Abe and T. Ito, *Bull. Chem. Soc. Jpn.,* 1968, **41,** 2366.
28. T. Ohsaka, M. Abe, Y. Kanzaki, T. Kotani and S. Awano, *Mat. Res. Bull.,* 1999, **34,** in press.

# SYNTHESIS OF NEW ILMENITE-TYPE OXIDES, AgMO$_3$ (M=Sb,Bi) BY ION-EXCHANGE REACTION

N. Kumada, N. Takahashi and N. Kinomura
Faculty of Engineering, Yamanashi University
Miyamae-cho 7, Kofu 400-8511 Japan

A. W. Sleight
Department of Chemistry, Oregon State University
Corvallis, Oregon 97331-4003 USA

## 1 ABSTRACT

New ilmenite-type oxides, AgMO$_3$ (M=Sb,Bi) were prepared by ion-exchange reaction of NaBiO$_3$·nH$_2$O and KSbO$_3$. The hexagonal lattice parameters are $a$=5.323(1) and $c$=16.698(5)Å for AgSbO$_3$, and $a$=5.641(1) and $c$=16.118(2)Å for AgBiO$_3$. The crystal structure of AgBiO$_3$ was refined using neutron powder diffraction data and the final R factors were R$_{WP}$= 7.96 and R$_p$=6.12%. AgSbO$_3$ changed from the ilmenite-type to the defect pyrochlore-type structure at 770°C and AgBiO$_3$ decomposed to Ag and Bi$_2$O$_3$ above 200°C.

## 2 INTRODUCTION

We have previously reported several new bismuth oxides with pentavalent bismuth prepared by low temperature hydrothermal reaction using hydrated sodium bismuth oxide, NaBiO$_3$·nH$_2$O as a starting material[1-8]. The crystal structures for the new bismuth oxides are closely related to those for antimony oxides; for example, Bi$_2$O$_4$[5] is isostructural with β-Sb$_2$O$_4$, LiBiO$_3$[6] has the LiSbO$_2$-related structure and ABi$_2$O$_6$ (A=Mg,Zn)[7] has the trirutile-type structure. During this investigation we found that the new ilmenite-type oxide, AgBiO$_3$, was prepared from NaBiO$_3$·nH$_2$O by an ion-exchange reaction. AgBiO$_3$ with the cubic KBiO$_3$-type structure was prepared by ion-exchange of K$^+$ ion in KBiO$_3$·nH$_2$O[9], but the ilmenite-type AgBiO$_3$ has not been prevously reported. On the other hand AgSbO$_3$ has been known to have two types of crystal structure; one is the defect pyrochlore-type structure[10,11] and the other is the cubic KBiO$_3$-type structure. The compound with the latter structural type was prepared under high pressure or ion-exchange of the K$^+$ ion in KSbO$_3$ with the Ag$^+$ ion[12]. Taking into account that the ionic radius of Ag$^+$ ion is smaller than that of K$^+$ ion, AgSbO$_3$ can be expected to adopt the ilmenite-type structure, but this form has not been reported yet. The preparation of the ilmenite-type AgSbO$_3$ was attempted by an ion-exchange reaction using ilmenite-type KSbO$_3$.

The synthesis, crystal structure and thermal stability of these new ilmenite-type oxides are described in this paper.

## 3 EXPERIMENTAL

### 3.1 Sample preparation and characterization

Ilmenite-type AgSbO$_3$ was prepared by the ion-exchange reaction of the K$^+$ ion in ilmenite-type KSbO$_3$ with the Ag$^+$ ion. The starting compound, KSbO$_3$ was prepared by heating the stoichiometric mixture of K$_2$CO$_3$ and Sb$_2$O$_3$ at 600°C for 12h, then grinding and heating again at 800°C for 12h. The ion-exchange reaction was carried out in molten

AgNO$_3$ at 220~300°C. After cooling, AgNO$_3$ and KNO$_3$ were dissolved with distilled water and the product was filtrated and dried at 50°C.

Ilmenite-type AgBiO$_3$ was prepared by the ion-exchange reaction of the Na$^+$ ion in NaBiO$_3$·nH$_2$O with the Ag$^+$ ion. The starting material of NaBiO$_3$·nH$_2$O (Nacalai Tesque Inc.) was placed in an autoclave (70mL) with a teflon lining with AgNO$_3$ and H$_2$O (30mL) and was heated at 70~120°C. The molar ratio of the starting material to silver nitrate was varied from 1 to 4. The solid products were separated by centrifuging, washed with distilled water, and dried at 50°C.

The products were identified by X-ray powder diffraction using CuK$\alpha$ radiation. The thermal stability was investigated by TG-DTA with a heating rate of 10K/min. The gases evolved during TG-DTA measurements in flowing He were analyzed by mass spectrometry.

To determine chemical composition, the products were dissolved completely in concentrated HNO$_3$ for AgBiO$_3$ and by a mixture of HF and HNO$_3$ under hydrothermal condition for AgSbO$_3$. The amount of alkaline metals, silver, antimony and bismuth in the solution was analyzed by atomic absorption spectroscopy or induced coupled plasma spectroscopy.

## 3.2 Crystal structure refinement

For the structural refinement of AgBiO$_3$ by neutron diffraction, intensity data were collected at 295K on a high resolution powder diffractometer at the National Institute of Standards and Technology. Data were obtained at intervals of 0.05° 2θ from 5.00 to 140.00° 2θ using a wave length of 1.8857Å. The data were refined using the Rietveld method with the program RIETAN[13].

## 4 RESULTS AND DISCUSSION

### 4.1 Preparation

AgSbO$_3$ was obtained after 3 h at 250°C and the colour of this product was yellow, though a small amount of unidentified phase was observed from the X-ray powder pattern. The X-ray powder pattern of AgSbO$_3$ can be indexed with a hexagonal cell with $a$=5.323(1) and $c$=16.698(5)Å as shown in Fig 1(a). From the chemical analysis the chemical composition was found to be Ag$_{0.87}$K$_{0.13}$SbO$_3$, and the partial ion-exchange corresponded to the fact that its c-axis was longer than that of AgBiO$_3$ in which the ion-exchange was completed.

Single phase AgBiO$_3$ was obtained after 1 day at 70°C using a molar ratio of Ag/Bi=2~4, and the colour of product was black. When the reaction temperature was above 100°C, unknown phases coexisted with ilmenite-type AgBiO$_3$. The X-ray powder pattern of AgBiO$_3$ can be indexed with the hexagonal lattice parameters of $a$=5.641(1) and $c$=16.118(2)Å as shown in Fig. 1(b). The ratio of Ag/Bi was found to be 1.02, and no sodium was detected by the chemical analysis.

### 4.2 Thermal Behavior

No mass loss was observed in the TG curve for AgSbO$_3$ and an exothermic peak at 770°C was observed in the DTA curve as shown in Fig.2. The X-ray powder pattern of the sample heated above the exothermic peak was identical with that of the defect pyrochlore-type AgSbO$_3$.

Figure 3 shows the TG-DTA curves and temperature dependence of gas evolution during TG-DTA measurements for AgBiO$_3$. The TG curve indicates mass losses above 200°C, and only oxygen was evolved during the course of pyrolysis. From the X-ray powder pattern, the sample heated up to 500°C was identified as a mixture of Ag and Bi$_2$O$_3$. The observed total mass loss (6.60wt%) agrees well with the value (6.58wt%) calculated on the assumption that AgBiO$_3$ decomposed to Ag and Bi$_2$O$_3$ by release of oxygen due to complete reduction of Ag$^+$ to Ag and Bi$^{5+}$ to Bi$^{3+}$.

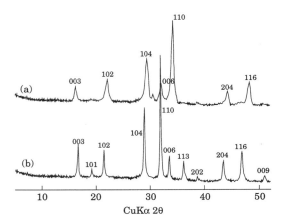

**Figure 1** *X-ray powder patterns of AgSbO₃ (a) and AgBiO₃ (b)*

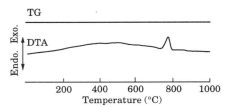

**Figure 2** *TG-DTA curves of AgSbO₃*

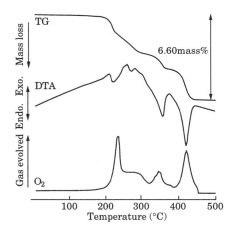

**Figure 3** *TG-DTA curves and gas evolution during TG-DTA measurement for AgBiO₃*

## 4.3 Crystal Structure

The crystal structure refinement for $AgBiO_3$ using neutron diffraction data led to $R_{WP}= 7.96$ and $R_P=6.12\%$. Figure 4 shows observed and calculated neutron diffraction patterns. The crystallographic data is summarized in Table 1. Tables 2 and 3 show the atomic parameters, and selected interatomic distances, respectively.   The crystal structure can be described as a layer structure where $BiO_6$ and $AgO_6$ octahedra are stacked alternatively along the c-axis as shown in Fig. 5(a)  Environments around silver and bismuth atoms are shown in Fig. 5(b). The mean Bi-O distance (2.12Å) is in agreement with 2.10Å in $KBiO_3$[14], 2.101Å in $Bi_2O_4$[5] and 2.10Å in $MgBi_2O_6$[7] .  The mean Ag-O distance (2.52Å) is somewhat longer than 2.458Å and 2.498Å for two distorted octahedra in $Ag_2Cr_2O_7$[15] .

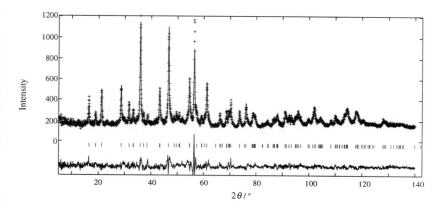

**Figure 4** *Observed and calculated neutron diffraction patterns for AgBiO₃*

**Table 1** *Crystal data and intensity collection for AgBiO₃*

| Colour | Black | Crystal system | Trigonal |
|---|---|---|---|
| Space group | R$\bar{3}$ (No.148) | Z | 6 |
| Lattice parameters | a=5.641(1),c=16.118(2)Å | Formula weight | 364.83 |
| Volume | 444.2(1)Å$^3$ | Calculated density | 8.18 g/cm$^3$ |
| Wavelength | λ=1.8857Å | Temperature | 23°C |
| 2θ scan range | 5.00-140.00° | | |
| $R_{WP}$ | 7.96% | $R_P$ | 6.12% |
| $R_I$ | 7.36% | $R_F$ | 3.62% |

**Table 2** *Positional and thermal parameters (Å2) for AgBiO₃*

| Atom | Site | x | y | z | Biso |
|---|---|---|---|---|---|
| Ag | 6c | 0 | 0 | 0.1312(8) | 1.8(3) |
| Bi | 6c | 0 | 0 | 0.3404(6) | 1.1(2) |
| O | 18f | 0.277(1) | -0.042(2) | 0.2605(6) | 1.5(2) |

**Table 3** *Selected interatomic distances (Å) for AgBiO₃*

| AgO₆ octahedron | | | | BiO₆ octahedron | | | |
|---|---|---|---|---|---|---|---|
| Ag | - | O | 2.356(9) x3 | Bi | - | O | 2.106(9) x3 |
|  |  | O | 2.69(1) x3 |  |  | O | 2.13(1) x3 |
| mean |  |  | 2.52 | mean |  |  | 2.12 |
| O | - | O | 2.70(2) x3 | O | - | O | 2.94(1) x3 |
|  |  |  | 2.94(1) x3 |  |  |  | 3.15(1) x3 |
|  |  |  | 3.53(2) x3 |  |  |  | 3.19(2) x3 |
|  |  |  | 3.74(1) x3 |  |  |  | 3.69(2) x3 |
| mean |  |  | 3.23 | mean |  |  | 3.24 |

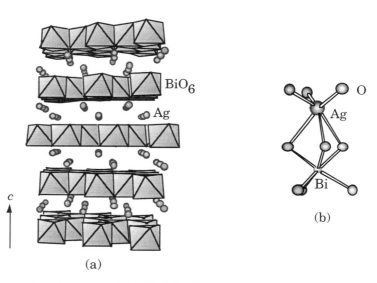

**Figure 5** *Crystal structure of AgBiO₃ (a) and environments around silver and bismuth atoms (b)*

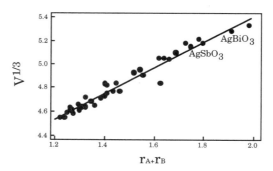

**Figure 6** $V^{1/3}$ *vs. sum of ionic radii for ilmenite type oxides*

According to Goodenough and Kafalas[9] , in the $ABO_3$ oxide compounds including the $Ag^+$ ion with a polarizable $4d^{10}$ core as the A ion, the structure prefers to have corner sharing linkages of $BO_6$ octahedra to make B-O-B angles as high as possible, such as in the pyrochlore and perovskite type structures, rather than to have edge sharing linkages of $BO_6$ octahedra with low B-O-B angle, as in the ilmenite and $KSbO_3$ type structure. Actually $AgSbO_3$ adopts the defect pyrochlore type structure, and the $KSbO_3$ type $AgSbO_3$ can be prepared only by ion-exchange reaction. The ion-exchange reaction is topotactic which often enables the preparation of metastable compounds. The ilmenite type $AgMO_3$ (M=Sb,Bi) prepared in this study is considered to be also metastable, whereas these contradict the criterion mentioned above. The unit cell volume of $AgMO_3$ falls on the straight line given by the plot of (unit cell volume)$^{1/3}$ vs. sum of ionic radii for ilmenite type oxides as shown in Fig. 6.

## 5 CONCLUSION

New ilmenite type oxides, $AgMO_3$ (M=Sb,Bi), were prepared for the first time by ion-exchange reaction. $AgSbO_3$ changed to the defect pyrochlore-type structure at 770°C and $AgBiO_3$ decomposed to Ag and $Bi_2O_3$ above 200°C.

## Acknowledgment

We acknowledge the support of the National Institute of Standards and Technology, U. S. Department of Commerce, in providing the neutron research facilities used in this work.

## References

1. N. Kumada, M. Hosoda and N. Kinomura, *J. Solid State Chem.*, 1993, **106**, 476.
2. N. Kumada, N. Kinomura, S. Kodialam and A. W. Sleight, *Mater. Res. Bull.*, 1994, **29**, 497.
3. S. Kodialam, N. Kumada, R. Mackey and A. W. Sleight, *European J. Solid State and Inorg. Chem.*, 1994, **31**, 739.
4. N. Kinomura and N. Kumada, *Mater. Res. Bull.*, 1995, **30**, 129.
5. N. Kumada, N. Kinomura, P. M. Woodward and A. W. Sleight, *J. Solid State Chem.*, 1995, **116**, 281.
6. N. Kumada, N. Kinomura, N. Takahashi and A. W. Sleight, *J. Solid State Chem.*, 1996, **126**, 121.
7. N. Kumada, N. Kinomura, N. Takahashi and A. W. Sleight, *Mater. Res. Bull.*, 1997, **32**, 1003.
8. N. Kumada, N. Kinomura, N. Takahashi and A. W. Sleight, *J. Solid State Chem.*, in press.
9. J. B. Goodenough and J. A. Kafalas, *J. Solid State Chem.*, 1973, **6**, 493.
10. N. Schrewelius, *Z. Anorg. Allg. Chem.*, 1938, **238**, 241.
11. A. W. Sleight, *Mater. Res. Bull.*, 1969, **4**, 377.
12. H. Y-P. Hong, J. A. Kafalas and J. B. Goodenough, *J. Solid State Chem.*, 1974, **9**,345.
13. F. Izumi, *Kobutsugaku Zasshi*, 1985, **17**, 37.
14. S. Kodialam, V. C. Korthius, R. D. Hoffmann and A. W. Sleight, *Mater. Res. Bull.*, 1992, **27**, 1379.
15. P. A. Durif and M. T. A-Pouchot, *Acta Cryst.*, 1978, **C34**, 3335.

**Novel Organic Ion Exchangers**

# THE DEVELOPMENT AND APPLICATION OF TYPE III STRONG

# BASE ANION EXCHANGE RESINS (PUROLITE A-555)

By Jim Irving, James A Dale, Purolite International Ltd. Wales, UK.CF 72 8YL

## Abstract

Purolite A-555 is a novel type of strong base anion exchange resin that has a combination of properties which offers unique performance under certain conditions. In comparison with Type-II resins, its special functional group based on N-propanolamine, enhances thermal stability, and silica removal in water treatment processes, while preserving good regenerability to the hydroxide form. It has also been found to be less susceptible to silica fouling as the resin type ages. In comparison with Type-I resins it offers improved operating capacity, and greater resistance to organic fouling, while offering good performance in terms of silica removal. It offers improved thermal stability, more stable rinse characteristics and superior handling of waters with high silica to total anions ratio.

## I LABORATORY EVALUATION

### I.i Functional Group

Figure 1 compares the chemical structure of the Type-III resin with those of the commonly available, conventional strong base anion (SBA) exchange resins. Briefly the original Type-I structure has a quaternary ammonium functionality, made up of methyl groups fixed to the nitrogen atom and thence to the styrene polymer backbone. This is one of the most thermally stable of the structures shown, and the most thermally stable resin commonly available. Even for this resin, the maximum recommended operating temperature is 60°C. Shortly after the invention of Type-I SBA resins, it was discovered that replacement of one of the methyl groups by an ethanolamine group improved regenerability[1]. The new structure was called Type-II. This resin type offered savings on plant size and in caustic soda usage (of up to 50%) while still achieving the same removal performance with respect to mineral acids[2], depending, of course, on the regeneration levels chosen and the analysis of the feed water.

It was realised soon after the launch of the Type-II SBA resins, that the thermal stability was poor compared with the Type I structure. The maximum recommended temperature for this resin type was finally set at 35-40°C (depending on the exact usage). Despite this restriction, the resin has been commonly used in the Middle East at higher temperatures. However the expected life-time is halved when compared with that in colder climates. Later, when removal of silica (necessary for water supplied to high-pressure boilers) became an essential requirement for boiler feed water treatment, the need for the Type-I resins increased further. In fact it can be said that the excellent silica removal properties of the Type-I resins was quite fortunate, as the Type-I resin was invented before efficient silica removal was a requirement. Nevertheless the complement in properties has since resulted in generally successful water treatment processes provided the correct resin is

FIGURE 1    STRONG BASE ANION EXCHANGE RESIN STRUCTURES

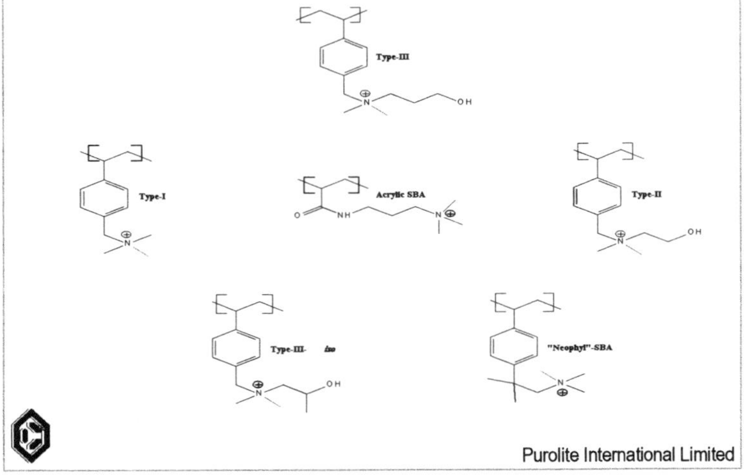

Purolite International Limited

matched to the process, and appropriate cleaning treatments applied to keep the resins in good condition.

The main exception to this statement is where SBA resins are treating water high in organic content. It was found, approximately 20 years ago, that SBA resins with an acrylic structure had a similar regenerability to that of the Type-II structure. This slightly more expensive resin also has good silica removal properties. However, the thermal stability is, if anything, marginally worse than that of the Type-II resin. Nevertheless its use has continued to increase worldwide, from its beginnings in the West Coast of Scotland and in Ireland, where organic fouling was a particular problem. On the other hand the neophile resin (also illustrated) has never found favour commercially because of cost. It was proposed initially as having a more thermally stable structure[3]. However results were fairly disappointing. This is not so for similar structure proposed by H.Kubota et al[4]

Studies on the mechanisms of thermal stability, and resin regenerability, which are discussed in previous papers on these subjects[5-7] led Purolite to investigate the N-propanolamine and iso-propanolamine structures. Some examples of these had been prepared previously[8], but were not fully evaluated.

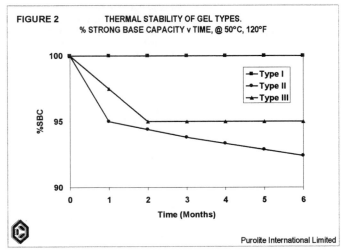

FIGURE 2    THERMAL STABILITY OF GEL TYPES.
% STRONG BASE CAPACITY v TIME, @ 50°C, 120°F

Purolite International Limited

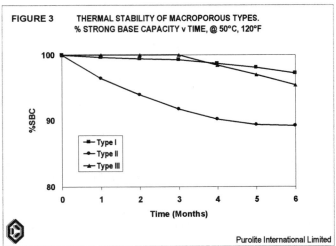

FIGURE 3    THERMAL STABILITY OF MACROPOROUS TYPES.
% STRONG BASE CAPACITY v TIME, @ 50°C, 120°F

Purolite International Limited

## I.ii Thermal Stability

Ion exchange resins with the novel functional groups were prepared[5] using both gel-type[9] and macroporous type[5] co-polymers. The thermal stability of the N-propanolamine based resin has been evaluated both at 80°C and 50°C. The iso-propanolamine degraded more rapidly than any other product ever tested at 80°C. The details were very useful for the understanding of the mechanism of thermal degradation. Naturally however, further investigations were limited to N-propanolamine. Figure 2 shows the loss of hydroxide (strong base groups) for the gel type at 50°C. Comparison with Figure 3 shows that the macroporous material is almost as stable as the Type-I resin at this temperature. At 80°C for the more stable macroporous resin, the loss of hydroxide is much higher for all three types as expected, see Figure 4. The macroporous Type-III lies more or less mid-way between Type-I and Type-II, Thus the initial results showed that potential for satisfactory operation in countries with hot ambient climates was promising, because the properties more closely approached those of Type-I resin at 50°C.

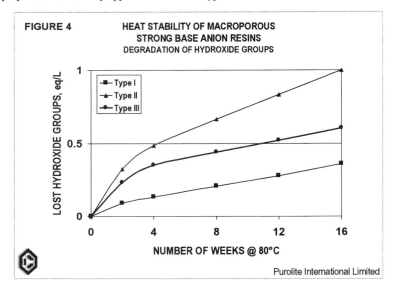

**FIGURE 4**          **HEAT STABILITY OF MACROPOROUS STRONG BASE ANION RESINS**
DEGRADATION OF HYDROXIDE GROUPS

LOST HYDROXIDE GROUPS, eq/L

— Type I
— Type II
— Type III

NUMBER OF WEEKS @ 80°C

Purolite International Limited

## I.iii Regeneration Capacity

Figure 5 compares the regeneration capacity for Type-I, Type-II and, both the iso and normal Type-III. The curves for displacing both the chloride and sulphate ion with hydroxide ion show that the capacity is between Type-I and Type-II. Interestingly the iso is better, but this is of no benefit because of the poor thermal stability. The Figure 6 compares the sulphate and chloride displacement more carefully. The higher release of sulphate is as expected[2]. The mechanism of the regeneration is dependent on selectivity differences between the hydroxide and chloride or sulphate. This was investigated previously[9].

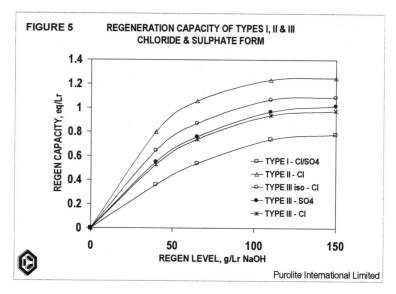

FIGURE 5    REGENERATION CAPACITY OF TYPES I, II & III
            CHLORIDE & SULPHATE FORM

Purolite International Limited

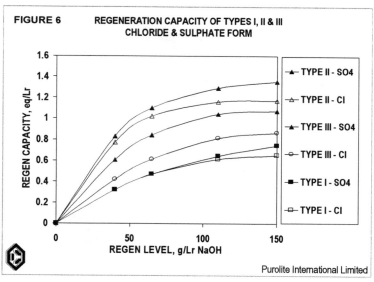

FIGURE 6    REGENERATION CAPACITY OF TYPES I, II & III
            CHLORIDE & SULPHATE FORM

Purolite International Limited

## I.iv Operating Capacity

Anion exchange in the loading phase is a neutralisation process, so the regeneration capacity virtually controls the operating capacity. Figure 7 for co-flow regeneration shows that the operating capacity mirrors the regeneration capacity closely. For counter-flow regeneration, however, the capacity of the gel Type-III, is actually higher than that of the macroporous Type-II, and the macroporous Type-III appears only a little better than the Gel Type-I, which was a little disappointing, see Figure 8.

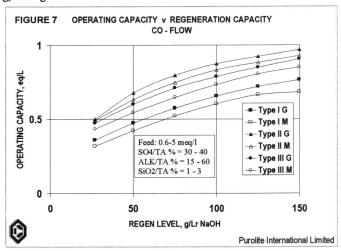

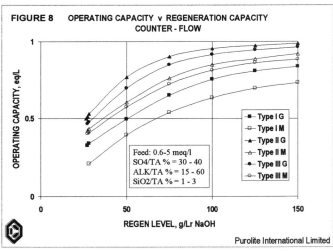

## a) Silica Removal

Figure 9a gives a silica leakage and conductivity profile for a particular feed water available for co-flow operation. Figure 9b gives the data for counter-flow operation. Values for silica are below 10 ppb co-flow and below 5 ppb counter-flow, very close to Type-I performance. Table I compares values obtained for the resin types. This shows how close the values are to those for a Type-I.

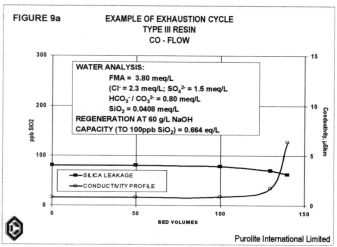

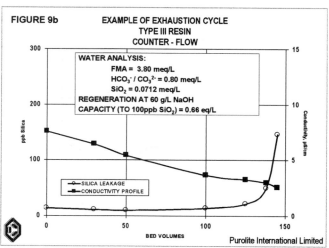

TABLE 1                    SILICA LEAKAGE OF TYPE III
                        SODIUM HYDROXIDE REGENERATION
                      TO 100mg/m³ SILICA LEAKAGE END-POINT

| REGENERATION LEVEL, MODE | % SILICA / TOTAL ANIONS | TYPE III SiO₂ mg/m³ | PUBLISHED DATA, SiO₂ mg/m³ | |
|---|---|---|---|---|
| | | | TYPE-I | TYPE-II |
| CO-FLOW | | | | |
| 60 g/L | 1% | 18 | 15 | 40 |
| 150 g/L | 3% | 14 | 10 | 20 |
| COUNTER-FLOW | | | | |
| 60 g/L | 1% | 12 | 8 | 20 |
| 150 g/L | 3% | 10 | 6 | 15 |

Purolite International Limited

TABLE 2        SUMMARY OF PROPERTIES FOR RESIN TYPES I, II & III

| CHARACTERISTIC | TYPE-I | TYPE-II | TYPE-III | TYPE-I ACRYLIC |
|---|---|---|---|---|
| OPERATING CAPACITY | ✗ | ✓ | ✓ | ✓ |
| SILICA REMOVAL | ✓ | ✗ | ½ | ✓ |
| HEAT STABILITY | ✓ | ✗ | ½ | ✗ |
| TOTAL | 2 | 1 | 2 | 2 |

KEY TO TABLE :   ✓ - GOOD    ✗ - POOR    ½ - INTERMEDIATE

Purolite International Limited

## b) Organics Removal

Accurate data for organics removal needs a lengthy programme of work that can only be achieved with a thorough field trial. Those organic matters loaded during the experimental determination of operating capacity, were easily removed, but insufficient data was available to make authoritative predictions.

## Summary of Laboratory Evaluation

Table 2 gives an overall comparison of the differences in the properties of the resin types. The advantage of the Type-III over the others is that it does not have any major weakness in this rather crude summary. What it shows is that this promising structure really merits extended field trials.

## II PLANT OPERATING PERFORMANCE

A site for a field trail was identified at New Plymouth Power Station in New Zealand. A detailed report of the performance over the first 15 months has been presented [10] in 1996. The paper gives details of the first 10 months operation.

### a) Capacity

Quite surprisingly and contrary to what was expected from the laboratory data, the capacity obtained for the Type-III is higher than that for the Type-II resin over the first 10 months (See Figure 10). Results are quite variable because of changes in the cation limitation of some runs; but overall, the trends are clear. The reason for the higher capacity is that the resins are run to a silica end-point.

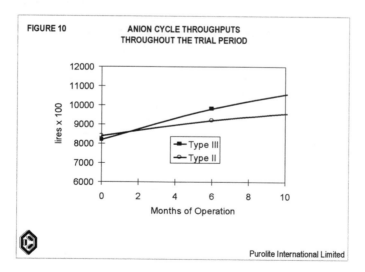

FIGURE 10    ANION CYCLE THROUGHPUTS
THROUGHOUT THE TRIAL PERIOD

Purolite International Limited

## b) Silica Removal

The Type-III has a better silica removal profile: thus to a silica end point, the Type-III has an added advantage. This is particularly surprising because the end point is very much higher than usual at 4.5 ppm. It would therefore be expected that the advantage would be even greater for an end point of 0.5 ppm or less. Figure 10A shows the silica and conductivity profile for the Type-II resin after start-up. Figure 10B gives the same data for the Type-III resin.

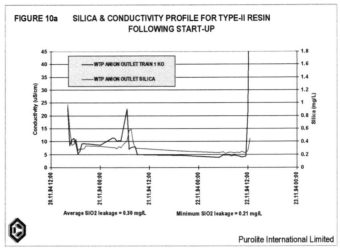

FIGURE 10a    SILICA & CONDUCTIVITY PROFILE FOR TYPE-II RESIN FOLLOWING START-UP

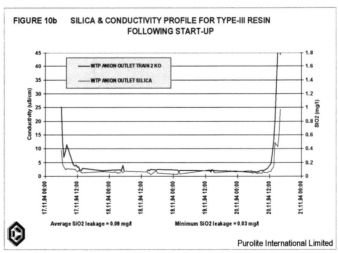

FIGURE 10b    SILICA & CONDUCTIVITY PROFILE FOR TYPE-III RESIN FOLLOWING START-UP

**FIGURE 11**

**SILICA LEAKAGE**
**FIRST 12 MONTHS OF OPERATION**

| Anion 1, Type-II | SiO2 (mg/L) | | Anion 2, Type-III | SiO2 (mg/L) | |
|---|---|---|---|---|---|
| | Average | Min | | Average | Min |
| October '94 | 0.23 | 0.14 | October '94 | 0.23 | 0.16 |
| November ' 94 | 0.27 | 0.16 | November ' 94 | 0.09 | 0.02 |
| December '94 | 0.35 | 0.29 | December '94 | 0.17 | 0.12 |
| January '95 * | 0.31 | 0.21 | January '95 * | - | - |
| February '95 * | - | - | February '95 * | - | - |
| March '95 * | 0.35 | 0.20 | March '95 * | 0.25 | 0.10 |
| April '95 * | 0.30 | 0.20 | April '95 * | 0.13 | 0.10 |
| May '95 * | 0.23 | 0.20 | May '95 * | - | - |
| June '95 | 0.34 | 0.24 | June '95 | 0.26 | 0.09 |
| July '95 | 0.28 | 0.21 | July '95 | 0.19 | 0.13 |
| August '95 | 0.34 | 0.28 | August '95 | 0.13 | 0.10 |
| September '95 | 0.44 | 0.29 | September '95 | 0.13 | 0.09 |
| October '95 | 0.45 | 0.17 | October '95 | 0.12 | 0.06 |

- Minimal or no information collected due to lack of plant operation or due to cation limited service cycles

Purolite International Limited

Figure 11 gives the average silica leakage over the first 12 months of operation. It is particularly interesting that the Type III is improving slightly, where the Type-II is deteriorating. The resin with higher capacity is clearly regenerating more easily. This means that the conversion to the hydroxide form is improving for the Type-III. The deterioration in the Type-II resin is explicable by the degradation of the strong base groups to weak base (see Figure 12).

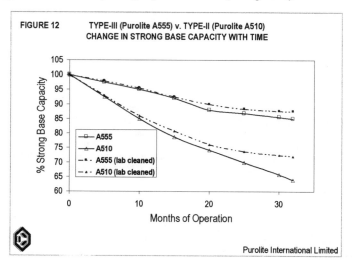

**FIGURE 12**     **TYPE-III (Purolite A555) v. TYPE-II (Purolite A510)**
**CHANGE IN STRONG BASE CAPACITY WITH TIME**

Purolite International Limited

## c) Resin Life

As the lifetimes of these resins evolve, the trends in capacity initially seen are reversed. The Type-II resin performance is deteriorating significantly (See Figure 13). The reason for the initial increase in capacity for the Type-III resin is that the introduction of weak base groups increases the regeneration efficiency, and yet there are sufficient strong base groups to remove all the weak acids (organics, silica, and carbonates). As the ratio of strong base groups to weak base groups further decreases, the concentration of strong base groups is not sufficient to maintain quality. So the quality or capacity falls off, depending on the water analysis and the treated water specification.

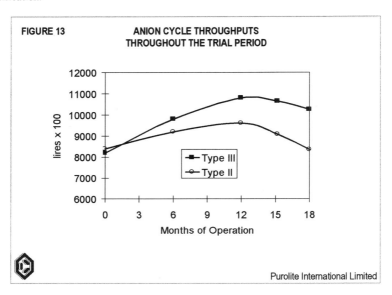

FIGURE 13      ANION CYCLE THROUGHPUTS THROUGHOUT THE TRIAL PERIOD

Purolite International Limited

## d) Thermal Stability

One further point : the ratiod loss of SBC, for the two resin types is predicted quite closely by the 50° C data from the laboratory thermal stability tests described earlier. This prediction is much closer than for 80° C tests. This is very useful information for evaluating thermal stability of resins, to guide further developments. It should be born in mind that this plant is operating at only 25-30°C, so the resin is not severely stressed thermally. The fact that, at this temperature, the model ratio is still accurate, suggests good performance at temperatures up to 50°C.

## e) Silica Fouling

One very interesting result of the degradation process from strong base to weak base described above is the change of the pH inside the resin. As the degraded resin containing weak base groups exhausts, the mineral acids are taken up by addition to the free base sites. These free base sites are typically at a pH of less than three. This means that, as the resin approaches exhaustion, the resin

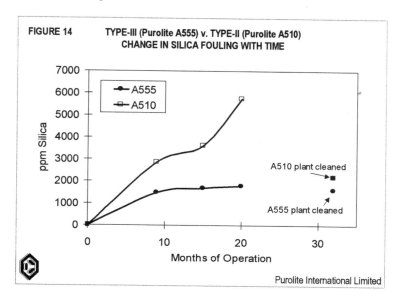

FIGURE 14    TYPE-III (Purolite A555) v. TYPE-II (Purolite A510)
CHANGE IN SILICA FOULING WITH TIME

Purolite International Limited

internal environment is more acidic than that of a resin containing a higher proportion of strong base sites which essentially moves from alkaline to neutral pH. This is because the Donnan exclusion applies. In dilute solution, the cations (positively charged) are excluded from within the anion beads (positively charged functional groups). For the Type-II resin the proportion of weak base groups is seen to increase more rapidly. Hence it becomes more prone to silica fouling as the silica precipitates in the increasingly acidic environment. Figure 14 shows exactly what has happened to the silica content of the resin over the first 20 months. As the weak base content increases to over 25%, so the silica fouling has escalated rapidly. It is true that the same problem is occurring with the Type-III resin, but at a much slower rate. The situation is compounded further for the Type-II resin. The standard method for silica removal is to soak the resin in hot caustic soda. This dissolves the silica. However this process increases the percentage of weak base groups which of course makes the resin more prone to silica fouling in the future. It should also be pointed out that previous experience with acrylic resins has indicated that they suffer when treating waters with a high ratio of silica to total anions. No such problems are anticipated with this thermally stable polystyrenic type.

## f) Organics

The resins have been subjected to a reasonable load of organics, which has built up over 20 months. However this does not appear to have affected performance, certainly not of the Type-III resin, though it might be a contributary factor to the decrease in the performance of the Type-II resin. Organic load has been a problem with certain resins in the past, which made this a good site for the trial. Further operating cycles may provide more significant data. The profile of organics elution during regeneration also indicates a wider spectra of organics removal, see Fig.15. In all probability this arises from more efficient organics removal. Certainly, deterioration of performance has been seen from time to time during the three years of operation; however, the use

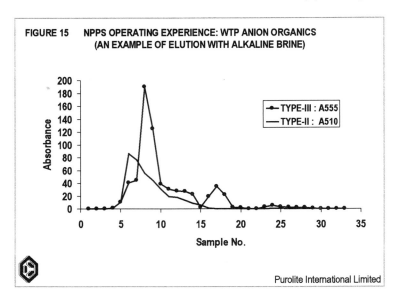

FIGURE 15    NPPS OPERATING EXPERIENCE: WTP ANION ORGANICS
(AN EXAMPLE OF ELUTION WITH ALKALINE BRINE)

Purolite International Limited

of alkaline brine treatments has restored performance, indicating that any fouling has been reversible.

## Commercial

Results over the last two years have indicated that the Type-III resin has fulfilled its potential. Comparing with Type-I resin: for every £10,000 plant equipment costs the resin cost is approximately £1000. If Type-I resin is replaced by Type-III the plant size can be reduced by 33%, operating counter-flow at 50g/L, Thus there is a saving of £3,300 on capital cost as well as a 33% saving on regenerant. The capital cost alone justifies a resin increase of £1000 pounds sterling 3.3 times over. The extra price of the resin is insignificant compared with the savings.

The comparison against Type-II is more difficult. The plant size will be approximately the same, although a saving of 10% only in improved capacity would justify the increased resin cost twice over: for capital savings on equipment and regeneration costs, not to mention the improved performance. However the real gain is in the resin life. When operating in warm climates, the operating life of the Type-II can be as low as two years, with part of that time giving poor performance. If the Type-III resin lasts 3.5-4 years, this represents a similar overall cost except that the cost of reordering, replacement and commissioning are avoided. Inflationary rises, if any, over the intermediate two year period do not figure. In addition if the improved performance is cut back to an equivalent performance, regenerant savings of 10-20% may be possible.  Thus the overall package demonstrates that, in certain situations, the Type-III resin is commercially attractive.

## Overall Rating

Table 2 assessed an overall comparison of the various strong base resin types, based on the laboratory tests. In Table 3 this is extended to cover the areas of silica fouling, and performance when treating waters containing organics. The review also takes a re-look at the basic parameters covered in Table 2. This overall appraisal clearly shows that there are significant advantages to be gained by choosing this resin where appropriate despite its increased cost.

TABLE 3    SUMMARY OF PROPERTIES FOR RESIN TYPES I, II & III
- PLANT EXPERIENCE

| CHARACTERISTIC | TYPE-I | TYPE-II | TYPE-III | TYPE-I ACRYLIC |
|---|---|---|---|---|
| OPERATING CAPACITY | ✗ | ✓ | ✓✓ | ✓ |
| SILICA REMOVAL (TREATED WATER) | ✓ | ✗ | ½ | ✓ |
| SILICA FOULING | (✓) | ✗ | ✓ | ½ |
| HEAT STABILITY | ✓ | ✗ | ½ | ✗ |
| ORGANIC FOULING RESISTANCE | (½) | ✓ | ✓✓ | (✓✓) |
| RATING | 3½ | 2 | 6 | 4½ |

KEY TO TABLE :   ✓ - GOOD    ✗ - POOR    ½ - INTERMEDIATE

Purolite International Limited

## Conclusion

The point is rapidly being approached where the justification of cost of manufacture is being confirmed. The demand for this product has escalated sharply, though its use is still limited.

## Acknowledgements

To Wayne Williamson of New Plymouth Power Station in New Zealand who generated the performance data. To Elsa Davies, Phillip Warn, Andreas Gotthardt, Graham Crooks, and Monica Forster for the hard work to obtain and process the laboratory Data.

## References

1. R.M.Wheaton & W.C.Bauman Ind.& Eng. Chem 1951 **43** 1088.
2. J.D. Nolan and J.Irving "Ion Exchange Technology" 1984 Ed. D. Naden & M.Streat, Ellis Horwood Publishers, 1984, p.160.
3. M.J.Hatch, & W.D.Lloyd, J.App Poly Science, 1964 **8** 1659
4. H.Kubota, K.Yano, S.Sawada, Y.Aosaki, J.Watanabe, T.Usui, S.Ono, T.Shoda, K.Okazaki, M.Tomoi, M.Shindo, and T. Onozuka, "Ion Exchange Developments and Applications" Ed. J.A.Greig SCI, London 1996 p182.
5. J.Irving & J.A.Dale, IWC-92-34 Pittsburgh Water Conf, 1996 p302.
6. B. Chu, D.C.Whitney, R.M. Diamond, J.Inorg Nuclear Chem 1962 **24** 1405-1415.
7. B.A.Soldano, & G.E.Boyd, J.Am Chem Soc. 1953 **75** 6099
8. W.C.Bauman & R.Mckellar, U.S.Pat 2614099 1952
9. J.A.Dale & J.Irving, "Ion Exchange Advances" Ed. M.J.Slater, Elsevier 1992 p33
10. W.S. Williamson & J.Irving, "Ion Exchange Developments and Applications" Ed. J.A.Greig, SCI London 1996 p.43.
11. F.G. Donnan, Z. Electrochem 1911 **17,** 572.

# SOME PROPERTIES OF THE STRONG BASE ANION EXCHANGERS BASED ON 4-VINYLPYRIDINE - DIVINYLBENZENE COPOLYMER

C. Luca[*], V. Neagu[*], B. C. Simionescu[**], R. Mocanu[***], P. Jitaru[***] and R. Buhaceanu[***]

[*] "Petru Poni" Institute of Macromolecular Chemistry, 6600, Iasi, Romania
[**] "Gh. Asachi" Technical University, 6600, Iasi, Romania
[***] "Al. Ioan-Cuza" University, 6600, Iasi, Romania

## 1 INTRODUCTION

Among the strong base anion exchangers, those based on quaternized 4-vinylpyridine - divinylbenzene (DVB) copolymers can be an alternative of the above-mentioned exchangers, in certain applications. Anion exchangers with 1-alkyl-4-vinylpyridinium structural units are used in recovery and purification processes of plutonium from scrap materials. The process consists of treatment of these materials with nitric acid leading to the anionic Pu (IV) nitrato complex, it being retained by means of the strong base anion exchangers. The advantage of the quaternized 4-vinylpyridine based exchangers compared to classical ones, based on styrene - DVB copolymers, is due to the much higher resistance to nitric acid of the pyridine aromatic rings than of those of styrene which are much more susceptible to the nitration reaction[1].

It is known that quaternized 4-vinylpyridine - DVB structures are made by the free-radical copolymerization of the two monomers, by means of the aqueous suspension technique, followed by the alkylation reaction of the copolymer. The most common alkylation is the nucleophilic substitution of the copolymer with different reactive halogenated derivatives[2]. Another alkylation method is by the nucleophilic addition reaction of the protonated copolymer to ethylenic electrophilic compounds[3].

Our previous study showed some characteristics of the strong base anion exchangers made by alkylation reaction such as: ion exchange and Fe(III) ion retention capacities, as well as chemical stabilities in aqueous HCl solutions of various concentrations[4]. The Fe(III) cation retention was from $Fe_2(SO_4)_3$ aqueous solution at pH=2.0 as the complex anion of the $[Fe_3(SO_4)_2(OH)_6]^-$.

The present study deals with the analysis of the characteristics of a larger range of 4-vinylpyridine - DVB copolymer based strong base anion exchangers made both by substitution and addition reactions. The paper shows chemical stabilities in the acid and alkaline media of these types of anion exchangers as well as, in brief, retention capacities for six metal cations, as anion their chlorocomplexes.

## 2 EXPERIMENTAL

The synthesis and characterization of the 4-vinylpyridine - DVB starting copolymer, as well as of the pyridine strong base anion exchangers, have been described elsewhere[4]. The chemical stabilities were determined as shown elsewhere[4].

The retention capacities of the M(II) cations were determined as follows: anion exchanger samples of 0.2 g with the known humidities were contacted for 24 hours at $25^0$C with 12 cm$^3$ of aqueous solutions of chlorocomplexes of known metal quantity. After contact, the exchanger samples were separated by filtration and the metal concentrations in the aqueous phase were determined by means of atomic absorption spectroscopy on a 3300 Perkin-Elmer spectroscope.

The difference between the initial metal concentration (from the volume of 12 cm$^3$ of aqueous chlorocomplex solutions) and the metal concentration into the aqueous phase after 24 hours, gave the retained metal concentration expressed as mmols M(II)/g dry ion exchanger. The chlorocomplexes for six cations as Hg(II), Zn(II), Pb(II), Cu(II), Mn(II) and Ni(II) were studied.

## 3 RESULTS AND DISCUSSION

Scheme 1 shows the synthesis, and chemical structure, for all the strong base anion exchangers used in this study.

The treatment with NaCl of the $A_1$ - $A_3$ samples meant the passing of aqueous NaCl solution of 5% concentration through the anion exchangers packed into columns, up to the absence of I⁻ ion in the effluent solution.

Scheme 1 shows that the seven structures can be separated in two types, namely those carried out by well-known nucleophilic substitutions ($A_1$-$A_4$ samples) and $A_5$-$A_7$ samples synthesized by an uncommon method; the nucleophilic addition of the 4-vinylpyridine - DVB copolymer to appropriate reagents (acrylamide, acrylonitrile and methylvinylketone). A copolymer with a content of 8% DVB was used as a starting compound for all the reactions.

### 3.1 Chemical stabilities

Chemical stabilities in the acid and alkaline media are very important properties of ion exchangers because these products have a very great advantage due to their participation in more regeneration - reuse cycles. Treatments with acid, especially HCl, and NaOH of the strong base anion exchangers intervene both in the elution of the retained metals and the regeneration processes. The Cl⁻ and OH⁻ forms of these exchangers are the most common when strong base anion exchangers are used.

Chemical stabilities in alkaline media of the $A_1$ - $A_7$ samples were examined and compared to those of Amberlite IRA - 400 and IRA - 410 samples. This aspect was studied for the possible use of the pyridine type strong base anion exchangers where the regeneration with NaOH is required, because this has not been reported yet in the literature. Amberlite IRA - 400 and IRA - 410 are commercially available strong base anion exchangers with the structural units of benzyltrimethylammonium chloride and benzyldimethyl-2-hydroxyethylammonium chloride, respectively, these samples being, so-called, strong base anion exchangers of I and II types. The starting compounds for these exchangers are gel type copolymers. The chemical stabilities in alkaline media are shown in Table 1.

From the data in Table 1 one can see that the treatment with aqueous 1M NaOH solution of the strong base anion exchangers containing pyridine structures leads to a much greater decrease of their ion exchange capacities compared to those with styrene structures (Amberlite IRA - 400 and IRA - 410). The decreases of the ion exchange capacities are due to the well-known Hofmann degradation of the quaternary ammonium hydroxides[5]. It is observed that the presence of the quaternary nitrogen atom in the aromatic ring determines the much stronger degradation compared to the situation when this atom is outside to the ring (samples $A_1$ - $A_7$ compared to the two Amberlite samples).

**Table 1** *Chemical stabilities in the alkaline media for the strong base anion exchangers studied*

| Sample | Nature and concentration of aqueous alkaline solution | | Total ion exchange capacity | | Loss of ion exchange capacity % | |
|---|---|---|---|---|---|---|
| | | | meq/g | meq/cm$^3$ | weight capacity | volume capacity |
| $A_1$ | | 0 | 5.07 | 2.03 | | |
| | NaOH | 1N | 3.05 | 1.21 | 39.84 | 40.20 |
| $A_2$ | | 0 | 4.90 | 1.90 | | |
| | NaOH | 1N | 2.15 | 0.84 | 56.12 | 55.79 |
| | NaHCO$_3$ | 1N | 4.90 | 1.90 | 0 | 0 |
| $A_3$ | | 0 | 3.77 | 1.58 | | |
| | NaOH | 1N | 0.96 | 0.43 | 74.53 | 73.20 |
| $A_4$ | | 0 | 3.00 | 1.40 | | |
| | NaOH | 1N | 1.77 | 0.84 | 41.00 | 40.00 |
| $A_5$ | | 0 | 3.38 | 1.42 | | |
| | NaOH | 1N | 0.013 | 0.006 | 99.61 | 99.57 |
| $A_6$ | | 0 | 4.00 | 1.78 | | |
| | NaOH | 1N | 0.617 | 0.277 | 84.57 | 84.44 |
| $A_7$ | | 0 | 2.23 | 1.30 | | |
| | NaOH | 1N | 0.158 | 0.056 | 92.91 | 95.69 |
| Amberlite IRA-400 | | 0 | 4.98 | 1.47 | | |
| | NaOH | 1N | 4.90 | 1.40 | 1.60 | 4.76 |
| Amberlite IRA-410 | | 0 | 3.45 | 1.44 | | |
| | NaOH | 1N | 3.40 | 1.40 | 1.44 | 2.77 |

Among the pyridine quaternary structures, those which contain the alkyl substitutents on quaternary nitrogen atoms ($A_1$ - $A_3$ samples) display decreases of the ion exchange capacities with increases of alkyl substitute length. The $A_1$ and $A_4$ samples, with

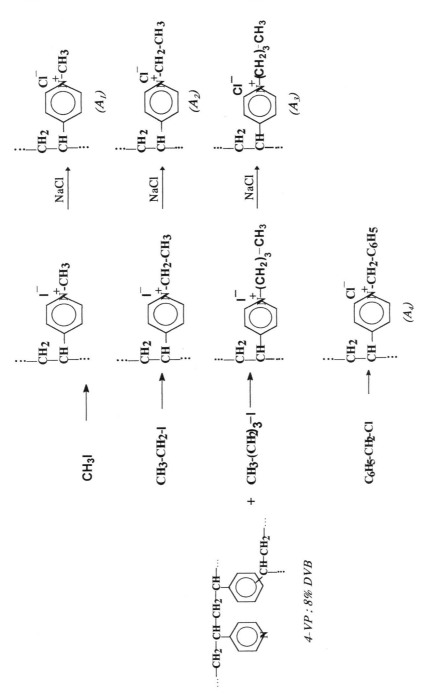

**Scheme 1**

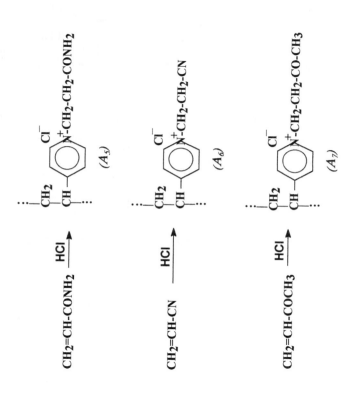

Scheme 1 (continued)

methyl and benzyl substitutents respectively, have the same behaviour in the NaOH solution.

The loss of the strong base exchange capacities become much more marked when the substitutents on the quaternary nitrogen atoms contain electron-attracting groups, such as amide, nitrile, and ketone ($A_5$ - $A_7$ samples). It is clear that in the behaviour of the pyridine strong base anion exchangers in the NaOH aqueous medium, the chemical nature of the substitutent on the quaternary nitrogen atom is of great importance.

It must be pointed out that a mild alkaline medium, such as sodium bicarbonate, does not modify the ion exchange capacities. The situation of $A_2$ sample is shown in Table 1 to illustrate this.

The good stabilities of the pyridine strong base anion exchangers in the HCl media, especially of those with methyl and ethyl as substitutents on quaternary nitrogen atoms, can also be seen. Also, Table 2 shows the effect of more severe treatment with HCl at 90°C. From Table 2 one can see that the $A_1$ and $A_2$ samples have higher chemical stabilities than the two commercially available strong base anion exchangers, whose loss of ion exchange capacities is about of 20%.

**Table 2** *Chemical stability in acid medium for the strong base anion exchangers studied*

| Sample | Concentration of aqueous HCl solution | Total ion exchange capacity meq/g | meq/cm$^3$ | Reference |
|--------|--------|--------|--------|--------|
| $A_1$ | 0 | 5.07 | 2.03 | present study |
|  | $1N^*$ | 5.07 | 2.03 |  |
|  | $2N^*$ | 5.07 | 2.03 |  |
|  | $3N^*$ | 5.04 | 2.00 |  |
|  | $1N^{**}$ | 5.04 | 2.00 |  |
| $A_2$ | 0 | 4.90 | 1.90 | present study |
|  | $1N^*$ | 4.90 | 1.90 |  |
|  | $2N^*$ | 4.90 | 1.90 |  |
|  | $3N^*$ | 4.90 | 1.90 |  |
|  | $1N^{**}$ | 4.90 | 1.90 |  |
| $A_3$ | 0 | 3.77 | 1.58 | present study |
|  | $1N^*$ | 3.77 | 1.58 |  |
|  | $2N^*$ | 3.77 | 1.58 |  |
|  | $3N^*$ | 3.67 | 1.46 |  |
|  | $1N^{**}$ | 3.60 | 1.40 |  |
| $A_4$ | 0 | 3.00 | 1.40 | present study |
|  | $1N^*$ | 2.72 | 1.29 |  |
|  | $2N^*$ | 2.72 | 1.29 |  |
|  | $3N^*$ | 2.51 | 1.19 |  |
|  | $1N^{**}$ | 2.57 | 1.20 |  |
| $A_5$ | 0 | 3.38 | 1.42 | 4 |
|  | $1N^*$ | 3.28 | 1.40 |  |
|  | $2N^{**}$ | 3.18 | 1.38 |  |
|  | $3N^*$ | 3.00 | 1.37 |  |
|  | $1N^{**}$ | 2.90 | 1.17 |  |

**Table 2** (continued)

| | | | | |
|---|---|---|---|---|
| $A_6$ | 0 | 4.00 | 1.78 | 4 |
| | $1N^*$ | 4.00 | 1.78 | |
| | $2N^*$ | 3.70 | 1.75 | |
| | $3N^*$ | 3.70 | 1.75 | |
| | $1N^{**}$ | 3.68 | 1.68 | |
| $A_7$ | 0 | 3.45 | 1.38 | 4 |
| | $1N^*$ | 3.35 | 1.36 | |
| | $2N^*$ | 3.30 | 1.35 | |
| | $3N^*$ | 3.20 | 1.34 | |
| | $1N^{**}$ | 3.00 | 1.20 | |
| Amberlite IRA - 400 | 0 | 4.98 | 1.47 | present study |
| | $1N^*$ | 4.98 | 1.47 | |
| | $2N^*$ | 4.90 | 1.40 | |
| | $3N^*$ | 4.90 | 1.40 | |
| Amberlite IRA - 410 | 0 | 3.45 | 1.44 | present study |
| | $1N^*$ | 2.90 | 1.20 | |
| | $2N^*$ | 2.90 | 1.20 | |
| | $3N^*$ | 2.90 | 1.20 | |
| | $1N^{**}$ | 2.75 | 1.15 | |

\* all the experiments were made at $25^0C$ for 24 hours
\*\* all the experiments were made at $90^0C$ for 10 minutes

## 3.2 Retention of M(II) cations as chlorocomplexes

The retention of cations of transition, heavy and noble metals can be performed by the ion exchange process using strong and weak acid cation exchangers as well as strong base ones, the latter exchangers being used if the metal cations there are in the conditions to form complex anions.

In the presence of an excess of $X^-$ anion, a M(II) cation may occur in solution in the following forms: M(II), $MX^+$, $MX_2$, $[MX_3]^-$ and $[MX_4]^{2-}$. The two last species, because of their anionic character, may be retained by the strong base anion exchangers by an ion exchange process. More than that, the $MX_2$ species may be also retained by the exchangers due to an Donnan type equilibrium[6]. In the present study the M(II) cations were Hg(II), Zn(II), Pb(II), Cu(II), Mn(II) and Ni(II) and the pyridine strong base anion exchangers were $A_1$, $A_2$, $A_5$ and $A_7$. The two commercially available strong base anion exchangers, Amberlite IRA - 400 and Amberlite IRA - 410 also were studied.

The preliminary retention studies showed that the above-mentioned strong base anion exchangers show preference only for Hg(II), Zn(II) and Pb(II) cations.

The metal quantities retained are listed in Table 3.

**Table 3** *Retention capacities of the samples studied*

| Sample | Retention capacities (mmols M/g dry ion exchanger) | | |
|---|---|---|---|
| | Hg(II) | Zn(II) | Pb(II) |
| A$_1$ | 2.290 | 1.125 | 1.090 |
| A$_2$ | 4.420 | 1.625 | 1.251 |
| A$_5$ | 2.675 | 1.345 | 1.120 |
| A$_7$ | 2.045 | 1.170 | 0.935 |
| Amberlite IRA-400 | 2.645 | 1.080 | 0.585 |
| Amberlite IRA-410 | 2.365 | 1.460 | 0.490 |

From Table 3 one can see that, for all the samples, the retentions follow the sequence: Hg(II) > Zn(II) > Pb(II). Generally, the pyridine strong base anion exchanger with the ethyl substituent on the quaternary nitrogen atoms (sample A$_2$) had the highest retention values for all the three metal cations. The largest difference between this sample and the others was for the Hg(II) cation. Probably, the more hydrophobic structure of the A$_2$ sample determined the larger preference for the chlorocomplex of Hg(II). The smallest differences in retention values between the A$_2$ sample and the others were in the case of the Zn(II) cation. It was clear that the pyridine strong base anion exchangers have a higher preference than the styrene ones for the chlorocomplex of the Pb(II) cation.

CONCLUSIONS

The alkylation reactions of the 4-vinylpyridine-DVB copolymers, both by nucleophilic substitutions with reactive halogenated compounds and nucleophilic additions to electrophilic compounds, lead to strong base anion exchangers. The values of their ion exchange capacities are of the same magnitude as those of strong base anion exchangers based on styrene-DVB copolymers.

A disadvantage of the strong base anion exchangers with pyridine structure is their instability in NaOH aqueous solution which leads to the loss of strong base capacity.

However, the pyridine strong base anion exchangers have good stabilities in mildly alkaline (sodium bicarbonate) and HCl media.

Strong base anion exchangers, with the pyridine structure, show higher retention capacities for some metal cations, as complex anions, than the commercially available strong base anion exchangers based on styrene-DVB copolymers.

## References

1. J. Haggin, *Chemical and Eng. News,* 1989, **67**, 27.
2. J. M. J. Frechet and M. V. de Meftahi, *British Polymer J.*, 1984, **16**, 193.
3. C. Luca, V. Neagu, G. Grigoriu and B. C. Simionescu, *Rev. Roum. Chim.*, 1995, **40(7-8)**, 717.
4. C. Luca, V. Neagu, G. Grigoriu and B. C. Simionescu, "Processes in Ion Exchange: Advances and Applications", A. Dyer, M. J. Hudson, P. A. Williams (Eds.), The Royal Society of Chemistry, Special Publication No. 196, Cambridge, U. K.,1997, p. 70.
5. S. Patai, "The chemistry of the amino groups", Interscience Publishes, John Wiley and Sons, New York, 1967.
6. T. Onofrei and S. Fisel, "Polimeri chelatizanti in chimia analitica anorganica", Tehnopress, Iasi, 1998.

# PROPERTIES OF PHOSPHORIC ACID RESINS DERIVED FROM POLY(GLYCIDYL METHACRYLATE)S CROSSLINKED WITH DIMETHACRYLATES OF OLIGO(ETHYLENE GLYCOL)S

A. Jyo, K. Yamabe, and G. Kimura

Department of Applied Chemistry and Biochemistry
Kumamoto University
Kumamoto 860-8555, Japan

## 1 INTRODUCTION

Recently, we have reported properties of phosphoric acid resins (RGP), which were prepared by addition of phosphoric acid to epoxy groups in macroreticular poly(glycidyl methacrylate)s crosslinked with divinylbenzene.[1,2] These resins have larger acid capacities compared with phosphoric acid resins so far reported.[3-9] For example, even the RGP crosslinked with 10 nominal mol% of divinylbenzene has an acid capacity as large as 7 meq/g. Furthermore, the metal ion selectivity of RGP is quite different from that of conventional cation exchange resins having sulfonic acid groups. For instance, RGP exhibits extremely high selectivity toward Fe(III), U(VI), and Ti(IV) even in strongly acidic media (1 - 6 M of mineral acids), which is not observed in the case of the sulfonic acid resins.[1,2]

As reviewed by Helfferich,[10] the metal ion selectivity of crosslinked poly(styrene) based sulfonic acid resins is enhanced with an increase in content of divinylbenzene (or degree of crosslinking). Thus, of interest is the effect of crosslinking species on properties of phosphoric acid resins derived from crosslinked poly(glycidyl methacrylate)s. In this work, we have prepared phosphoric acid resins by addition of phosphoric acid to epoxy groups in poly(glycidyl methacrylate)s crosslinked with dimethacrylates of four kinds of oligo(ethylene glycol)s (Scheme 1), and properties of the resulting resins were studied.

## 2 EXPERIMENTAL SECTION

### 2.1 Preparation of Resins (Scheme 1)

*2.1.1 Materials.* Glycidyl methacrylate was provided from Nippon Oil and Fats Co. and was purified by distillation under reduced pressure. Crosslinkers used were ethylene dimethacrylte (1G), diethylene glycol dimethacrylate (2G), triethylene glycol dimethacrylate (3G), and tetraethylene glycol dimethacrylate (4G). These crosslinkers were supplied from Shin Nakamura Chemical Co. and were washed with cooled 1 M NaOH and then with cooled water in a separating funnel. Other reagents used were of reagent grade unless otherwise noted.

*2.1.2 Suspension Polymerization.* Glycidyl methacrylate and a crosslinker (1G, 2G, 3G or 4G) were polymerized by means of the suspension polymerization technique in the presence of isobutyl acetate as diluent. An aqueous solution (500 mL, act as dispersion bath) containing gelatin (0.5 g) and sodium sulfate (6 g) was taken into a 1000 mL autoclave equipped with stirrer and heater, and nitrogen was bubbled into this solution for 60 mim. A monomer solution (50 mL) consisting of glycidyl methacrylate and each

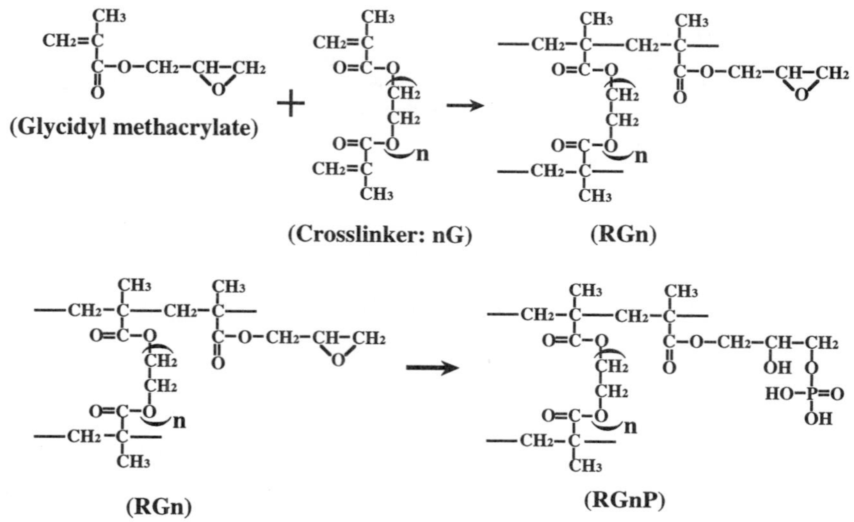

**Scheme 1** *Preparation of phosphoric acid resins. The number n is 1, 2, 3 or 4.*

crosslinker (its content = 10 mol%) was prepared, and then isobutyl acetate (50 mL) was added to the monomer solution. After nitrogen was bubbled into the resulting organic solution for 60 min, azobisisobutyronitrile (0.5 g) as initiator was dissolved. The monomer solution was added to the dispersion bath in the autoclave, and the resulting mixture was stirred (260 - 270 rpm). Then, the temperature of the stirring mixture was raised to 70 °C for 1 h and was maintained at the same temperature for more 1 h. In order to complete polymerization, the temperature of the mixture was raised to 90 °C for 1.5 h and maintained at this temperature for 2 h more. After cooling the mixture, the resulting copolymer beads were collected by filtration, and washed with a large amount of deionized water. The beads were boiled in a beaker for 30 min to eliminate gelatin, and they were immersed in methanol overnight to eliminate the diluent. After filtration, the copolymer beads were air-dried and beads of 32 - 60 mesh were collected by meshing. Spherical copolymer particles were selected as described,[1] and dried in vacuum at 40 °C for more than 24 h prior to their functionalization. Hereafter, copolymer beads crosslinked with 1G, 2G, 3G, and 4G are denoted by the symbols RG1, RG2, RG3, and RG4, respectively.

    *2.1.3 Functionalization of Copolymer Beads.* Copolymer beads (5 g of RG1, RG2, RG3, or RG4) and commercially available phosphoric acid (100 mL, 85 %) were heated in a beaker on an oil bath at 140 °C for 2 h. The resulting phosphoric acid resin was washed with water until the washing became acid-free. After air-drying, they were dried in vacuum at 40 °C for more than 24 h. In a smaller scale preparation, 1 g of copolymer beads and 20 mL of the phosphoric acid were used. The phosphoric acid resins derived from RG1, RG2, RG3, and RG4 are denoted by the symbols RG1P, RG2P, RG3P, and RG4P, respectively.

## 2.2 Characterization of the Resulting Resins

Phosphorus contents, acid capacities, water contents and swelling ratios of the resulting resins were measured according to reported methods.[1,7,10,11] Infrared spectra of resins were recorded by means of a KBr disk method.

## 2.3 Capacity for Metal Ion Uptake

Into a 100 mL Erlenmeyer flask, a metal ion solution (50 mL, 0.01 M) and a resin (0.125 g) were taken, and the flask was shaken at 30 °C for 24 h. The concentration of metal ion in the aqueous phase then was determined by means of EDTA titration. From the decrease in the metal ion concentration in the aqueous phase, a capacity for the metal ion in mmol/g-dry resin was calculated. The pH of the solutions was  adjusted with hydrochloric acid,  acetic acid and/or sodium acetate.

## 2.4 Distribution Ratios

Into a 50 mL Erlenmeyer flask, a metal ion solution (25 mL, 0.0001 M) and a resin (40 mg) were taken, and the flask was shaken at 30 °C for 24 h. Then, the concentration of metal ion in the aqueous phase was determined by means of inductively coupled plasma atomic emission spectrometry. From the decrease in the metal ion concentration in the aqueous phase, a distribution ratio (D) designated by Equation (1) was  calculated.

$$D = \frac{\text{amount of metal ion in resin at equilibrium [mmol/g]}}{\text{amount of metal ion in solution at equilibrium [mmol/mL]}} \quad (1)$$

The pH of  metal ion solutions was adjusted with hydrochloric acid.

## 3 RESULTS AND DISCUSSION

### 3.1 Effect of Temperature and Reaction Time on Functionalization

Effect of temperature and reaction time on the functionalization reaction was examined by using the precursory copolymer beads RG1. First, by changing the reaction temperature but fixing the reaction time for 2 h, the phosphoric acid resins RG1P were prepared. Figure 1a shows acid capacities observed and ones estimated from phosphorus contents of the resulting resins. Up to 140 °C, both parameters increase with an increase in the reaction temperature and coincide well with each other. At 160 °C deviation between both parameters became rather marked. These results  indicate that the optimum synthesis temperature is ca. 140 °C. Next a series of the RG1P resins were prepared by changing the reaction time but fixing the temperature at 140 °C. Figure 1b shows capacities observed and ones estimated from phosphorus contents as a function of the reaction time. Until 2 h, both parameters increase with an increase in the reaction time and coincide well with each other. However, the deviation between both parameters becomes significant with a further increase in the reaction time; namely, acid capacities estimated from phosphorus contents decrease but observed ones are essentially constant. The results shown in Figures 1a and 1b clearly suggest that addition of phosphoric acid and acid catalyzed ester bond cleavage simultaneously occur. The former reaction leads to introduction of phosphoric acid groups but the latter does to the formation of carboxylic groups and decrease in phosphorus contents. The formation of carboxylic groups compensates the loss of acid capacities  resulting  from the cleavage of ester bonds  which connect  phosphoric acid groups to resins. Indeed, IR adsorption around 1550 cm⁻¹ of the

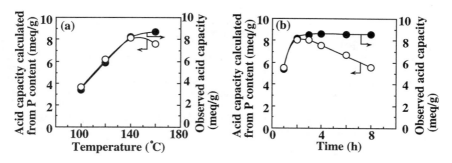

**Figure 1** *Effect of temperature (a) and reaction time (b) on functionalization reaction. Precursory copolymer RG1 (1 g), phosphoric acid (20 mL). Reaction time in (a) and reaction temperature in (b) are 2 h and 140 °C, respectively.*

RG1P resins in sodium form becomes stronger with an increase in the reaction time suggesting that too high temperature, and prolonged heating, brings about cleavage of the ester bonds. From the results shown in Figures 1a and 1b, the optimized functionalization is achieved by heating at 140 °C for 2 h as in the case of the macroreticular poly(glycidyl methacrylate) crosslinked with 10 nominal mol% of divinylbenzene.[1]

## 3.2 Characterizations of RG1P, RG2P, RG3P, and RG4P

Referring to the results given in the preceding section, four kinds of phosphoric acid resins, RG1P, RG2P, RG3P, and RG4P, were prepared by heating RG1, RG2, RG3, and RG4 in commercially available phosphoric acid at 140 °C for 2 hr. Table 1 summarizes their properties with those of the phosphoric acid resin RGP, which was derived from macroreticular poly(glycidyl methacrylate) crosslinked with 10 nominal mol% of divinylbenzene.

**Table 1** *Properties of Phosphoric Acid Resins*

| Resin | PC [1] | AC [2] | DV[3] | WV [4] | SR [5] | WC[6] |
|-------|------|------|------|------|------|------|
| RG1P | 3.6 | 7.4 | 1.3 | 2.8 | 2.2 | 43.7 |
| RG2P | 3.9 | 7.9 | 1.2 | 3.7 | 3.1 | 60.7 |
| RG3P | 3.9 | 7.7 | 1.2 | 4.2 | 3.5 | 64.5 |
| RG4P | 4.0 | 8.4 | 1.2 | 4.2 | 3.5 | 66.4 |
| RGP | 3.5 | 7.0 | 1.9 | 3.2 | 1.7 | 55.2 |

[1] Phosphorus content [mmol/g]. [2] Acid capacity [meq/g].
[3] Dry volume [mL/g]. [4] Wet volume [mL/g].
[5] Swelling ratio [-]. [6] Water content [wt%].

Properties of RG1P are significantly different from those of RG2P. However, properties of RG3P and RG4P are not so greatly different from those of RG2P. In particular, parameters concerning hydrophilicity of RG1P are largely different from those of the other three resins. Water content of RG1P is only 43.7 % but those of the other three resins are ranged from 60.7 to 66.4 %. In addition, swelling ratio of RG1P is 2.2, and this is much less than those of other resins (3.1 - 3.5). These data indicate RG1P is more hydrophobic than the other three resins.

Dry volumes of the four resins are nearly equal (1.2 - 1.3 mL/g) but they are much less than that of RGP (1.9 mL/g). This indicates that highly stable porous structures comparable to those of RGP are difficult to achieve in the case of the phosphoric acid resins derived from poly(glycidyl methacrylate)s crosslinked with oligo(ethylene glycol)s. Acid capacities and phosphorus contents increase in the order of RG1P < RG2P ≈ RG3P ≤ RG4P. This probably means that the efficiency of the functionalization increases with an increase in length of the crosslinking bridge.

### 3.3 Capacities for Uptake of Metal Ions

Capacities for uptake of Cu(II), Pb(II), and U(VI) by the four phosphoric acid resins were studied in acidic pH regions (0 < pH < 5). Figure 2 shows the uptake of these metal ions as a function of pH. The metal ion affinity of each resin increases in the order of Cu(II) < Pb(II) < U(VI). This order of the metal ion affinity is characteristic to phosphoric acid resins.[1,2]

Capacities of RG1P for Pb(II) are slightly smaller than those of RG2P, RG3P, and RG4P, but it seems that all four resins give nearly equal capacities for Cu(II). Uptake of

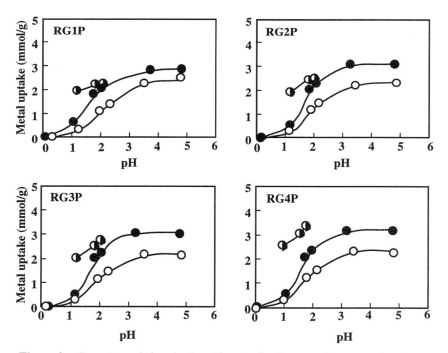

**Figure 2**   *Capacities of phosphoric acid resins for three metal ions as a function of pH.   Resin 0.125 g,   solution 50 mL, initial concentration of each metal ion 0.01 M.  ◑ :U(VI),  ●: Pb(II),  O:  Cu(II).*

Pb(II) closely corresponds to their phosphorus contents (Table 1). In uptake of U(VI), however, the situation is somewhat different from the uptake of Cu(II) and Pb(II). Although there is no marked difference in phosphorus contents between RG2P, RG3P, and RG4P, RG4P can take up U(VI) more effectively than RG2P and RG3P, suggesting that, the longer the crosslinking bridge, the larger the uptake of U(VI).   Since U(VI) tends to form bulky multi-nuclear hydrolyzed species, diffusion of U(VI) in the resin phases is strongly affected by the structure of the polymer frameworks. It seems that RG4P, having the longest crosslinking bridge, is suitable for rapid diffusion of bulky hydrolyzed species of U(VI). Thus, the suitable selection of crosslinker species is important in the effective adsorption of specific metal ions, such as U(VI).

## 3.4 Distribution of Metal Ions from Hydrochloric Acid Media

Effect of pH on distribution of metal ions was tested by using RG1P, RG2P, RG3P, and RG4P. The tested metal ions were Pb(II), Mn(II), Zn(II), Cu(II), Ni(II), Lu(III), Gd(III), and La(III). In this work, -log[HCl] was used as pH above 1 M of hydrochloric acid. Here, results of RG1P and RG4P are mainly discussed, since the results  of RG2P and RG3P are rather similar to those of RG4P. Figure 3 compares log D vs. pH curves between RG1P and RG4P. Both resins exhibit almost the same metal ion selectivity sequence: Lu(III) $\geq$ Gd(III) > La(III) > Pb(II)  $\approx$ Mn(II) > Zn(II) $\approx$ Cu(II) $\approx$ Ni(II). This selectivity sequence qualitatively explains why capacities of the phosphoric acid resins for Pb(II) are larger than those for Cu(II) (Figure 2). Clearly,  RG1P more strongly adsorbs lanthanides than does RG4P. Furthermore the mutual selectivity of RG1P to the divalent metal ions is higher than that of RG4P. These results mean that the metal ion selectivity of phosphoric acid resins is affected by crosslinking species.

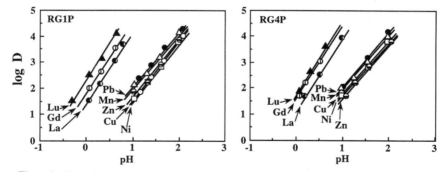

**Figure 3**  *Dependence of partition coefficients on pH.   Resin 0.04 g,   solution 25 mL .  Initial concentration of each metal ion  is 0.0001 M.  The pH was adjusted  with hydrochloric acid except for Pb(II) (nitric acid).*

As is well known, the metal ion selectivity of sulfonic acid resins based on poly(styrene)s crosslinked with divinylbenzene is enhanced with an increase in divinylbenzene contents (degree of crosslinking). In these kinds of resins, water contents of lightly crosslinked resins are usually much larger than those of highly crosslinked ones. Thus, the metal ion selectivity of cation exchangers is closely related to hydrophilicity of the resin phases; in general, the higher the hydrophilicity of resin phase the lower the metal ion selectivity. Thus, the lower inter-cationic selectivity of RG4P can be ascribed to the enhanced hydrophilicity of RG4P. Indeed, the hydrophilicity of RG1P is much less than that of RG4P as estimated from swelling ratios as well as water contents of the phosphoric acid resins studied (Table 1). Although the four phosphoric acid resins prepared in this work have the same degree of crosslinking (10 mol% of each crosslinker), their metal ion selectivity is lowered with an increase in number of ethylene glycol moiety in the crosslinkers. In conclusion, the present work clarified the important role of crosslinking species in overall properties of phosphoric acid resins.

## Acknowledgment

The authors are grateful to Nippon Oil and Fats Co. for gift of glycidyl methacrylate as well as to Shin Nakamura Chemical Co. for gift of ethylene dimethacryalte, diethylene glycol dimethacrylate, triethylene glycol dimethacrylate, and tetraethylene glycol dimethacrylate.

## References

1.  A. Jyo, A. Matsufune, H. Ono, and H. Egawa, *J. Appl. Polym. Sci.,* 1997, **63**, 1327.
2.  A. Jyo, X. Zhu, 'Chemistry for the Protection of the Environment 3' , Plenum Press, New York, in press.
3.  J. D. Reid and L. W. Mazzeno, Jr., *Ind. Eng. Chem.,* 1949, **41**, 2828.
4.  G. C. Daul, J. D. Reid, and R. M. Reinhardt , *Ind. Eng. Chem.,* 1954, **46**, 1042.
5.  G. M. Blackburn, M. J. Brown, M. R. Harris, and D. Shire, *J. Chem., Soc. C,* **1969**, 676.
6.  E. Bayer, P.-A. Grathwohl, and K. Geckeler, *Angew. Makromol. Chem.,* 1983, **113**, 141.
7.  H. Egawa, T. Nonaka, and H. Maeda, *J. Appl. Polym. Sci.,* 1985, **30**, 3239.
8.  R. B. Selzer and D. G. Howery, *Macromolecules,* 1986, **19**, 2673.
9.  R. B. Selzer and D. G. Howery, *Macromolecules,* 1986, **19**, 2974.
10. F. Helfferich, 'Ion Exchange', McGraw Hill, New York, 1962.
11. H. Egawa, T. Nonaka, and M. Ikari, *J. Appl. Polym. Sci.,* 1984, **29** 2045.

# PREPARATION OF ION EXCHANGE FIBRES FROM LOOFAH (NATURAL FIBRE) AND THEIR PROPERTIES

T. Nonaka, E. Noda and S. Kurihra

Department of Applied Chemistry and Biochemistry, Faculty of Engineering,
Kumamoto University,
Kumamoto 860-8555, JAPAN

## 1 INTRODUCTION

Cellulose is a naturally occurring polymer and is biodegradable. The chemical modification of cellulose has been investigated by many researchers[1,2,3] and modified celluloses are used for many purposes. Loofah fibre is a natural fibre and it has a strong network structure. The graft copolymerization of vinyl monomers onto loofah fibres has been achieved and their use as adsorbents or antibacterial fibres studied.

## 2 EXPERIMENTAL

### 2.1 Materials

Dried bleached loofahs obtained commercially were refluxed with benzene/ethanol (1/1 volume ratio) for 6 h and then dried. The dried loofahs were boiled for 12 h in a 1% NaOH solution under a nitrogen atmosphere followed by treating three times with 1% acetic acid solution, then washed with water until the smell of acetic acid disappeared. The loofah fibres obtained were used for experiments after complete drying.

Tributyl-4-vinylbenzylphosphonium chloride(TBVB) and methacryloyloxyethyl-trimethylammonium chloride(METAC) were kindly supplied by Nippon Kagaku Kogyo Co. and Nitto Riken Kogyo Co., respectively. They were used without further purification. 2,3-Epithiopropyl methacrylate(ETMA) was prepared by the method reported earlier[4]. Other chemicals were reagent grade.

### 2.2 Graft copolymerization of vinyl monomers onto loofah fibres

Pretreated loofah fibres(about 50 mg) were soaked in deionized water (or deionized water-dimethylsulfoxide(DMSO) mixed solution) and water-soluble monomers(METAC and TBVB) (or water-insoluble ETMA) were added. Nitrogen was introduced into this solution for 1 h in an ice/water bath followed by addition of a desired amount of initiator (ammonium cerium nitrate for METAC, $H_2O_2$ for TBVB and ETMA) and by further introduction of nitrogen for 30 min. The polymerization was carried out in a sealed glass vessel. The grafted fibres obtained were purified by extraction of unreacted monomers and homopolymers with solvents (boiled water for METAC, methanol for TBVB, and DMSO for ETMA). The grafted loofah fibres obtained were abbreviated as L-g-MATAC, L-g-TBVB, and L-g-ETMA (LE), respectively.(Fig.1)

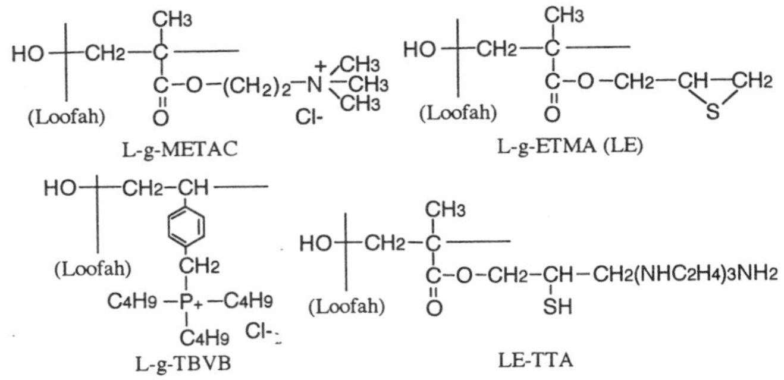

Fig. 1        Structure of grafted loofah fibres

## 2.3 Amination of LE

LE was aminated with triethylenetetramine (TTA) in 1,4-dioxane at 90 °C for 8h under occasional shaking.    First LE-TTA (Fig. 1) was washed with methanol and deionized water, then washed three times with 1 mol $dm^{-3}$ HCl solution to remove unreacted TTA, and followed by washing several times with 1 mol $dm^{-3}$ NaOH solution, and deionized water, until washings became neutral, and then dried.

The polymer add-on(%) was calculated by use of equation (1).

Polymer add-on(%) =100 x (wt of grafted loofah - wt of loofah)/wt of loofah    (1)

## 2.4 Measurement of adsorption capacity of the grafted loofah fibres for $Ag^+$ ions

Weighed grafted loofah fibres (about 100mg) and 50 $cm^3$ of 0.01 mol $dm^{-3}$ $AgNO_3$ solution were placed in a 100 $cm^3$ glass-stoppered Erlenmeyer flask.    The mixture was shaken at 30 °C for 24 h.    The adsorption capacity was calculated by determining the concentration of metal ions in the supernatant with inductively coupled argon plasma atomic emission spectrophotometry (Shimadzu ICPS-5000).

## 2.5    Measurement of antibacterial activity

Antibacterial activity was estimated as follows: After contacting the gels with a bacteria (*E. coli* or *S. aureus*) suspension for a prescribed time, 1 $cm^3$ of bacteria suspension was pipetted from the flask, and 9 $cm^3$ of sterile water was added to the bacteria suspension. The suspension was diluted several times, and 0.1 $cm^3$ of the diluted suspension was spread on an agar plate which was made of nutrient agar.    After the plate was kept at 30℃ for 15-24 h, the numbers of viable cells were calculated by counting the numbers of the colonies formed on the plate.

## 3 RESULTS AND DISCUSSION

## 3.1 Graft copolymerization of vinyl monomers onto loofah fibres

First graft copolymerization of METAC on loofah fibres was carried out by varying the

Table 1 Optimum condition for copolymerization of each monomer

| Monomer | Solvent | Maximum polymer add-on (%) | Monomer concentration (mol/dm³) | Initiator | Initiator concentration (mol/dm³) | Temperature (°C) |
|---------|---------|------|------|------|------|------|
| **METAC** | H₂O | 25 | 0.75 | Ce(IV) | 0.016 | 30-40 |
| **TBVB** | H₂O | 120 | 0.4 | H₂O₂ | 0.1 | 40 |
| **ETMA** | H₂O/DMSO(4/1) | 500 | 0.5 | H₂O₂ | 0.02 | 60 |

amount of ammonium cerium nitrate as an initiator. The maximum polymer add-on (%) was obtained at the concentration of 0.01-0.016 mol dm⁻³ Ce(IV). In the case of TBVB and ETMA, the maximum polymer add-on (%) was obtained at the concentration of 0.1 and 0.02 mol dm⁻³ of H₂O₂, respectively. The optimum condition to get the highest polymer add-on(%) for each monomer was investigated by changing temperature, time, monomer concentration, and solvent. Table 1 shows the optimum conditions and the maximum polymer add-on(%) for each copolymerization. The copolymerization of TBVB and ETMA could be initiated using H₂O₂, while METAC could be initiated using ammonium cerium nitrate. It was found that ETMA shoud be copolymerized in water/DMSO mixed solution, because ETMA was insoluble in water.

### 3.2 Confirmation of introduction of vinyl monomers into loofah fibres

The graft copolymerization was confirmed by elemental analysis and infrared spectra. The introduction of ETMA and TTA was confirmed from the absorption peaks at 1720 cm⁻¹ due to carboxyl groups and at 1550 cm⁻¹ due to imino groups, respectively.

### 3.3 Adsorption of Ag⁺ on LE and LE-TTA

Adsorption of Ag⁺ on LE and LE-TTA with different polymer add-on(%) was investigated. The results are shown in Fig.2. The LE and LE-TTA with about 100 % of polymer add-on(%) exhibited the maximum adsorption capacity (meq/g). The adsorption capacity of LE for Ag⁺ was higher than that of LE-TTA.

### 3.4 Antibacterial activity of LE-METAC and LE-TBVB

First antibacterial activity of L-g-METAC with different polymer add-on(%) against *E.*

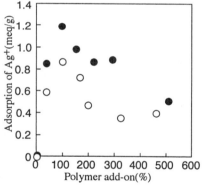

Fig. 2 Effect of polymer add-on on the adsorption of Ag⁺

Initial Ag⁺ concentration : 0.01 M
Shaking at 30 °C, 24 h
● : LE, ○ : LE-TTA

*coli* and *S. aureus* was investigated. L-g- METAC exhibited high antibacterial activity both against *E. coli* and *S. aureus* and the antibacterial activity increased with increasing polymer add-on(%).

Next antibacterial activity of L-g-TBVB with different polymer add-on(%) against *E. coli* and *S. aureus* was investigated (Fig.3). L-g-TBVB also exhibited high antibacterial activity against *E. coli* and *S. aureus* and the antibacterial activity was higher than that of LE-METAC. This may be due to the higher polymer add-on (%) of L-g-TBVB than that

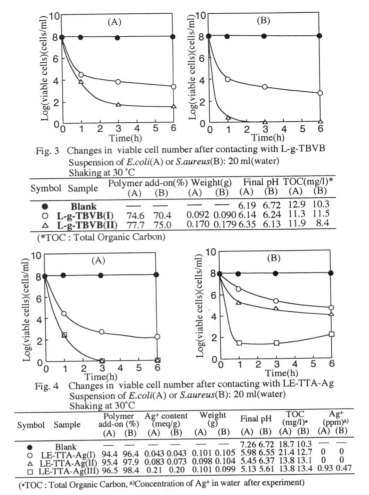

Fig. 3  Changes in viable cell number after contacting with L-g-TBVB
Suspension of *E.coli*(A) or *S.aureus*(B): 20 ml(water)
Shaking at 30 °C

| Symbol | Sample | Polymer add-on(%) (A) | (B) | Weight(g) (A) | (B) | Final pH (A) | (B) | TOC(mg/l)* (A) | (B) |
|---|---|---|---|---|---|---|---|---|---|
| ● | Blank | — | — | — | — | 6.19 | 6.72 | 12.9 | 10.3 |
| ○ | L-g-TBVB(I) | 74.6 | 70.4 | 0.092 | 0.090 | 6.14 | 6.24 | 11.3 | 11.5 |
| △ | L-g-TBVB(II) | 77.7 | 75.0 | 0.170 | 0.179 | 6.35 | 6.13 | 11.9 | 8.4 |

(*TOC : Total Organic Carbon)

Fig. 4  Changes in viable cell number after contacting with LE-TTA-Ag
Suspension of *E.coli*(A) or *S.aureus*(B): 20 ml(water)
Shaking at 30°C

| Symbol | Sample | Polymer add-on (%) (A) | (B) | Ag+ content (meq/g) (A) | (B) | Weight (g) (A) | (B) | Final pH (A) | (B) | TOC (mg/l)* (A) | (B) | Ag+ (ppm)a) (A) | (B) |
|---|---|---|---|---|---|---|---|---|---|---|---|---|---|
| ● | Blank | — | — | — | — | — | — | 7.26 | 6.72 | 18.7 | 10.3 | — | — |
| ○ | LE-TTA-Ag(I) | 94.4 | 96.4 | 0.043 | 0.043 | 0.101 | 0.105 | 5.98 | 6.55 | 21.4 | 12.7 | 0 | 0 |
| △ | LE-TTA-Ag(II) | 95.4 | 97.9 | 0.083 | 0.073 | 0.098 | 0.104 | 5.45 | 6.37 | 13.8 | 13.1 | 0 | 0 |
| □ | LE-TTA-Ag(III) | 96.5 | 98.4 | 0.21 | 0.20 | 0.101 | 0.099 | 5.13 | 5.61 | 13.8 | 13.4 | 0.93 | 0.47 |

(*TOC : Total Organic Carbon, a)Concentration of Ag+ in water after experiment)

of L-g-METAC.  Furthermore they exhibited higher antibacterial activity against *S. aureus* than *E. coli*.

### 3.5  Antibacterial activity of LE-Ag and LE-TTA-Ag

First antibacterial activity of LE which adsorbed different amounts of Ag+ was investigated against *E. coli* and *S. aureus*.  These LE-Ag (Ag+ adsorbed : 0.044-0.079 meq/g) exhibited no antibacterial activity against both bacteria except LE-Ag having the highest Ag+ adsorption (0.20 meq/g) against *S. aureus*.

The antibacterial activity of LE-TTA which adsorbed different amounts of Ag+ was also investigated against *E. coli* and *S. aureus* (Fig.4).  These LE-TTA-Ag (Ag+ adsorbed : 0.043-0.21 meq/g) exhibited high antibacterial activity against both bacteria.  They exhibited higher antibacterial activity against *E. coli* than *S. aureus*.

The high antibacterial activity of LE-TTA-Ag is due to the release of $Ag^+$ from LE-TTA and subsequent penetration into the bacteria. Therefore, the higher antibacterial activity of LE-TTA-Ag is thought to be due to easier release of $Ag^+$ from the LE-TTA than the LE.

## 3.6 Adsorption of bacteria on L-g-METAC, L-g-TBVB and LE-TTA-Ag

The surface on the grafted loofah fibres after contacting with bacteria was observed by scanning electron microscopy. It was found that many bacteria were adsorbed on L-g-METAC and L-g-TBVB, but not on LE-TTA-Ag. This indicates that the grafted loofah fibres with quaternary ammonium or phosphonium groups can adsorb bacteria such as *E. coli* which is known to have negative charges on the surface.

## REFERENCES

1) G.F.Fanta, R.C.Burr, and W.M.Doane, *J.Appl.Polym.Sci.*, **27**, 2731(1982).
2) H.Kubota and Y.Shigehisa, *J.Appl.Polym.Sci.*, **56**, 147(1995).
3) H.Hebeish, A.Waly, F.A.Abel-Mohdy, A.S.Aly, *J.Appl.Polym.Sci.*, **66**, 1029 (1997).
4) H.Egawa and T.Nonaka, *Kobunshi Ronbunshu*, **35**, 21(1978).

# Removal of Isotopes from Aqueous Solutions

# Treatment of Radioactive Waste using a Titanosilicate Analogue of the Mineral Zorite

A. Dyer and M. Pillinger

Chemistry Division, Science Research Institute, Cockcroft Building,
University of Salford, Salford M5 4WT, U.K.

## Introduction

Large volumes of aqueous nuclear waste arise from several different sources including nuclear power stations, reprocessing plants, military sites and research centres. One preferred technology involved in the treatment of such waste is ion exchange, specifically the use of inorganic ion exchangers for the separation of actinide elements, fission products and activated corrosion products. Column operation enables large reductions in volumes of liquid waste together with high decontamination factors. After use, the ion exchanger can be vitrified or encapsulated in cement for long-term disposal as high-level waste, while the larger volume of remaining material can be more easily disposed of as low-level waste. Inorganic ion exchangers have a number of advantages over organic ion exchangers including superior thermal, chemical and radiolytic stability, together with better compatibility with final waste forms. They also have high capacities and selectivities for certain monovalent and divalent cations.

One promising group of inorganic ion exchangers are framework titanosilicates, the structures of which are built up from silicon in tetrahedral coordination and titanium in octahedral and/or semi-octahedral coordination. The negatively charged frameworks are balanced by interstitial cations located in tunnels or cavities. Pore sizes are generally in the range 3-7.5 Å. Some of these materials have novel structures while others are structural analogues of naturally occurring minerals. To date we have prepared a large number of variants and screened them for their ability to remove certain radionuclides from aqueous solution. The performance of the more promising materials has been further evaluated in the presence of the most common macro-components present in nuclear waste ($H^+$, $Na^+$, $K^+$, $Mg^{2+}$, $Ca^{2+}$, EDTA, citrate and borate). Tests have then been carried out in simulated feedstreams. This paper describes results for one of these materials, a titanosilicate analogue of the mineral zorite.

The crystal structure of zorite was first reported in 1979 by Sandomiriskii and Belov.[1] It has a highly disordered framework with ostensibly a two-dimensional channel system.[2] The two orthogonal sets of channels are defined by 12-Si/Ti atom and 8-Si atom rings but disorder results in the larger 12-ring channels becoming partitioned into sections. A cation diffusing into the 12-ring channel would have to make detours through the 8-Si ring channel system in order to pass freely. The ideal unit cell chemical formula for zorite is $Na_6[Ti_5Si_{12}O_{35}(OH)_4]11H_2O$.

## Experimental

### Materials and methods

All reagents were reagent grade or better and were used without further purification. The sources of titanium, niobium, antimony and silicon used were titanium(IV) isobutoxide (Aldrich), niobium(V) ethoxide (99.95 %, Aldrich), antimony(V) butoxide (95 wt.% solution in butanol and toluene, Aldrich) and Cab-O-Sil M5 fumed silica (BDH) respectively. Powder X-ray diffraction

(XRD) patterns were collected on a Siemens D5000 diffractometer with Cu Kα radiation (step size 0.02 °2θ, 1 s/step). Thermogravimetric analyses were performed on a Mettler TA3000 system at a heating rate of 5 K/min under a nitrogen atmosphere. Elemental analyses for titanium, antimony, silicon and sodium were conducted by dissolving each solid in HF, after which the solutions were analysed by atomic absorption/emission spectroscopy (AAS/AES). Filtrates from all ion-exchange experiments were measured using an Orion model 720A pH meter fitted with an Accumet semi micro calomel electrode.

*Synthesis*

Gels with the composition $SiO_2$ : 1.92NaOH : 0.15TBABr : 25 $H_2O$ : 1.8 $H_2O_2$ : 0.1$TiO_2$ were heated in teflon-lined stainless steel autoclaves at 180-200 °C for 4-5 days.[3] The crystalline products were isolated by centrifugation and washed before air-drying at 50-60 °C. For the sample synthesised at 200 °C TGA mass loss showed 10% water loss up to 200 °C followed by a mass loss of 5.8% between 200 and 300 °C characterised by a strong sharp peak in the DTG profile at 223 °C. The structure probably collapses at this point. Elemental analysis yielded the composition 12.0% Na, 16.5% Ti, 24.5% Si (Na:Ti:Si = 1.5:1:2.5, 5.2 meq Na/g hydrous basis). The Si:Ti ratio of 2.5 is close to that for the ideal unit cell chemical formula for zorite (Si:Ti = 2.4). There was a good match between the powder XRD pattern of this sample and the observed d-spacings for zorite (Fig. 1).

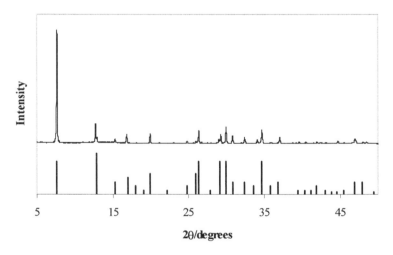

**Figure 1** Powder XRD pattern of zorite analogue (top, 180 °C synthesis) compared with observed d-spacings and intensities for the natural mineral (JCPDS 25-1298) ($\lambda$ = 1.5406 Å)

Attempts were also made to substitute antimony or niobium for titanium. Gels with the composition $SiO_2$ : 1.92NaOH : 0.15TBABr : 25 $H_2O$ : 1.8 $H_2O_2$ : 0.0978$TiO_2$ : 0.0011$M_2O_5$ were autoclaved at 200 °C for 5 days. Elemental analysis of the antimony-containing product yielded the composition 10.8% Na, 17.6% Ti, 21.1% Si, 0.6% Sb (Na:Ti:Si:Sb = 1.3:1:2.0:0.01, 4.7 meq Na/g hydrous basis). The powder XRD patterns of both materials were identical to that of the pure titanium derivative.

Maximally-exchanged K-form, Ca-form and Mg-form samples were prepared by equilibrating as-synthesised Na-form samples several times at room temperature with 1 mol/L solutions of the nitrates (solution to exchanger ratio = 100).

## Ion exchange experiments

*Determination of cation exchange capacity by isotope dilution analysis*

0.01 g of Na-form exchanger synthesised at 200 °C was equilibrated at room temperature for 1 day with 10 ml 0.025 mol/L $NaNO_3/NaOH$ ($NO_3^-$ :$OH^-$ = 1:1) to which [22]Na tracer had been added. The solid phase was separated from the solution by centrifugation (4000 rpm, 20 min) followed by filtration through a 0.2 μm PVDF membrane. [22]Na in solution was measured using liquid scintillation counting. The exchangeable sodium content in the solid corresponded to 5.0 meq/g (hydrous basis), in good agreement with the value obtained from AAS/AES.

*Distribution coefficient measurements*

Batch distribution coefficients ($K_d$) of tracer quantities of appropriate ions on Na-form samples were determined initially in 0.1 mol/L $NaNO_3$, 0.1 mol/L $NaNO_3$/0.1 mol/L NaOH and 0.1 mol/L $HNO_3$. Exchangers were equilibrated with the solutions for 1-2 days (rolling) and the solid phase was separated from the solution as described above (no filtration in the case of [236]Pu). $K_d$ was calculated from the activity measurements of radio-tracers using the equation

$$K_d = \frac{(Ai - A)}{A} \frac{V}{m} = \frac{\text{Concentration of ion in exchanger}}{\text{Concentration of ion in solution}}$$

where V/m is the solution to exchanger ratio (batch factor, L/kg, normally 100 or 200). [57]Co, [65]Zn, [54]Mn and [59]Fe in solutions were measured by gamma spectrometry; [236]Pu, [89]Sr and [137]Cs were measured by liquid scintillation counting.

The performance of the material was also evaluated in the presence of certain macro-components. Distribution coefficients of trace caesium and strontium were measured on the Na-form sample as a function of pH in 0.1 mol/L $NaNO_3$ and as a function of sodium ion concentration (pH adjusted with concentrated nitric acid). The concentration of strontium was 2.7→5.7E-6 mol/L and the concentration of caesium was 7.7E→8-1.5E-6 mol/L. Similarly distribution coefficients of caesium and strontium were measured on a K-form sample as a function of potassium ion concentration. Distribution coefficients of strontium were also measured on Ca-form and Mg-form samples as a function of calcium ion and magnesium ion concentration respectively. Finally distribution coefficients of caesium and strontium were measured as a function of caesium ion and strontium ion concentration respectively.

*Tests in simulated wastes*

Distribution coefficients of caesium and strontium on a Na-form sample synthesised at 200 °C were determined in five simulated waste feed-streams as a function of pH: Granitic ground-water (GGW), Sellafield ground-water (SGW), Site Exchange Effluent Plant (SIXEP), Nuclear Power Plant (NPP) and Neutralised Current Acid Waste (NCAW, Hanford Tank 241-AZ-102) (Table 1). A sample of the natural zeolite clinoptilolite as used in the BNF plc Sellafield SIXEP process was used as a reference material. The equilibration time was one week and a relatively large batch factor of 1000 was used in order to minimise changes in solution composition.

## Ion Exchange Results and Discussion

*Screening test results*

The test results indicated that the material was very efficient for removing trace caesium and strontium from neutral to mildly alkaline sodium salt solutions (Table 2). A very high $K_d$ of 175000 L/kg was also obtained for [137]Cs in acidic solution (pH = 1.4, 99.94 % sorption). However, $K_d$s for the other radionuclides were very low in 0.1 mol/L $HNO_3$.

**Table 1** Compositions of the simulated waste feedstreams (mmol/L)

|        | GGW   | SGW   | SIXEP | NPP    | NCAW  |
|--------|-------|-------|-------|--------|-------|
| Na     | 2.838 | 360   | 4.345 | 8.874  | 4987  |
| K      |       | 3.8   |       | 0.515  | 120   |
| Rb     |       |       |       |        | 0.05  |
| Cs     | Trace | Trace | Trace | Trace  | 0.5   |
| Mg     | 0.177 | 5.9   | 0.0615| 0.494  |       |
| Ca     | 0.45  | 27.0  | 0.0375| 1.00   |       |
| Sr     | Trace | 0.28  | Trace | Trace  | Trace |
| Al     |       |       |       |        | 430   |
| $SiO_2$| 0.206 |       | 0.167 |        |       |
| $NO_2$ |       |       |       |        | 430   |
| $NO_3$ | 1.56  |       | 0.198 | 10.883 | 1.669 |
| $CO_3$ |       |       |       |        | 230   |
| $HCO_3$| 0.2   | 2.0   | 4.345 |        |       |
| $SO_4$ |       | 11.0  | 0.1   |        |       |
| $PO_4$ |       |       |       |        | 25    |
| F      |       |       |       |        | 0.089 |
| Cl     | 1.975 | 406.2 |       |        |       |
| OH     |       |       |       |        | 3400  |

**Table 2** Distribution coefficients (L/kg) on zorite analogue synthesised at 180 °C (V:m = 100)

|                                    | $^{57}$Co | $^{65}$Zn | $^{54}$Mn | $^{59}$Fe | $^{236}$Pu | $^{137}$Cs | $^{89}$Sr |
|------------------------------------|-----------|-----------|-----------|-----------|------------|------------|-----------|
| 0.1 M NaNO$_3$ (pH 10.5)           | 620       | 174700    | 60100     | >200000   | 2630       | >200000    | >200000   |
| 0.1 M NaNO$_3$/0.1 M NaOH (pH 12.9)| 0         | 5310      | 19070     | 63800     | 1100       | 83100      | >200000   |
| 0.1 M HNO$_3$ (pH 1.4)             | 2         | 0         | 0         | 0         | 0          | 175000     | 13        |

*Distribution coefficients of trace caesium and strontium as a function of pH in 0.1 mol/L NaNO₃*

In general the situation is favourable for the removal of trace caesium ion across a wide pH range (Fig. 2). Effective removal of trace strontium is limited to the pH range 7-13; the distribution coefficient falls off sharply once the pH decreases below neutral. Very similar behaviour was obtained for two batches differing only in the synthesis temperature (180 °C and 200 °C).

As a result of incorporation of antimony or niobium in the synthesis mixture there was a significant improvement in sorption of $^{137}$Cs in the pH range 1-9 compared to the pure titanium zorite analogue (Fig. 3). For example, at pH 1.2, the $K_d$ of $^{137}$Cs on the antimony-containing material (final Ti:Sb in the product ≈ 80) was 25000 L/kg (99.2 % sorption) compared to 2500 L/kg (92.6 % sorption) on the pure titanium analogue. The magnitudes of the distribution coefficients on both metal-doped materials were very similar at any given pH. An unexplained dip was observed in the distribution coefficient of $^{137}$Cs at about pH 7 on all three materials. Niobium/antimony substitution had less of an effect on $^{89}$Sr uptake although both materials showed slightly better performance in the pH range 1-9.

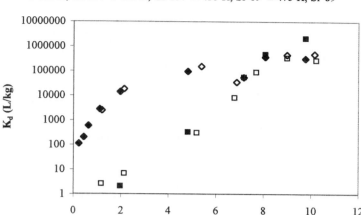

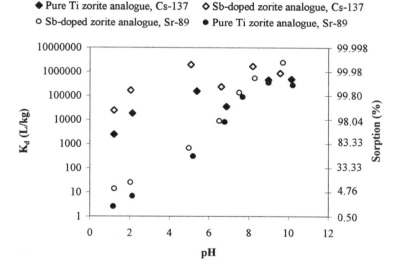

**Figure 2** Distribution coefficients of trace caesium and strontium on zorite analogues synthesised at different temperatures as a function of pH in 0.1 mol/L NaNO₃ (V:m = 100 for 180 °C synthesis, 200 for 200 °C synthesis)

**Figure 3** Distribution coefficients of trace caesium and strontium on titanium and titanium/antimony (Ti:Sb ≈ 80 in the product) zorite analogue as a function of pH in 0.1 mol/L NaNO₃ (V:m = 200)

*Distribution coefficients of trace caesium and strontium as a function of sodium ion concentration*

The distribution coefficients of both trace caesium and strontium on the zorite analogue increase by greater than one order of magnitude between sodium ion concentrations of 0.001 and 0.1 mol/L (Fig. 4). For example, the distribution coefficients in 0.001 mol/L $NaNO_3$ were 7000 L/kg for $^{137}Cs$ (98.6 % sorption) and 5840 L/kg for $^{89}Sr$ (98.3 % sorption). In 0.1 mol/L $NaNO_3$ the activity of both radionuclides in the equilibration solutions was below the detection limit ($K_d > 250000$ L/Kg). At low sodium ion concentrations hydronium ion exchange is probably significant resulting in an alkaline equilibrium pH and an equilibrium sodium ion concentration that is greater than the initial value. With higher initial sodium ion concentrations the equilibrium is shifted in favour of the exchanger in the sodium-form and the equilibrium pH is lower. Since uptake of caesium or strontium is more favourable on the sodium-ion form than the hydronium-ion form the distribution coefficients increase. Above a sodium ion concentration of 0.1 mol/L there is a linear decrease in the distribution coefficient on a logarithmic scale (slope -1.26 caesium and -1.69 for strontium). According to theory, expressing $LogK_d$ for a trace ion $M^{z+}$ and an ion exchanger AR, where $A^{y+}$ is the exchangeable ion, as a function of $A^{y+}$ concentration in the solution, should give a straight line with a slope of $-z/y$, in the case that the process is pure ion exchange.[4] The more positive value obtained for strontium ion exchange may be due to a contribution from $Sr(OH)^+$.

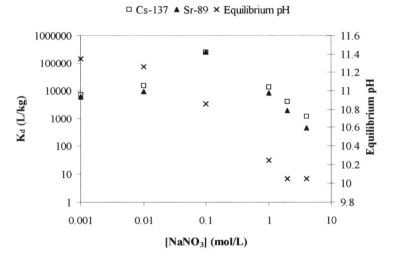

□ Cs-137   ▲ Sr-89   × Equilibrium pH

**Figure 4** Distribution coefficients of trace caesium and strontium on zorite analogue synthesised at 180 °C as a function of sodium ion concentration (V:m = 100)

*Distribution coefficients of trace caesium and strontium on maximally-exchanged potassium-form zorite analogue as a function of potassium ion concentration*

The distribution coefficient of $^{137}Cs$ decreases approximately linearly on a logarithmic scale with a slope of -1.22 (Fig. 5). The distribution coefficient of $^{89}Sr$ initially increases slightly for the same reasons described above and then decreases linearly with a slope of -1.1. In contrast to the effect of sodium ions, potassium ions interfere much more strongly in the sorption efficiency of trace caesium compared to strontium. The distribution coefficients of $^{137}Cs$ are two to three orders of magnitude lower than that of $^{89}Sr$ at any particular potassium ion concentration. For example, in 1 mol/L $KNO_3$, the Kd of $^{137}Cs$ was 57 L/kg (22 % sorption) and the $K_d$ of $^{89}Sr$ was 42200 L/kg (99.5 % sorption).

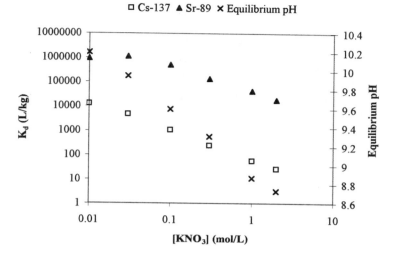

**Figure 5** Distribution coefficients of trace caesium and strontium on maximally-exchanged potassium-form zorite analogue synthesised at 180 °C as a function of potassium ion concentration (V:m = 200)

*Distribution coefficients of trace strontium on maximally-exchanged calcium-form zorite analogue as a function of calcium ion concentration*

Calcium ions interfere much more strongly on the uptake of $^{89}$Sr compared to magnesium ions (Fig. 6). In the concentration range 0.01 to 1 mol/L there is a linear decrease in the distribution coefficient on a logarithmic scale with slope -1.54 for $Mg^{2+}$ and -1.70 for $Ca^{2+}$.

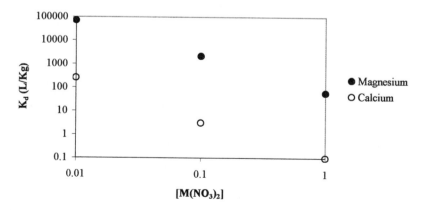

**Figure 6** Distribution coefficients of trace strontium on maximally-exchanged calcium-form and magnesium-form zorite analogues as a function of calcium ion and magnesium ion concentration respectively (V:m = 100, equilibrium pH 9.0-9.1)

*Distribution coefficients of caesium and strontium as a function of caesium and strontium ion concentration*

The distribution coefficients of caesium as a function of caesium ion concentration or strontium as a function of strontium ion concentration are of a similar magnitude and are constant or increase slightly up to a concentration of 0.001 mol/L for the same reasons as described above (Fig. 7). Sorption of both trace ions was still at least 99 % at this concentration (V:m = 100). Towards higher concentrations the distribution coefficients decrease smoothly as the saturation capacity is reached. With an initial caesium ion concentration of 0.1 mol/L the $K_d$ was 25 L/kg which corresponds to 20 % sorption, i.e., 2 meq/g which is about 40% of the total cation exchange capacity.

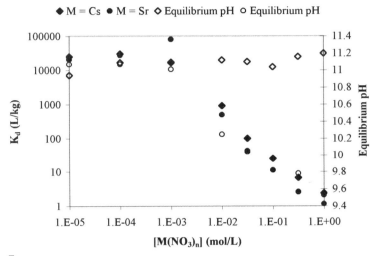

◆ M = Cs   ● M = Sr   ◇ Equilibrium pH   ○ Equilibrium pH

**Figure 7**

*Distribution coefficients of caesium and strontium in the simulated waste solutions*

The batch distribution coefficient is a measure of the maximum processing capacity, i.e., the number of litres of solution treatable with 1 kg of material. Comparatively good caesium capacities of at least 10000 L/kg were obtained on the zorite sample in all of the simulants except the very high salt strongly alkaline neutralised current acid waste ($K_d$ 17 L/kg, Table 3). The distribution coefficients of caesium on both the clinoptilolite and zorite samples were about the same in the SIXEP simulant (64000 L/kg). The sorption efficiency of strontium on the zorite sample was not as good as that on the clinoptilolite sample in the low salt AGW and NPP simulants (pH less than 8). In contrast a distribution coefficient of at least 1000000 L/kg was obtained on the zorite sample in the SIXEP simulant at pH 9. Neither material was effective for strontium in the SGW simulant due to interference from calcium ions.

**Conclusion**

The zorite analogue shows promise as a material for the removal of both [137]Cs and [89]Sr from effluent solutions. It is effective over a wide pH range for caesium but is restricted to pH greater than seven for strontium. High $K_d$s were obtained in concentrated sodium salt solutions but potassium and calcium ions interfere very strongly on the sorption efficiency of caesium and strontium ion respectively. The material also shows promise as a material for scavenging other radionuclides such as $^{65}$Zn and $^{54}$Mn from mildly alkaline sodium salt solutions. Doping of antimony or niobium in the synthesis gel resulted in a significant improvement in performance especially for caesium in acidic solution. It is known that niobium proxies for titanium in the natural mineral, a typical chemical formula being $Na_6Ti(Ti_{0.9}Nb_{0.1})_4(Si_6O_{17})_2(OH)_5.10.5H_2O$. It is feasible, therefore, that isomorphous substitution of niobium or antimony for titanium occurs in the synthetic analogue but further structural characterisation will be required to confirm this.

**Table 3** Distribution coefficients (L/kg) of caesium and strontium on zorite analogue synthesised at 200 °C and clinoptilolite in five simulants (V:m = 1000, pH in parenthesis, pH adjusted with $HNO_3$)

|        |    | AGW    | SGW    | SIXEP     | NPP     | NCAW   |
|--------|----|--------|--------|-----------|---------|--------|
| Zorite | Cs | 545600 | 9426   | 63830     | 129500  | 17     |
|        |    | (8.01) | (7.09) | (9.04)    | 6.08)   | (14.5) |
| Clino  | Cs | 59870  | 615    | 63490     | 20630   | 15     |
|        |    | (7.63) | (7.79) | (8.79)    | (6.15)  | (14.5) |
| Zorite | Sr | 889    | 5      | >1000000  | 156     | 110    |
|        |    | (7.97) | (7.22) | (8.95)    | (5.84)  | (14.5) |
| Clino  | Sr | 11410  | 23     | 48860     | 2332    | 35     |
|        |    | (7.63) | (7.80) | (8.79)    | (6.45)  | (14.5) |

### Acknowledgement

The work reported herein was funded by the European Commission Framework 4 Nuclear Fission Safety Program under contract number FI4W-CT95-0016.

### References

1.     P. A. Sandomiriskii and N. V. Belov, *Sov. Phys. Crystallogr.*, 1979, **24**, 686.
2.     A. Philippou and M. W. Anderson, *Zeolites*, 1996, **16**, 98
3.     H. Du, F. Zhou, W. Pang and Y. Yue, *Microporous Mater.*, 1996, **7**, 73
4.     J. Lehto and R. Harjula, *Reactive & Functional Polymers*, 1995, **27**, 121

# FRACTIONATION OF LITHIUM ISOTOPES ON INORGANIC ION EXCHANGERS

Y. Makita, H. Kanoh, A. Sonoda, T. Hirotsu and K. Ooi

Shikoku National Industrial Research Institute,
2217-14 Hayashi, Takamatsu, 761-0395,
JAPAN

## 1 INTRODUCTION

In its natural state, the lithium element consists of two stable isotopes , $^6$Li and $^7$Li. The natural abundance of lithium isotopes is given, on average, as approximately 7% of $^6$Li and 93% of $^7$Li. The study of lithium isotopic effects is important, not only for industrial nuclear science, but also for fundamental material and geological sciences.

There are three typical methods for lithium isotope separation, i.e. amalgam, electrophoresis and ion-exchange methods.[1-3] The amalgam method makes use of isotope fractionation in the exchange reaction between aqueous lithium ions and lithium amalgam. It has a high separation factor for lithium isotopes, but there is a serious industrial problem of disposal of mercury waste. Electrophoresis makes use of the difference in electrical mobility between the lithium isotopes in a lithium molten salt. It has a relatively high separation factor, but it consumes a large amount of energy since the process must be carried out at a high temperature. Ion-exchange makes use of isotope fractionation in the ion-exchange reaction between the solution and the exchanger. Although this method is the simplest, its separation factor is very low compared with the other methods. We expect that the ion-exchange method will become of practical use when novel exchangers with a high separation factor comparable to that of the amalgam method are developed. Crown ether compounds may be one of the promising ion exchangers.[4]

Lithium isotope fractionation by ion-exchange reactions has been studied extensively since the first work by Tayler and Urey.[5] However, the fractionation mechanism of ion-exchange has not yet been clearly elucidated because the fractionation depends on the kind of ion-exchanger used as well as on the conditions of ion-exchange. This paper summarizes our results of lithium isotope fractionation using inorganic ion exchangers, and discusses the fractionation properties in order to elucidate the isotope fractionation mechanism, with a view to developing novel isotope separation materials.

## 2 EXPERIMENTAL

### 2.1 Inorganic Ion Exchangers

*2.1.1 Oxide Adsorbents*    The spinel-type manganese oxides of the redox type

(SpiMO-R) and the ion-exchange type (SpiMO-I), the hollandite-type (HolMO) and the birnessite-type (BirMO) were prepared. Cubic antimonic acid (CSbA) was purchased from Toagosei Co. Ltd. All the oxide adsorbents show different ion-sieve properties depending strongly on the pore size.[6-9]

*2.1.2 Phosphate Adsorbents*   α-titanium phosphate ($Ti(HPO_4)_2 \cdot nH_2O$), α-zirconium phosphate ($Zr(HPO_4)_2 \cdot nH_2O$), α-tin phosphate ($Sn(HPO_4)_2 \cdot nH_2O$) were supplied by Toagosei Co. Ltd., and zirconium oxonium phosphate ($Zr_2(H_3O)(PO_4)_3 \cdot nH_2O$) by Daiichikigenso Co. Ltd. We named these phosphates TiP, ZrP, SnP and ZOP, respectively. All the phosphate adsorbents have layered structures, except for ZOP, and they have different ion-sieve properties.[10,11] Since we assumed that this property depends on the basal spacing between a neighbouring pair of layers, we prepared the heat-treated samples by heating at 200, 400 or 600 °C in air for 4 h. These samples are denoted TiP(200), TiP(400), etc., with temperatures in parentheses.

X-ray diffraction (XRD) analysis of the inorganic ion exchangers was carried out using a Rigaku RINT-1200 diffractometer and DTA-TG using a Macscience TG-DTA2000 analyzer.

## 2.2 Lithium Ion Exchange and Isotope Separation Factor

The fractionation experiment of lithium isotopes was carried out batchwise. An inorganic ion exchanger was added to an aqueous lithium solution (0.1M LiOH or 0.03M LiCl + 0.07M LiOH) and shaken for about 2 weeks. After equilibrium, the ion exchanger was filtered off with a membrane filter. The $Li^+$ concentration in the filtrate was determined with a Shimadzu model AA-670 atomic absorption spectrometer. The $Li^+$ uptake by the exchanger was calculated by subtracting the final concentration of $Li^+$ in the filtrate from the initial concentration.

The lithium isotope ratio in the filtrate was measured using a surface ionization mass spectrometer (VG336 with a wide flight tube, V.G. Co. Ltd.). The filtrate converted to LiOH solution by passing through an anion-exchange resin and to LiI solution by the following addition of HI solution. A droplet containing 1 μg of $Li^+$ was loaded in a side filament of a Ta-Re-Ta triple filament ion source and dried. The filament was mounted into the specimen chamber and the lithium isotope ratio was measured by a positive ion mode. The magnitudes of both lithium isotopes were determined with double Faraday collectors simultaneously. The relative standard deviation of $^7Li/^6Li$ measurement was less than 0.02%.

The separation factor ($S$) of the lithium isotope was calculated using the following equation:[12]

$$S = (^7Li/^6Li)_{solution} / (^7Li/^6Li)_{exchanger} \qquad (1)$$

where the lithium isotope ratio in the exchanger, $(^7Li/^6Li)_{exchanger}$, was calculated from the lithium isotope ratios in the solution and the lithium concentrations of the solution phase before and after the ion-exchange reaction.

## 2.3 Distribution Coefficient

Distribution coefficients, $Kd$, were measured by a batch method using a $NH_3$-$NH_4Cl$ buffer solution (pH8.2) containing 1mM LiCl, 1mM NaCl and 1mM KCl. The Kd values were calculated using the following equation:[13]

$$Kd= \frac{\text{amount of the ion uptake per 1g of the adsorbent}}{\text{amount of ions remaining per 1cm}^3 \text{ in the solution phase}} \qquad (2)$$

The *Kd* value shows an affinity for each alkali metal ion on an exchanger site without metal-metal interactions, since the ion-exchange was carried out in very dilute solution.


### 3 RESULTS AND DISCUSSION

In this section, we show the fractionation properties of lithium isotopes using inorganic ion exchangers and discuss the factors affecting the isotopic fractionation taking into account the ion selectivity of the exchange site and the increase in water content of the exchanger accompanying the ion-exchange.

### 3.1 Separation Factor for Lithium Isotopes

Lithium uptakes and isotope separation factors with the oxide adsorbents are given in Table 1.[12] Separation factors larger than 1 mean that the lighter isotope, $^6Li$, is preferentially fractionated into the solid phase, and the heavier isotope, $^7Li$, is enriched in the solution phase. CSbA shows the highest separation factor (1.024) for lithium isotopes.[12,13] It has, to the best of our knowledge, the highest lithium isotope separation factor among the inorganic ion exchangers studied so far. The separation factor increases in the order BirMO, HolMO and SpiMO-I, SpiMO-R, CSbA in the LiOH solution. SpiMO, CSbA, HolMO and BirMO have ion sieve properties for $Li^+$, $Na^+$, $K^+$ and $Cs^+$, respectively.[6-9] This indicates that the fractionation of lithium isotopes does not necessarily correlate with the degree of $Li^+$ selectivity.

Lithium uptake and isotope separation factors with phosphate adsorbents are given in Table 2. This table shows that the isotope fractionation of phosphates with a layered structure is influenced by the chemical species of the central metal atoms as well as by the heat-treatment temperature. TiP, TiP(200) and TiP(400) show high isotope separation factors of 1.017, 1.018 and 1.019 respectively, whereas TiP(600) shows a lower isotope separation factor of 1.010. ZrP(600) gives an isotope separation factor of 1.010, although ZrP, ZrP(200) and ZrP(400) give low separation factors in spite of the fact that they have a layered structure similar to TiP. SnP(400) and ZOP have relatively high separation

Table 1 *Lithium Isotope Fractionation with Oxide Adsorbents*

| Sample | 0.1M LiOH | | 0.03M LiCl + 0.07M LiOH | |
|---|---|---|---|---|
| | *$Li^+$ uptake / mmol g$^{-1}$* | *Separation factor* | *$Li^+$ uptake / mmol g$^{-1}$* | *Separation factor* |
| SpiMO-R | 3.44 | 1.013 | 1.07 | 1.002 |
| SpiMO-I | 5.64 | 1.006 | 1.10 | 0.997 |
| CSbA | 5.47 | 1.024 | 4.27 | 1.024 |
| HolMO | 3.73 | 1.006 | 0.85 | 1.006 |
| BirMO | 3.66 | 1.004 | 1.06 | 1.006 |

(From Table 1 of reference 12)

Table 2 *Lithium Isotope Fractionation with Phosphate Adsorbents and Distribution Coefficients for Alkali Metal Ions*

| Sample | 0.03M LiCl + 0.07M LiOH | | $Kd$ / cm$^3$ g$^{-1}$ | | | $\Delta H_2O/Li$ |
| | Li$^+$ uptake / mmol g$^{-1}$ | Separation factor | Li | Na | K | |
|---|---|---|---|---|---|---|
| TiP | 2.6 | 1.017 | 1600 | 530 | <5 | 0.28 |
| TiP(200) | 3.7 | 1.018 | 650 | 750 | 14 | 0.07 |
| TiP(400) | 3.6 | 1.019 | 580 | 7000 | 29 | n.d. |
| TiP(600) | 3.0 | 1.010 | 8 | 20 | <5 | n.d. |
| ZrP | 4.1 | 1.001 | <5 | <5 | <5 | 0.70 |
| ZrP(200) | 4.3 | 1.003 | <5 | <5 | <5 | 0.69 |
| ZrP(400) | 3.3 | 1.007 | <5 | <5 | <5 | n.d. |
| ZrP(600) | 2.1 | 1.010 | 16 | 140 | <5 | n.d. |
| SnP | 3.2 | 1.003 | 370 | 27 | 20 | 0.25 |
| SnP(200) | - | - | 970 | 1060 | 14 | - |
| SnP(400) | 4.0 | 1.015 | 510 | 2000 | 32 | 0.34 |
| SnP(600) | - | - | <5 | <5 | <5 | - |
| ZOP | 1.4 | 1.014 | >10000 | >10000 | <5 | 0.48 |

n.d. : not determined owing to the collapse of the crystalline exchangers under high temperature.

factors of 1.015 and 1.014, respectively, although the unheated sample SnP gives a low separation factor.

## 3.2 Influence of Crystal Structure and Ion Selectivity on Isotope Fractionation

Crystal structure and ion selectivity were investigated for the phosphate adsorbents. XRD patterns of TiP and its heat-treated samples are shown in Figure 1. Interlayer distance slightly decreases with an increase in temperature. As can be seen from the decrease of the intensity of (002) in Figure 1, the layered crystalline structure partially collapses when heated at 400 °C and even more at 600 °C. The structural changes of ZrP and SnP by heating were similar to that of TiP in Figure 1.

Distribution coefficients, $Kd$, of phosphate adsorbents are summarized in Table 2. TiP, TiP(200) and TiP(400) give high $Kd$ values for both Li$^+$ and Na$^+$. These results agree well with those in the literature where TiP shows a sodium ion-sieve property (exchangeable for Li$^+$ and Na$^+$ but not for the other alkali metal ions with ionic radii larger than Na$^+$).[10] ZrP, ZrP(200) and ZrP(400) give low $Kd$ values for both Li$^+$ and Na$^+$, although they have large ion-exchange capacities.[11] Since they can intercalate cations unselectively due to the expansion of basal spacing, most of the sites are probably exchanged by NH$_4^+$, the major cation in the buffered solution. SnP shows a lithium ion-sieve property ($Kd(Li) > Kd(Na)$), while SnP(200) and SnP(400) show sodium ion-sieve properties. TiP(600), ZrP(600) and SnP(600) give low $Kd$ values for all alkali metal ions. This is due to the low crystallinity caused by excessive heating of the exchanger, which is presumed from XRD patterns (Figure 1(d)). Comparing the $Kd$ value with the separation factors, we can see a tendency that the samples with a high selectivity for sodium ions give high separation factors for lithium isotopes.

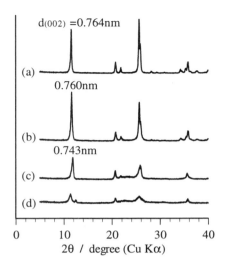

**Figure 1** *XRD patterns of TiP (a), TiP(200) (b), TiP(400) (c) and TiP(600) (d)*

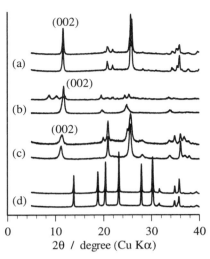

**Figure 2** *XRD patterns of TiP (a), ZrP (b), SnP (c) and ZOP (d). The lower and upper lines show XRD patterns before and after the Li⁺ exchange, respectively.*

### 3.3 Factors Affecting Lithium Isotope Fractionation

In our previous paper,[12] we proposed two factors for the origin of lithium isotope fractionation: the degree of $Li^+$ dehydration in the solid phase and the affinity of the ion-exchange site toward $Li^+$. We will discuss the lithium isotope fractionation described in Tables 1 and 2 from the change of water content in the exchanger and the distribution coefficient for alkali metal ions.

The water content in the sample was calculated from the weight loss during heating up to 380 °C on a TG curve, [14,15] assuming that the crystal structure of the sample was not changed by the heat-treatment. Increase in water content by ion-exchange reaction per mole of Li ($\Delta H_2O/Li$) is given in Table 2. The samples with low $\Delta H_2O/Li$ value tend to have higher separation factors for lithium isotopes, except for SnP. The low $\Delta H_2O/Li$ value corresponds to the large dehydration of $Li^+$ in the exchanger, since $Li^+$ has a hydration number of about 4 in an aqueous solution. The low separation factors of ZrP and ZrP(200) can be explained by insufficient dehydration of $Li^+$ in the solid phase, since their $\Delta H_2O/Li$ values are larger than those of other samples with high separation factors.

The XRD patterns of unheated samples before and after the ion-exchange are given in Figure 2. The basal spacings are nearly constant regardless of $Li^+$ exchange for TiP, SnP and ZOP, while ZrP shows a marked increase in the basal spacing with $Li^+$ exchange. This indicates that ZrP expands in an aqueous medium. This property is disadvantageous for lithium isotope fractionation since $Li^+$ can exchange smoothly in the solid phase without dehydration.

A theoretical study of isotope fractionation has shown that the heavier isotope is more stable in the molecule, while the lighter isotope prefers the isolated form of atoms.[16] This suggests that $^6$Li is enriched in the solid phase when the degree of Li$^+$ dehydration becomes higher, but that $^7$Li is more fractionated to the solid phase when the lithium ions are strongly bound on the ion-exchange sites. The exchangers with an ion-sieve property for sodium ions can make full use of the former isotopic effect of dehydration, since they have a low selectivity for lithium ions and adsorb lithium ions in a less hydrated form. The low separation factor of SnP in spite of its low $\Delta H_2O/Li$ value may be due to its high affinity for lithium ions as shown in Table 2. The former isotopic effect of dehydration may be cancelled by the latter reverse isotopic effect due to Li$^+$ stabilization.

## 4 CONCLUSION

Antimonic acid, with a cubic structure, has the highest fractionation property for lithium isotopes among the inorganic ion exchangers studied, and titanium phosphate, heat-treated tin phosphate and zirconium oxonium phosphate have relatively high fractionation properties. Our results demonstrate that the separation factor of the inorganic ion exchangers is principally determined by the isotopic effect of dehydration. There is a close correlation between the isotope separation factors and the affinity for Na$^+$. The present results give a new approach for the design of inorganic ion exchangers effective for lithium isotope separation.

### References

1. A. A. Palko, J. S. Drury and G. M. Begun, *J. Chem. Phys.*, 1976, **64**, 1828.
2. K. Okuyama, I. Okada and N. Saito, *J. Inorg. Nucl. Chem.*, 1973, **35**, 2883.
3. T. Oi, H. Ogino, N. Izawa, H. Kakihana, K. Ooi, Y. Miyai and S. Katoh, 'New Developments in Ion Exchange (M. Abe, T. Kataoka and T. Suzuki. Eds)', Kodansha Press, Tokyo, 1991, p. 613.
4. D. W. Kim, C. S. Kim, Y. K. Jeong, J. K. Lee, C. P. Hong and Y. S. Jeon, *J. Radioanal. Nucl. Chem.*, 1997, **220**, 229.
5. T. I. Tayler and H. C. Urey, *J. Chem. Phys.*, 1937, **5**, 597.
6. K. Ooi, Y. Miyai and S. Katoh, *Solv. Extr. Ion Exch.*, 1987, **5**, 561.
7. M. Tsuji and M. Abe, *Solv. Extr. Ion Exch.*, 1984, **2**, 253.
8. M. Tsuji, S. Komarneni, Y. Tamaura and M. Abe, *Mater. Res. Bull.*, 1992, **27**, 741.
9. M. Abe, *J. Inorg. Nucl. Chem.*, 1979, **41**, 85.
10. G. Alberti, P. Cardini-Galli, U. Costantino and N. Tomassini, *J. Inorg. Nucl. Chem.*, 1967, **29**, 571.
11. A. Clearfield and J. A. Stynes, *J. Inorg. Nucl. Chem.*, 1964, **26**, 117.
12. K. Ooi, Q. Feng, H. Kanoh, T, Hirotsu and T. Oi, *Sep. Sci. Technol.*, 1995, **30**(20), 3761.
13. T. Oi, 'Proc. Int. Conf. Ion Exch., ICIE'95', Takamatsu, Japan Association of Ion Exchange, 1995, p. 147.
14. A. Clearfield, 'Inorganic Ion Exchange Materials', CRC Press, 1982, Chapter 2, pp. 81-82.
15. Y. Makita, H. Kanoh, T. Hirotsu and K. Ooi, *Chem. Lett.*, 1998, pp. 77-78.
16. J. Bigeleisen and M. G. Mayer, *J. Chem. Phys.*, 1947, **15**, 261.

# UPTAKE OF RADIONUCLIDES BY METALLOSILICATE ION EXCHANGERS

Teresia Möller and Risto Harjula

*Laboratory of Radiochemistry*
*Department of Chemistry*
*P.O. Box 55*
*FIN-00014 University of Helsinki*
*Helsinki, Finland*

## 1 INTRODUCTION

Selective separation of radionuclides from nuclear waste solutions is a highly effective method for the minimisation of nuclear waste volumes for final disposal. There are numerous selective inorganic ion exchangers (zeolites, sodium titanates, titanosilicates, hexacyanoferrates) available commercially and the number of their applications in nuclear sites is increasing. Almost all of these exchangers have been developed for the separation of $^{137}Cs$ and $^{90}Sr$, and they often have limited selectivity for other radionuclides. Because separation of $^{137}Cs$ and $^{90}Sr$ has been historically the main emphasis in the developmental work, little information is available for the selectivity of inorganic exchangers for other radionuclides. There is an increasing demand especially for the selective removal of activated corrosion product nuclides (e.g. $^{63}Ni$, $^{58,60}Co$, $^{51}Cr$, $^{54}Mn$, $^{59}Fe$ and $^{65}Zn$) at nuclear power plants.

## 2 PURPOSE OF STUDY

A research programme is underway to prepare and evaluate a large number of different inorganic exchangers, traditional and novel ones, for removal of a wide range of radionuclides commonly found in nuclear waste solutions ($^{85}Sr$, $^{134}Cs$, $^{57}Co$, $^{65}Zn$, $^{59}Fe$, $^{54}Mn$, $^{51}Cr$, $^{110}Ag$, $^{236}Pu$ and $^{241}Am$). This paper will report some preliminary results obtained for a group of metallosilicate materials with framework or layered structures. Experiments were carried out by determining the distribution coefficients ($K_D$) of the radionuclides in several "model" solutions (0.1 M $HNO_3$, 0.1 M $NaNO_3$, 0.1 M $NaNO_3$/0.1 M NaOH). The effect of pH and concentrations of common cations ($Ca^{2+}$, $Mg^{2+}$, $Na^+$, $K^+$) present in nuclear waste solutions on $K_D$s were investigated as well. Several commercial ion exchangers (SrTreat®, clinoptilolite) were tested in parallel for reference.

### 3 EXPERIMENTAL SECTION
**3.1 Materials**
All reagents were of analytical grade (Fluka, Riedel-de Haën, Merck) and used without further purification. The radioactive tracers used in ion exchange experiments were obtained from Amersham International (UK).

**3.2 Syntheses**
All of the syntheses were carried out in Teflon lined pressure vessels under hydrothermal conditions at temperatures 423-503 K. After cooling the products were filtered, washed with distilled water at 298 K and dried at 373 K. The removal of the template was done by calcinating the material at 673-793 K when necessary. The following materials were prepared:

*3.2.1 Titanosilicate TS-1*
TS-1[1] with a MFI structure was synthesized by adding tetrapropylammonium hydroxide (TPAOH) [20% in water] to $Si(OEt)_4$ under stirring. To this an aqueous solution of $TiCl_3$ was added. The hydrothermal reaction was carried out in a 433 K oven for two days. A sample of the microcrystalline product used in the ion exchange experiments was calcined at 793 K. The product has an approximate $SiO_2$ : $TiO_2$ ratio of 55.

*3.2.2 Titanosilicate TS-2*
TS-2[2] with a MEL structure was synthesized tetrabutylammonium hydroxide (TBAOH) [40% in water] was added to a solution of tetraethylorthosilicate (TEOS). To this $Ti(OBu)_4$ in isopropyl alcohol was slowly added and finally water. The hydrothermal synthesis was carried out at 443 K for two days. A sample of the product was calcined at 793 K. The product has an approximate $SiO_2$ : $TiO_2$ ratio of 27.

*3.2.3 Sodium titanosilicate AM-4*
A novel layered sodium titanosilicate $Na_4Ti_2O_2(Si_2O_6)_2*2H_2O$ (AM-4)[3,4] was synthesized as in reference 3.

*3.2.4 Sodium titanosilicate TSi*
A sodium titanosilicate $Na_2Ti_2O_3SiO_4*2H_2O$ with a framework structure was synthesized according to reference 5. A sample was calcined at 673 K. The material is given a code of TSi in the ion exchange experiments.

*3.2.5 Titanosilicate TAM-5*
A sodium titanosilicate TAM-5 (ideal composition of $Na_2Ti_2O_3SiO_4*2H_2O$) was synthesized according to reference 6. The material has a structure consisting of tunnels and cavities but is less crystalline than the other sodium titanosilicate (TSi). A sample of the material was calcined at 673 K.

*3.2.6 Ferrosilicate*
A microporous crystalline ferrosilicate (FeSi) was synthesized according to example 1 in reference 7. This material has a product composition of 0.83 $Al_2O_3$ : $Fe_2O_3$ : 21.2 $SiO_2$ : 1.99 $Na_2O$ : 0.28 $R_2O$, where R is tetraethylammonium (TEA). A sample was calcined at 793 K.

*3.2.7 Cobaltsilicate*

A cobaltsilicate with structure similar to ZSM-5 was synthesized in fluoride conditions[8]. A mixture of HF, NaF, NaCl, $H_2O$ and a cobalt compound[9] was added to fumed $SiO_2$ in water and stirred vigorously. Tetrapropylammonium bromide (TPABr) was added as template and pH set to 5. The hydrothermal reaction was carried out at 443 K for three days. The obtained blue crystals were separated from the pink main product for the ion exchange experiments.

**3.3 Characterization**

Diffraction data of the materials were collected with a Philips PW1710 powder X-ray diffractometer (XRD) with Cu K$\alpha$ radiation (1.54 Å) using step counting method (step size [°2θ] 0.050 and time/step 1.25 - 2.00 s) in the range of °2θ 5-50°. The X-ray source was a rotating anode operating at 40 kV and 50 mA. A Philips PW1877 diffraction software program PC-APD version 3.5B was used to analyze the traces. Thermogravimetric analyses (TGA) were carried out on a Mettler-Toledo TA800 unit, at a heating rate of 10 K/min from 298 K to 873 K under a nitrogen atmosphere. Elemental studies for some of the materials were done by heating a mixture of the sample and $LiBO_2$ at 1273 K in a nickel crucible and dissolving the melt in HCl. The diluted samples were analyzed by ICP-AES or ICP-MS.

**3.4 Batch exchange studies**

Distribution coefficients ($K_D$) were determined in batch experiments. 0.06 g of each sample was equilibrated with 6 mL of acid or salt solution (0.1 M $HNO_3$, 0.1 M $NaNO_3$, 0.1 M $NaNO_3$/0.1 M NaOH) which was traced with a radioactive isotope in a 8 mL polypropylene centrifuge tube (Du Pont Sorvall). Depending on the selectivity of the ion exchanger the batch factor (volume/mass) varied between 100 and 1000. The equilibration time was 1 or 4 days at room temperature with constant rotary mixing. The samples were centrifuged using a Beckman L8-M Ultracentrifuge (Velocity of 30000 rpm and time 30 min) and filtered with 0.22 μm Minispike PVDF Bulk Acrodisk 13 filters. 5 mL aliquots of the filtered samples were measured for their activities (cpm) with a gamma-counter Wallac 1280 Ultrogamma in the cases of $^{134}$Cs, $^{85}$Sr, $^{57}$Co, $^{65}$Zn, $^{110}$Ag, $^{54}$Mn, $^{59}$Fe, $^{51}$Cr and $^{241}$Am and with a liquid scintillator Wallac LKB 1217 Rackbeta in the case of $^{236}$Pu. The distribution coefficients ($K_D$) values were calculated according to

$$K_D = \frac{\overline{A_i}}{A_i} = \frac{concentration\ in\ exchanger}{concentration\ in\ solution} = \frac{(A_0 - A_{eq})}{A_{eq}} \frac{V}{m}$$

(1)

where $A_0$ and $A_{eq}$ are the activities of the tracer initially and at equilibrium, respectively, and V/m is the solution to exchanger mass ratio (batch factor, mL/g).

The effects of macro components on the uptake of $^{85}$Sr, $^{134}$Cs and $^{57}$Co were studied in three different salt solution concentrations, 0.1 M, 0.5 M and 3 M $NaNO_3$ or $KNO_3$ and 0.01 M, 0.1 M and 1 M $Mg(NO_3)_2$ or $Ca(NO_3)_2$. The metallosilicate exchangers were first equilibrated with 1 M salt solutions in order to obtain the $K^+$-, $Na^+$-, $Mg^{2+}$- and $Ca^{2+}$-forms of the materials. The equilibration lasted for 3 days, a batch factor of 100-200 was used and the salt solution was changed once before washing with distilled water.

The effect of pH on the uptake of $^{85}$Sr and $^{57}$Co was studied in 0.1 M NaNO$_3$ solution by adding a varying amount of HNO$_3$ or NaOH into the solution. These experiments were carried out with materials obtained from the syntheses with no pretreatments.

### 4 RESULTS AND DISCUSSION

Powder X-ray traces of all of the materials corresponded to those in the references indicating that the obtained products were the desired ones. The removal of the organic templates were observed in the TGA measurements and based on this data the calcination temperatures were chosen.

### 4.1 Distribution coefficients (K$_D$) for several nuclides in 0.1 M HNO$_3$, 0.1 M NaNO$_3$ and 0.1 M NaNO$_3$/0.1 M NaOH

The distribution coefficients (K$_D$) determined for titanosilicates TS-1, TS-2, AM-4, TAM-5 and TSi and ferrosilicate and cobaltsilicate in the "model" solutions for several radionuclides are shown in Table 1. The titanosilicate TS-1 took up efficiently Fe-59, Zn-65, Cr-51 and Am-241 for which the K$_D$S are about 500000 mL/g, 18000 mL/g, 15000 mL/g and 82000 mL/g in 0.1 M NaNO$_3$, respectively. Titanosilicate TS-2 had a similar behaviour except that it appeared to take up Co-57 with a good efficiency (K$_D$ 25100 mL/g) as well. Of the other titanosilicates the framework materials TAM-5 and TSi were very selective for Cs-134, the K$_D$S being 33000 mL/g and 88000 mL/g in 0.1 M HNO$_3$, respectively. The Cs-134 K$_D$ for the layered titanosilicate AM-4 is one order of magnitude lower. All of the titanosilicates took up $^{134}$Cs better than clinoptilolite. Of the other metallosilicates, ferrosilicate showed good uptake for Co-57 only.

Table 1. Distribution coefficients (K$_D$) [mL/g] for metallosilicates

| NUCLIDE | 0.1 M HNO$_3$ | 0.1 M NaNO$_3$ | 0.1 M NaNO$_3$ / 0.1 M NaOH |
|---|---|---|---|
| **TS-1** | | | |
| Cs-134 | 10.4 | 53.3 | 9.47 |
| Sr-85 | 2.41 | 8.30 | 22350 |
| Co-57 | 4.17 | 180 | 60.2 |
| Zn-65 | 0.73 | 17670 | 4090 |
| Am-241 | 7.93 | 81870 | 4140 |
| Pu-236 | 1400 | 470 | 3590 |
| Mn-54 | 1.56 | 270 | 6690 |
| Fe-59 | 10.0 | 521900 ! | 490 |
| Cr-51 | 27.5 | 14880 | 510 |
| Ag-110m | 590 | 9350 | 3230 |
| **TS-2** | | | |
| Cs-134 | 4.47 | 50.2 | 67.2 |
| Sr-85 | ~ 0 | 360 | 3610 |
| Co-57 | 1.45 | 25100 | 47.6 |
| Zn-65 | ~ 0 | 17650 | 4230 |
| Am-241 | 0.61 | 71340 | --- |
| Mn-54 | 0.95 | 120 | 1130 |
| Fe-59 | 1.0 | 339700 | 57 |
| Cr-51 | 8.06 | 15490 | 480 |
| Ag-110m | 180 | 29800 | 3370 |

| AM-4 | | | |
|---|---|---|---|
| Cs-134 | 4250 | 3110 | 120 |
| Sr-85 | 4.84 | 43960 | 62520 |
| Co-57 | 3.80 | 111700 | 250 |
| Cr-51 | 20.7 | 8570 | 490 |
| **TSi** | | | |
| Cs-134 | 88430 | 5264000 | 113000 |
| Sr-85 | 1.85 | 7107200 | 23600000 |
| Co-57 | 0.81 | 11640 | 2870 |
| Zn-65 | --- | 20900 | 845600 |
| Am-241 | 89 | 80820! | 74920! |
| Pu-236 | 240 | 174200 | 15300 |
| Cr-51 | --- | 2900 | 130 |
| **TAM-5** | | | |
| Cs-134 | 32950 | 21500 | 4900 |
| Sr-85 | 4.86 | --- | 1610000! |
| Co-57 | ~ 0 | 31690 | 2470 |
| Zn-65 | ~ 0 | 961200 | 385700 |
| Pu-236 | 290 | 50100 | 19000 |
| **Ferrosilicate** | | | |
| Cs-134 | 1640 | 4220 | 2150 |
| Sr-85 | 1.02 | 2280 | 379500 |
| Co-57 | 0.20 | 15240 | 100 |
| Cr-51 | 1.73 | 650 | 66.7 |
| **Cobaltsilicate** | | | |
| Cs-134 | ~ 0 | 30.4 | 1880 |
| Sr-85 | ~ 0 | 30.3 | 1890 |
| Co-57 | 1.20 | 0.52 | 170 |
| **Clinoptilolite** (Mud Hills, USA) | | | |
| Cs-134 | 2520 | 1840 | 590 |
| Sr-85 | 34.0 | 420 | 350 |
| Co-57 | 4.05 | 5780 | 1310 |
| Pu-236 | 1020 | 2520 | 26700 |
| Am-241 | 3.78 | 10400 | 48900 ! |
| **SrTreat**[®] | | | |
| Cs-134 | 205 | 98.9 | 37.6 |
| Sr-85 | 25.0 | 6490000 | 8820000 |
| Am-241 | 60.0 | 23700 | 6100 |
| Pu-236 | 2950 | 2780000 | 21000 |

! = Close to detection limit, --- = not measured

## 4.2 Distribution coefficients ($K_D$) for $^{85}$Sr, $^{134}$Cs and $^{57}$Co as a function of macro components $K^+$, $Na^+$, $Mg^{2+}$ and $Ca^{2+}$

Some selected distribution coefficient ($K_D$) results for $^{85}$Sr, $^{134}$Cs and $^{57}$Co as a function of $K^+$, $Na^+$, $Mg^{2+}$ and $Ca^{2+}$ concentrations are shown in Figs. 1-8. For pure binary exchange reactions, the slopes of the logarithmic plots $\log K_D = f (\log C)$ should be theoretically equal to the ratio of the charges of the exchanging ions (radionuclide charge/macro-ion charge) with a minus sign. Because the solution pH also has an effect on the $K_D$-values and the equilibrium pH-value was dependent on the macro-ion concentration, the studied exchange reactions were not pure binary ones and the theoretical slopes were not observed in most cases. However, the general trend was in practically every case that $K_D$-values decreased as macro-ion concentration was increased.

The $K_D$-values of [85]Sr for the titanosilicates TAM-5 and TSi were very high in $NaNO_3$ solutions (Fig. 1), and 1-2 orders of magnitude higher than for the zeolite clinoptilolite. TSi had slightly higher $K_D$'s than TAM-5 and both behaved almost ideally (slope of $logK_D$ ~ -2). Potassium ions did not decrease [85]Sr uptake more than sodium ions in TAM-5 and TSi (Fig. 2) but for clinoptilolite there was a strong decrease in the uptake level of [85]Sr. Calcium ions, on the other hand, had a strong decreasing effect for TSi and clinoptilolite, while TAM-5 was less affected (Fig. 3).

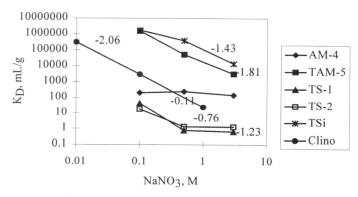

Figure 1. [85]Sr $K_D$ as a function of $NaNO_3$ concentration for titanosilicates and clinoptilolite

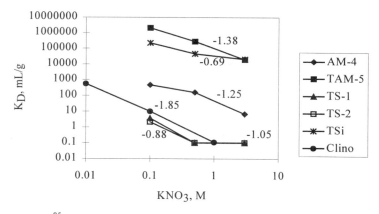

Figure 2. [85]Sr $K_D$ as a function of $KNO_3$ concentration for titanosilicates and clinoptilolite

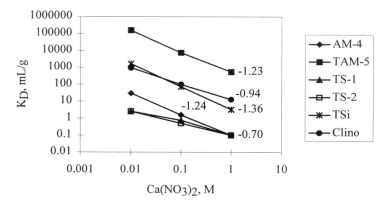

Figure 3. $^{85}$Sr $K_D$ as a function of $Ca(NO_3)_2$ concentration for titanosilicates and clinoptilolite

The $K_D$-values of $^{134}$Cs were also very high for TAM-5 and TSi in $NaNO_3$ and several orders of magnitude higher than for clinoptilolite (Fig. 4). Potassium ions decreased the level of $^{134}$Cs uptake strongly but TAM-5 was affected less strongly than other exchangers (Fig. 5). Calcium and magnesium ions did not decrease the level of $^{134}$Cs uptake noticeably for TAM-5 and TSi (Fig. 6). In the case of clinoptilolite, the uptake level increased compared to $NaNO_3$ solutions.

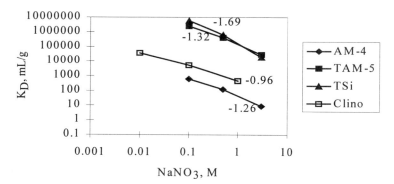

Figure 4. $^{134}$Cs $K_D$ as a function of $NaNO_3$ concentration for titanosilicates and clinoptilolite

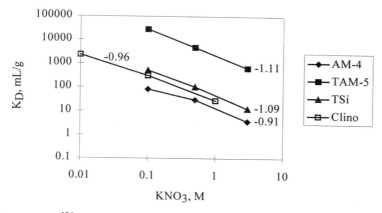

Figure 5. $^{134}$Cs $K_D$ as a function of KNO$_3$ concentration for titanosilicates and clinoptilolite

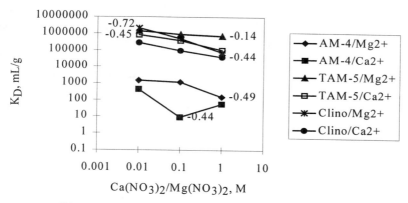

Figure 6. $^{134}$Cs $K_D$ as a function of Mg(NO$_3$)$_2$ and Ca(NO$_3$)$_2$ concentration for titanosilicates and clinoptilolite.

The $K_D$-values of $^{57}$Co were lower than those of $^{134}$Cs and $^{85}$Sr for titanosilicates TAM-5 and AM-4 (Figs. 7-8), but still at a reasonable level (10000-100000 mL/g) in the presence of all the macro-ions studied.

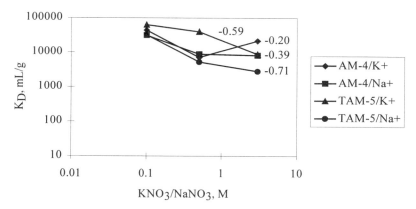

Figure 7. $^{57}$Co $K_D$ as a function of $KNO_3/NaNO_3$ concentration for titanosilicates

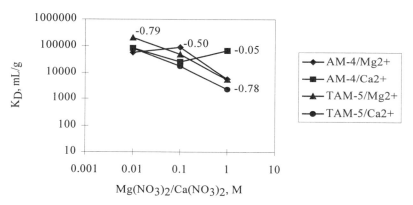

Figure 8. $^{57}$Co $K_D$ as a function of $Mg(NO_3)_2$ and $Ca(NO_3)_2$ concentration for titanosilicates

## 4.3 Distribution coefficients ($K_D$) for $^{85}$Sr and $^{57}$Co as a function of pH

Regarding the effect of pH, the titanosilicates behaved as weak-acid ion exchangers. The uptake of $^{57}$Co and $^{85}$Sr is strong in neutral and slightly alkaline solutions, but decreased with increasing pH (Figs. 9-10). Uptake of radionuclides was very low in slightly acidic solution. Sr-uptake in clinoptilolite appeared to be independent of pH over a wide pH-range (Fig. 9).

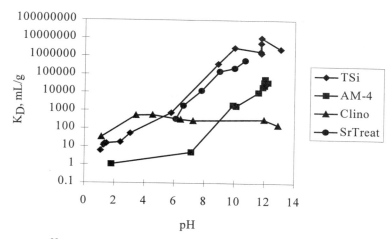

Figure 9. $^{85}$Sr $K_D$ as a function of pH in 0.1 M NaNO$_3$ for titanosilicates, clinoptilolite and SrTreat$^®$

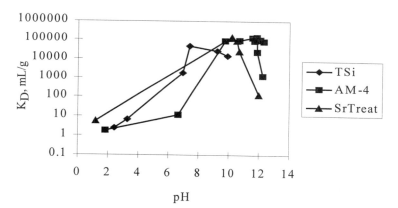

Figure 10. $^{57}$Co $K_D$ as a function of pH in 0.1 M NaNO$_3$ for titanosilicates and SrTreat$^®$

## 4.4 Largest d-spacings

A study of the relationship of the structure of crystalline titanosilicates to their selectivity for Cs$^+$ showed that when the value of the largest d-spacing measured by powder XRD approaches 0.78 nm the $K_D$(Cs) increases[10]. This was found to be true for TSi, TAM-5, TS-1 and TS-2 as seen in figure 11. The Cs-134 $K_D$s were measured in 0.1 M NaNO$_3$. A similar trend was found at the same conditions, for Sr-85 uptake (Figure 12).

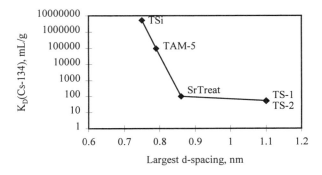

Figure 11. $^{134}$Cs $K_D$ in 0.1 M NaNO$_3$ as a function of the largest d-spacing

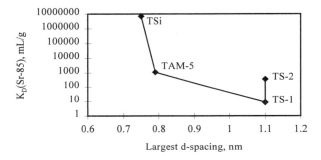

Figure 12. $^{85}$Sr $K_D$ in 0.1 M NaNO$_3$ as a function of the largest d-spacing

## 5 CONCLUSIONS

Titanosilicate materials appeared to be efficient for the removal of many key radionuclides commonly present in nuclear waste solutions. Different titanosilicates had very different selectivity patterns. TS-1 and TS-2 did not take up $^{134}$Cs or $^{85}$Sr appreciably, but had reasonable uptake for some corrosion product nuclides. Of the newer framework materials, TSi and TAM-5 took up efficiently both $^{134}$Cs and $^{85}$Sr. The layered titanosilicate AM-4 took up $^{85}$Sr reasonably but was not very selective for $^{134}$Cs. The interplanar d-spacings of the material appears to be an important factor in determining the selectivity. TSi and TAM-5 had also reasonable selectivities for several activated corrosion product nuclides (e.g. $^{57}$Co). Due to their weak-acid nature, the titanosilicates function efficiently for $^{85}$Sr and $^{57}$Co removal only in neutral or alkaline solution. In acidic and slightly acidic solutions the uptake of these radionuclides is very low.

6 REFERENCES

1. H. Gao, J. Suo and S. Li, *J. Chem. Soc., Chem. Commun.*, 1995, 835
2. J. S. Reddy and R. Kumar, *Zeolites*, 1992, **12**, 95-100
3. Z. Lin, J. Rocha, P. Brandão, A. Ferreira, A.P. Esculcas, J.D. Pedrosa de Jesus, A. Phillippou and M.W. Anderson, *J. Phys. Chem. B*, 1997, **101**, 7114-20
4. M.S. Dadachov, J. Rocha, A. Ferreira, Z. Lin and M.W. Anderson, *Chem. Commun.*, 1997, 2371- 2372
5. D.M. Poojary, R.A. Cahill and A. Clearfield, *Chem. Mater.* 1994, **6**, 2364-68
6. R.G. Anthony, R.G. Dosch and C.V. Philip, International Patent WO 94/19277, example 1
7. B. M-T. Lok, C.A. Messina, European Patent 0108271
8. C.I. Round, C.V.A. Duke and C.D. Williams, Private communication
9. N.S. Gill and F.B. Taylor, *Inorganic syntheses*, 1967, **9**, 136-142
10. R.G. Dosch, N.E. Brown, H.P, Stephens and R.G. Anthony, "Treatment of liquid nuclear wastes with advanced forms of titanate ion exchangers", Proc. Symp. Waste Management, Tuscon, AZ, USA, 1993 vol.2, 1751-54, ed. R.G. Post, M. E. Wacks

**Acknowledgements**

This work was funded by the European Commission within the Nuclear Fission Safety Programme (Contract No. FI4W-CT95-0016).

# INORGANIC ION EXCHANGE MATERIALS FOR THE REMOVAL OF STRONTIUM FROM SIMULATED HANFORD TANK WASTES

Paul Sylvester, Elizabeth A. Behrens, Gina M. Graziano and Abraham Clearfield

Department of Chemistry
Texas A&M University
PO Box 300012
College Station, TX 77842-3012
USA

## 1 INTRODUCTION

Over the last 50 years, large amounts of aqueous high level radioactive wastes (HLW) have been produced at the Hanford nuclear site in Washington State as a result of chemical extraction processes designed to separate $^{239}$Pu from irradiated uranium fuel. At present, the HLW is stored in 177 steel-lined tanks and has an estimated volume of approximately 65 million gallons. The composition of the waste varies considerably from tank to tank, but previous studies have shown that all of the tanks are believed to contain three distinct phases[1], which consist of a salt cake at the top of the tank, an alkaline supernate and a metal hydroxide-rich sludge at the bottom. In these wastes, strontium predominantly occurs in the sludges but is also found entrained within the salt cake and in the alkaline supernate.

In order to dispose of this waste in an economical manner, the Department of Energy (DOE) proposed to separate the majority of the radioactivity from the supernate and salt cake which enables the residual salts to be classed as low level waste (LLW), thus minimising the amount of HLW for vitrification. One method of attaining this objective is to use ion exchange to selectively remove heat-generating, high specific activity radioisotopes, such as $^{137}$Cs and $^{90}$Sr, from the supernate and dissolved salt cake. We have evaluated inorganic materials[2-5] because they have higher chemical, radiation and thermal stabilities than conventional organic resins. In addition, they typically exhibit greater capacities and selectivities for a wide range of monovalent and divalent ions[6-13] and desired properties can be 'fine-tuned' into the materials by varying synthetic parameters such as reaction times and temperatures.

In this paper we report the synthesis and evaluation of a titanosilicate pharmacosiderite material, a sodium nonatitanate and a sodium titanosilicate and we investigate their effectiveness at removing strontium from two simulated Hanford tank wastes using both batch techniques and column studies. All of these materials are synthesized in highly basic conditions and therefore will be stable in alkaline tank wastes. By contrast, many currently available ion exchange materials such as zeolites and hexacyanoferrates are unstable in alkaline wastes (pH > 14) and thus not appropriate for use in tank waste remediation.

2 EXPERIMENTAL

## 2.1 Materials

All reagents were of analytical grade and were used without further purification. $^{89}Sr$ (269 mCi/g total Sr) was used for NCAW experiments and $^{89}Sr$ (69.3 mCi/g total Sr) was used for 101SY-Cs5. Both batches of isotope were purchased from Isotope Products Laboratory, California. $^{89}Sr$ activities were measured using a Wallac 1410 liquid scintillation counter and Fisher Scientific Scintisafe Plus 50% scintillation cocktail. X-Ray powder diffraction (XRD) was performed on a Scintag Pad V diffractometer using CuKα radiation and a scan rate of 1°/min.

## 2.2 Synthesis of Ion Exchange Materials

Samples of the pharmacosiderite, sodium titanosilicate and sodium nonatitanate were prepared according to the following procedures which have been detailed in previous papers.[4, 14-16] Yields above 100 % suggest the presence of minor amounts of surface water in addition to the waters of crystallisation.

*2.2.1 Potassium Pharmacosiderite, $HK3(TiO)_4(SiO_4)_3 \cdot 4H_2O$.* 20 g of silica gel (454 mmol) (Aldrich, Grade 923) was combined with 80 mL of 10 M KOH and 140 mL of deionised water. In a separate beaker, 62 mL of titanium isopropoxide (203 mmol) (97%, Aldrich), 40 mL of 30 % $H_2O_2$, 200 mL of deionised water and 60 ml of 10 M KOH were added together with vigorous stirring. The two beakers were then combined, stirred to homogenize, placed in a 1 L Teflon-lined hydrothermal vessel and heated at 190°C for 4 days. The product, KTS-Ph, was filtered, washed with absolute ethanol and dried at 60°C. (Mass of product : 38.9 g, yield = 106 %).

*2.2.2 Sodium Titanosilicate, $Na_2Ti_2O_3SiO_4 \cdot 2H_2O$.* 33 g of tetraethoxyorthosilicate (158 mmol) (98 %, Aldrich) was combined with 45.6 g of titanium isopropoxide (156 mmol) and added to 260 mL of a 50 wt.% NaOH solution. The resultant white gel was stirred well to ensure homogeneity, placed in a 1 L Teflon-lined hydrothermal bomb and heated at 180°C for 4 days. The product, NaTS, was filtered, washed with absolute ethanol and dried at 60°C. (Mass of product : 25.2 g, yield = 102 %).

*2.2.3 Sodium Nonatitanate, $Na_4Ti_9O_{20} \cdot xH_2O$.* 77.5 g of titanium isopropoxide (273 mmol) was placed in a Teflon round-bottomed flask, and with constant stirring, 84.35 g of a 50 wt. % solution of NaOH was added resulting in a white gelatinous precipitate. The precipitate was subsequently stirred for one hour and then heated at approximately 108°C for a further 3 hours. The mixture was then transferred to a 1 L Teflon-lined hydrothermal vessel using 128 mL of deionised water and heated at 200°C for 21 hours. The product, NaTi, was then filtered through a 0.2 μm membrane, washed with 2 L of absolute ethanol and dried at 60°C for 48 hours. (Mass of product : 33.1 g, yield = 82 %).

The above materials were compared in column tests with IONSIV IE-911, a pelletised form of a crystalline silicotitanate (CST) manufactured and marketed by UOP. This material is believed to have a similar structure to NaTS and is marketed as a caesium and strontium selective ion exchanger and is being proposed for use in the treatment of alkaline tank wastes. A pelletised form of the sodium nonatitanate, manufactured at AlliedSignal according to the synthetic procedure developed at Texas A&M University, was also evaluated in column experiments for the removal of strontium from 101SY-Cs5 waste.

## 2.3 Batch Experiments

Two waste simulants, NCAW and 101SY-Cs5, were prepared according to information supplied by Pacific Northwest National Laboratory (PNNL). The compositions of these two wastes are given in Table 1. NCAW (Neutralised Current Acid Waste) is a relatively simple solution and contains no complexants and is representative of waste in tank 102-AZ diluted to 5 M $Na^+$. Hence, the strontium present is likely to occur as $Sr(OH)^+$ which should be readily extracted by ion exchange. 101SY-Cs5 is representative of tank waste 101-SY, also diluted to 5 M $Na^+$, and contains significant amounts of chelating agents such as EDTA and trisodium nitrilotriacetate which will complex any strontium present making removal by ion exchange far more difficult.

**Table 1** *The Composition of the Waste Simulants NCAW and 101SY-Cs5*

| Species | Concentration (M) | |
|---|---|---|
| | **NCAW** | **101SY-Cs5** |
| Al | 0.43 | 0.42 |
| Ca | 0 | 4.20E-03 |
| Cs | 5.00E-04 | 4.19E-05 |
| Fe | 0 | 1.96E-04 |
| K | 0.12 | 0.034 |
| Mo | 0 | 4.20E-04 |
| Ni | 0 | 2.50E-04 |
| Na | 4.99 | 5.1 |
| Rb | 5.00E-05 | 4.20E-06 |
| Sr (Total) | 2.70E-07 | 2.90E-07 |
| Zn | 0 | 5.00E-04 |
| Carbonate | 0.23 | 0.038 |
| Fluoride | 0.09 | 0.092 |
| Hydroxide | 3.4 | 3.78 |
| Hydroxide (Free) | 1.68 | 2.11 |
| Nitrate | 1.67 | 1.29 |
| Nitrite | 0.43 | 1.09 |
| Sulphate | 0.15 | 4.75E-03 |
| Phosphate | 0.025 | 0.02 |
| Citric acid | 0 | 5.00E-03 |
| $Na_4$EDTA | 0 | 5.00E-03 |
| Iminodiacetic acid | 0 | 0.031 |
| $Na_3$ nitrilotriacetate | 0 | 2.50E-04 |
| Sodium gluconate | 0 | 0.013 |
| Theoretical pH | 14.5 | 14.4 |

0.05 g or 0.01 g of ion exchanger was accurately weighed into a scintillation vial to give V:m ratios of 200 and 1000, respectively, and then contacted with 10 mL of the waste simulant which had been spiked with $^{89}$Sr. Sufficient $^{89}$Sr tracer was added to ensure that the residual activity in solution after contact with the exchanger could be measured with statistical accuracy on the liquid scintillation counter. (The total strontium concentration in the wastes for these batch tests was < 1 ppm, thus ensuring that no precipitation occurred). The vials were capped and placed on a rotary shaker for approximately 24 hours. The mixtures were then filtered using Whatman No.42 filter paper (which had previously been shown to be sufficient to retain any fines generated) and the residual activity in the solutions determined using liquid scintillation counting. Distribution coefficients ($K_d$s) were then calculated using Equation 1 below :

$$K_d = (C_o - C_f)/C_f \bullet (V/m) \hspace{2cm} (1)$$

where : $C_o$ = initial activity of solution (counts per minute, cpm/mL)
$C_f$ = final activity of solution (counts per minute, cpm/mL)
$V$ = volume of solution (mL)
$m$ = mass of exchanger (g)

All $K_d$ determinations were performed in duplicate and the $K_d$s quoted are the average of two or more separate determinations. Variation between individual $K_d$ determinations was generally less than 5 %. In cases where a greater variation was noted, the experiment was repeated until consistent $K_d$ values were obtained.

## 2.4 Column Experiments

Materials which showed high $K_d$s were further evaluated using column techniques. The exchangers typically consisted of fine powders (< 1 μm) which needed to be combined with an alkali-resistant inorganic binder to produce beads suitable for use in column experiments. This pelletizing was achieved using a titania-based binder and a proprietary method developed by AlliedSignal. The granular material produced was sieved to obtain 40-60 mesh particles (0.25-0.42 mm) and then placed in a Bio-Rad column (0.66 cm internal diameter, 5 cm length) to obtain a bed volume of between 1 and 1.5 mL. The column was connected to Tygon tubing and 1 M NaOH passed through to condition the column and ensure proper bed settling for columns containing NaTS, NaTi and IE-911, whilst 1 M KOH was used to ensure that KTS-Ph was in the $K^+$ form.

The tank waste simulant was spiked with $^{89}$Sr to give total strontium concentrations of 2.7 x $10^{-7}$ M (0.024 ppm) and 4.1 x $10^{-6}$ M (0.36 ppm) for the NCAW and 101SY simulants, respectively. (The strontium concentration in 101SY-Cs5 was considerably greater than the 2.90 x $10^{-7}$ M suggested in Table 1. This was necessary due to the lower specific activity of the batch of $^{89}$Sr isotope used for the 101SY-Cs5 work.) The simulants were then passed through the column at a rate of approximately 20 bed volumes (BVs) per hour for NCAW and 5 BVs per hour for the 101SY-Cs5 simulant. 20 BV/hr was chosen for NCAW due to the high $K_d$s and perceived long duration of the experiment. A more realistic 5 BV/hr[17] was selected for the 101SY-Cs5 work. Eluent fractions were collected twice a day for the NCAW and approximately once per hour for the 101SY-Cs5 simulant and analysed for $^{89}$Sr using liquid scintillation counting.

Percentage breakthrough (BT) was then calculated using Equation 2 below :

$$\% BT = (A_f / A_o) \bullet 100 \hspace{2cm} (2)$$

where : $A_f$ = activity of solution exiting column, cpm/mL
$A_o$ = activity of feed solution, cpm/mL

3  RESULTS

### 3.1 XRD

The XRD powder patterns of the sodium nonatitanate (NaTi), the sodium titanosilicate (NaTS) and the potassium pharmacosiderite (KTS-Ph) corresponded to those described in the literature.[2-4,9,14,15] Both the titanate and the pharmacosiderite exhibited low crystallinities whilst the titanosilicate was highly crystalline. Powder diffraction patterns of all three materials used in this study have been published earlier.[14]

### 3.2 Ion Exchange

The $K_d$s of the three ion exchange materials in the two Hanford tank waste simulants NCAW and 101SY-Cs5 are given in Tables 2 and 3, respectively.

In NCAW, both the titanosilicate and the titanate performed well at a V:m ratio of 200 and exhibited $K_d$s over 200,000 mL/g, which corresponds to the removal of over 99.9 % of the strontium present. The pharmacosiderite was less effective, with a $K_d$ of only 20,200 mL/g, but this still corresponds to the removal of over 98.8 % of the strontium. When the V:m ratio was increased to 1000, the observed $K_d$s for both the titanate and the pharmacosiderite dropped considerably whilst the $K_d$ for the titanosilicate changed little. Ion exchange theory dictates that for trace exchange, the $K_d$ should be independent of the V:m ratio if only the strontium is being removed from solution. This appears to be the case for the titanosilicate, but from the sharp reduction in the $K_d$s at the higher V:m ratio, it is apparent that other ions are being removed from solution by the titanate and the pharmacosiderite. The sodium titanate in particular is known to have some affinity for $K^+$ over $Na^+$[18] and it is likely that the uptake of $K^+$, which is present as a macro ion, is leading to saturation of the ion exchange sites and the reduction in the Sr $K_d$s.

**Table 2** *$K_d$s and % Sr Removal from NCAW Simulant*

|  | V:m = 200 | | V:m = 1000 | |
|---|---|---|---|---|
| Sample | $K_d$ / mL/g | % Removal | $K_d$ / mL/g | % Removal |
| KTS-Ph | 20,200 | 99.02 | 5,000 | 83.85 |
| NaTi | 235,100 | 99.92 | 39,600 | 97.52 |
| NaTS | 269,500 | 99.93 | 225,800 | 99.56 |
| IE-911 | 29,300 | 99.30 | 20,600 | 95.38 |

The removal of strontium from 101SY-Cs5 by the ion exchange materials was far less effective than from NCAW as can be seen in Table 3. The complexing agents present have made the strontium far more difficult to remove from solution and the highest $K_d$ recorded was only 295 mL/g which corresponds to the removal of only approximately 61 % of the strontium present. Both the titanosilicate and the titanate had similar $K_d$s and averaged 55 % and 61 % strontium removal, respectively. By contrast, the pharmacosiderite was much less effective and only extracted approximately 13 % of the strontium  from solution.

**Table 3** $K_d s$ *and % Sr Removal from 101SY-Cs5 Simulant at a V:m Ratio of 200.*

| Sample | $K_d$ / mL/g | % Removal |
|--------|--------------|-----------|
| KTS-Ph | 31 | 13.2 |
| NaTi | 295 | 61.1 |
| NaTS | 231 | 54.7 |
| IE-911 | 210 | 53.6 |

### 3.3 Column Experiments

The breakthrough curves obtained for the materials using the NCAW simulant are given in Figure 1. It was attempted to run all of the columns to 50 % breakthrough, but in some cases, this proved impossible due to the gradual precipitation of solids in both the column and the tubing. The precipitated solids also caused a gradual reduction in the flow rate from an initial 20 BV/hr to approximately 13 BV/hr over the duration of the experiment. Some precipitation was not unexpected since the columns were run for several weeks and both NCAW and 101SY-Cs5 contained large amounts of dissolved solids. Even storage of the simulants in sealed bottles resulted in the formation of crystals at the bottom of the solutions after only a few weeks of standing, so it was not surprising that some crystallisation or precipitation also occurred during the column experiments. This reduction in flow rate increased the bed residence time and consequently may have led to a marginally better performance than if the flow rate had been maintained at 20 BV/hr.

The binder used to produce pellets suitable for column studies was found to also remove some strontium from the NCAW solutions. However, the $K_d s$ for the binder were at least an order of magnitude lower than the ion exchange materials, so the performance of the beads will be related more to the ion exchanger than the binder. Consequently, as the percentage binder increases, the performance of the beads will decrease. All three of the samples prepared 'in-house' at Texas A&M contained approximately 85 wt.% active material and 15 wt.% binder allowing direct comparison of the exchangers.

The best material was the sodium titanosilicate, NaTS which exhibited less than a 10 % breakthrough after 3500 bed volumes. Both NaTi and KTS-Ph showed insignificant breakthrough up to approximately 1000 bed volumes. However, after this volume, both materials showed a relatively rapid breakthrough as the available ion exchange sites began to saturate with sodium and potassium ions. All three columns eventually blocked due to precipitation, but it can be clearly seen that the performance of the materials closely followed the $K_d s$ recorded in batch experiments with NaTS showing the least breakthrough and KTS the greatest.

Extrapolation of the curve for KTS-Ph to 50 % breakthrough indicated a total strontium capacity in NCAW of approximately $3.0 \times 10^{-3}$ meq/g. Strontium capacities for NaTi and particularly NaTS were clearly significantly greater than the capacity of KTS-Ph. The IE-911 performed very poorly in the column experiments and exhibited a [89]Sr breakthrough of almost 10 % immediately. Breakthrough continued to rapidly increase and reached almost 40 % after 2000 BVs. From the batch $K_d s$, it was expected that the IE-911 would breakthrough after KTS-Ph but before the sodium nonatitanate, NaTi. This unexpected early breakthrough may well be due to poor exchange kinetics in comparison with the other exchangers or due to the binder inhibiting access to the ion exchange sites within the IE-911.

**Figure 1** *Strontium Breakthrough Curves for the Pelletised KTS-Ph, NaTi, NaTS and IE-911 in NCAW Simulant.*

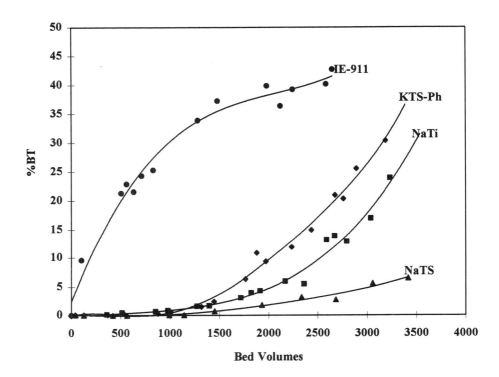

The breakthrough data for the 101SY-Cs5 simulant is given in Figure 2. As was expected from the far lower $K_d$s, 50 % breakthrough was reached in just a few days as opposed to the NCAW experiments which took several weeks to perform. In this simulant, strontium absorption by the binder was negligible allowing direct comparison of the materials. In these samples, there was some variation in the binder content which is outlined in Table 4. An industrially produced sample of pelletised sodium nonatitanate (Batch 8212-149) donated by AlliedSignal was also evaluated. This material was synthesized using the same method described for the sodium nonatitanate NaTi, but had been produced on a larger scale and pelletised by AlliedSignal. It performed better than any of the other exchangers.

**Figure 2** *Strontium Breakthrough Curves for the pelletised KTS-Ph, NaTi, NaTS, IE-911 and AlliedSignal Titanate in 101SY-Cs5 Simulant.*

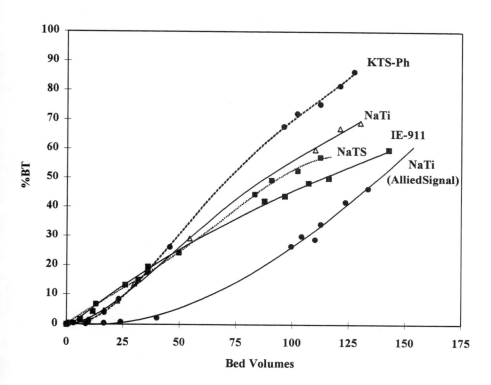

**Table 4** *Inorganic Binder Content of Pelletised Samples used in Column Studies with 101SY-Cs5 Simulant*

| Sample | % Binder |
|---|---|
| KTS-Ph | 23 |
| NaTi | 41 |
| NaTS | 19 |
| IE-911 | Not Known |
| AlliedSignal Titanate | Not Known |

The breakthrough characteristics followed the trend observed with the batch $K_ds$, and all of the materials except the AlliedSignal pelletised titanate showed immediate strontium breakthrough. The AlliedSignal material reached 50 % breakthrough after approximately 125 bed volumes, which was at least 50 bed volumes more than any of the other exchangers, including IE-911. The improved performance relative to NaTi may be due to either a reduced level of binder, an improved binder, better diffusion kinetics or a combination of all three.

### 4 CONCLUSIONS

Both the sodium nonatitanate, NaTi, and the sodium titanosilicate, NaTS, were highly efficient at removing strontium from NCAW, with greater than 99.9 % of the total strontium removed by both materials in batch experiments. Column studies confirmed the high affinity of both of these materials, particularly NaTS, for strontium in the presence of high alkalinity and high concentrations of inert salts. However, column blockage proved to be a problem due to precipitation of salts within both the column and the tubing which may cause problems in the application of these materials on a plant scale.

All materials performed poorly in removing strontium from 101SY-Cs5 due to the presence of high concentrations of complexants and $K_ds$ in batch experiments were reduced from > 200,000 mL/g for NCAW to < 250 mL/g for the 101SY-Cs5. All of the column experiments showed almost immediate breakthrough of strontium except for a sample of sodium nonatitanate provided by AlliedSignal which did not reach 50 % breakthrough until nearly 125 BVs. This materials contains essentially the same nonatitanate as the pelletised A&M titanate and it s superior performance must therefore be a result of an improved pelletising procedure or the incorporation of less inorganic binder.

The potassium pharmacosiderite, KTS-Ph, did not perform as well as either NaTS or NaTi in any of the waste simulants. This material has, however, been found to be more suitable for the removal of strontium and caesium from groundwaters[19] and other process wastes with a lower total ionic strength.

### 5 ACKNOWLEDGMENTS

We would like to thank Garrett N. Brown of PNNL for supplying data on the tank waste compositions and Steve Yates of AlliedSignal for supplying the sample of sodium nonatitanate pellets. We gratefully acknowledge the U.S. Department of Energy, Grant No. 198567-A-F1, through Pacific Northwest National Laboratory under DOE's Office of Science and Technology's Efficient Separations and Processing Crosscutting Program and the National Science Foundation, Grant No. DMR 9407899, for financial support.

### References

1. J.F. Keller and T.L.Stewart, 'First Hanford Separation Science Workshop' (Richland, WA.), 1993, p.I.35.
2. E.A. Behrens, D.M. Poojary and A. Clearfield, *Chem. Mater.*, 1996, **8**, 1236.
3. E.A. Behrens, D.M. Poojary and A. Clearfield, *Chem. Mater.*, (submitted).

4. D.M. Poojary, R.A. Cahill and A. Clearfield, *Chem. Mater.*, 1994, **6**, 2364.
5. D.M. Poojary, A.I. Bortun, L.N. Bortun and A. Clearfield, *Inorg. Chem.*, 1996, **35**, 6131.
6. L.P. Filina and F.A. Belinskaya, *Russ. J. Appl. Chem.*, 1995, **68**, 665.
7. A.I. Bortun, L.N. Bortun and A. Clearfield, *Solv. Extr. & Ion Exch.*,1996, **14**, 341.
8. L.H. Kullberg and A. Clearfield, *Solv. Extr. & Ion Exch.*, 1989, **7**, 527.
9. J. Lehto and A. Clearfield, *J. Radioanal. Nucl. Chem. Letters*, 1987, **118**, 1.
10. W.J. Paulus, S. Komarneni and R. Roy, *Nature*, 1992, **357**, 571.
11. T.M. Nenoff, J.E. Miller, S.G. Thoma and D.E. Trudell, *Environ. Sci. Technol.*, 1996, **30**, 3630.
12. C.B. Amphlett, 'Inorganic Ion Exchangers', Elsevier, New York, 1964.
13. F. Helfferich, 'Ion Exchange', McGraw Hill, New York, 1962.
14. E.A. Behrens, P. Sylvester and A. Clearfield, *Environ. Sci. Technol.*, 1998, **32**, 101.
15. E.A. Behrens, P. Sylvester, G.M. Graziano and A. Clearfield A in 'Science and Technology for Disposal of Radioactive Tank Wastes', N.J. Lombardo and W.W. Schulz (Eds.), Plenum Press, New York, 1998.
16. E.A. Behrens, PhD Thesis, Texas A&M University, 1997.
17. M. Johnson, BNFL Inc., Personal Communication.
18. G.M. Graziano, MS Dissertation, Texas A&M University, 1998.
19. P. Sylvester and A. Clearfield, *Solv. Extr. & Ion Exch.*, 1998, **16**(6), 1527.

# THE INFLUENCE OF THE PRESENCE OF ORGANIC ANIONS ON THE UPTAKE BY ZEOLITES OF TRANSITIONAL ELEMENT RADIOISOTOPES

A.Dyer,T.Shaheen and M.Zamin,

Chemistry Division, Science Research Institute,
Cockcroft Building,
University of Salford,
Salford, M5 4WT, U.K.

## 1 INTRODUCTION

One of the major problems associated with the removal of toxic elements from aqueous environments is the difficulty in defining the exact form of the species to be removed.This is particularly true when the species of concern is present in water at very low concentrations, as is the situation with radioisotopes in low-level aqueous nuclear waste streams.

A method employing a pragmatic approach is to use an ion-exchange material to scavenge the radioisotope and then to study the effect of trace concentrations on the ion-exchange process. Synthetic zeolites are appropriate exchangers because of their crystalline nature and availability as pure substances which are well characterised. These properties have prompted their use as models for cation exchange processes in many previous studies [1].

An earlier publication demonstrated that zeolites could be used to scavenge radioisotopes of Co, Ni and Zn from water and that uptake of these elements was sensitive to the presence of low concentrations of common anions [2]. In some cases knowledge of the structure and composition of the zeolite used helped to define the nature of the speciation of the isotope at low concentrations. In extreme cases the addition of an anion eliminated cation exchange,which could be explained on the basis of the metal ion forming an uncharged complex ion with the anion.

The elements studied here (Co,Fe,Ni,Zn) are constituents of nuclear wastes arising from neutron activation of constructional materials occurring in reactor operation, and decommissioning. Disposal of these elements, as part of low-level waste forms, will bring them into potential contact with organic moieties arising from the breakdown of cellulosic materials such as tissues and clothing. This contribution describes the influence of trace quantities of the citrate, oxalate and succinate anions on the uptake of $Co^{II}$, $Ni^{II}$, $Zn^{II}$ and $Fe^{III}$ onto several zeolites with a view to giving information on cation speciation and its potential effect on cation mobility in waste forms. The organic anions were chosen as model substances representative of the likely nature of cellulosic breakdown products [3].

## 2 EXPERIMENTAL

The synthetic zeolites used were A,X,Y, ZSM-5 and mordenite supplied in their sodium forms by Laporte Inorganics, Moorfield Rd, Widnes, Cheshire, UK (they are no longer available from this source but can be obtained from other suppliers). In addition a sample of the natural zeolite clinoptilolite was used. This was from the Mud Hills deposit in California (that used in the SIXEP facility at BNFL,Sellafield,UK, to clean-up aqueous nuclear waste). It contains Na,K and Ca as the most readily exchangeable cations. The composition of the zeolites is in Table1.

**Table1** *Composition of zeolites (suppliers data)*

| Zeolite | $SiO_2$ | $Al_2O_3$ | $Na_2O$ | $K_2O$ | $CaO$ | $MgO$ | $Fe_2O_3$ |
|---|---|---|---|---|---|---|---|
| A | 41.8 | 36.50 | 21.20 | 0.09 | 0.08 | 0.01 | 0.02 |
| X | 46.70 | 33.90 | 19.60 | 0.01 | 0.06 | 0.01 | 0.02 |
| Y | 61.90 | 23.70 | 12.50 | 0.20 | 0.37 | 0.11 | 0.51 |
| ZSM-5 | 97.90 | 0.62 | 0.67 | 0.01 | 0.02 | 0.01 | 0.04 |
| Mordenite | 80.20 | 15.90 | 2.00 | 0.90 | 0.20 | 0.10 | 0.30 |
| Clinoptilolite | 66.03 | 10.47 | 3.25 | 1.81 | 1.55 | 0.58 | 1.21 |

The radioisotopes, 59-Co, 63-Ni, 65-Zn and 59-Fe were supplied by Amersham International, UK, in the form of chlorides. Conversions to nitrate and sulphate forms were by use of a strong base anion exchange resin (Bayer Lewatit M500 KROH). The uptake of the radioisotopes onto the zeolites was monitored by loss of activity from the solution phase. 63-Ni and 65-Zn were conveniently measured by liquid scintillation counting as 1mL aliquots with 9mL PPO/POPOP toluene based scintillator mixed with Triton-X-100 detergent. The proportions of scintillator to detergent in the mixtures were 5:4 for 63-Ni and 6:3 for 65-Zn. Appropriate quench corrections were made. Both 58-Co and 59-Fe could be estimated via the Cerenkov effect using 10 mL of solution in polythene vials. Cerenkov light was detected in the tritium channel of a liquid scintillation spectrometer.
Ion-exchange experiments were performed as follows:
(a) isotopically labelled solutions of each cation were placed in contact with zeolites (maintaining the solid /liquid ratio at 200). The solutions were $10^{-2}$ M with respect to the metal ion. Each exchange was repeated using the metals as chlorides, nitrates and

sulphates. Experiments were carried out in polythene containers and stirred for one week.

(b) each experiment was repeated with additions of 30 ppb of citrate, oxalate and succinate anions as sodium salts- this being the concentration found in ground water near shallow land burial sites [4]. Chemicals were used in the purest form available; all organic materials were AnalaR grade and deionized water was used throughout.

## 3 RESULTS AND DISCUSSION

### 3.1 Zeolites X and Y

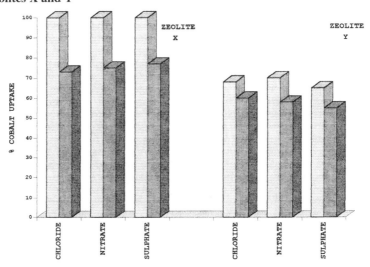

**Figure 1**. Uptake of cobalt onto zeolites X and Y.Right hand column with, and left hand column without, citrate anions in the presence of inorganic anions.

Zeolites X and Y are isostructural synthetic faujasites. They differ in Si/Al ratio (see Table1) and hence in the number of cations available for exchange. Details of their potential use to scavenge Co from aqueous nuclear wastes have been published [5]. An example of their ability to scavenge the Co cation can be seen in Fig.1. The capacity for Co is lower for Y than in X which reflects the general trend in zeolites that the X zeolite, with a lower Si/Al ratio, has a stronger preference for small hydrated ions than Y which has more Si in the zeolite framework

The presence of different inorganic anions has little effect on uptake which confirms previous work at this concentration ($10^{-2}$ M) [2].When additions of citrate anions are made the ability to take up Co drops in all cases, with changes of ~ 25% noted in X. Similar effects were seen in the presence of succinate in both X and Y zeolites and for oxalate in X. Reductions in the Co capacity of Y were small. Under similar experimental conditions uptakes of Zn and Fe were not greatly influenced by organic anions. Nickel uptake on X will be considered later.

The major opening ( ring of oxygen atoms) in the faujasite structure (see Figure 2) through which ions must pass in order to be exchanged has a diameter of 74 nm [6]. This dimension may be able to exclude a large complex ion but the 100% capacity for cobalt shown by zeolite X , in the absence of organics, is evidence that $[Co(H_2O)_6]^{2+}$ ions are not excluded. It is probable that the framework charge of X is able to strip co-ordinated water molecules from the $Co^{2+}$ ion to aid exchange. The capacity illustrated by Y, which has a much lower framework charge, suggests that hydrated $Co^{2+}$ ions can gain access to the internal cavities in both faujasites. The significance of the drop from 100% uptake in X to ~ 65% in Y may reflect relative differences in the availability of the sodium ions initially available for exchange. This trend again fits known zeolite exchange behaviour [1] in that the lower uptake in Y can be linked to a restriction of access to cation sites within the smaller cavities in the faujasite structure.

**Figure 2.** The aluminosilicate framework of synthetic faujasites X and Y.

Following this simple line of thought, reductions in Co uptake with an organic anion present could arise when the Co cation is contained in an uncharged complex species, such as $[Co^{II} L_2]$. Previous work noted an almost complete lack of exchange when uncharged species like $[Co^{II} Cl_2(H_2O)_4]$ were likely to be present [2]. Alternatively the complex species may be too large to enter the zeolite structure. Even the large framework charge on zeolite X may not be able to strip co-ordinated species from the cation if approach to the aluminosilicate surface is sterically hampered. An example of this has been observed by Rasqin *et al* [7] for a silver pyridine complex .

### 3.2 Zeolite A

Zeolite A has a similar framework charge to that of X, but access to cation sites is via a smaller oxygen window (40nm [6]). Figure 3 illustrates Co and Ni cation exchange on A with organic anions present in comparison to X. It seems that the A framework can strip ligands from the Ni cation but not always the Co cation. The same conclusion could be drawn from the equivalent experiments involving Zn and Fe cations (not shown). Lack of Ni uptake into the X zeolite seems to be influenced by steric restrictions with only about 60-70% of cations exchanged (as in the Co case discussed above - results repeated in Fig.3 for clarity).

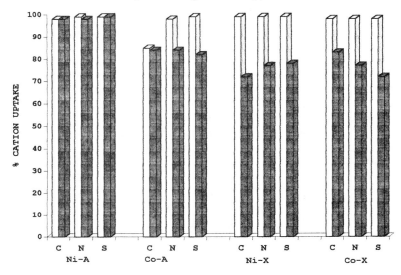

**Figure 3**. Uptake of cobalt and nickel onto zeolites A and X. Right hand column with, and left hand column without, citrate anions in the presence of chloride (C) nitrate(N) and sulphate (S).

### 3.3 Mordenite and ZSM-5

Table 1 shows that these zeolites have high Si/Al ratios and low cation contents , with ZSM-5 having very little ion-exchange capacity. Despite this low capacity, in most of the systems investigated the ZSM-5 cation uptakes were sensitive to the presence of organic species. An example of this for ZSM-5 with Zn as the cation of interest, when citrate and oxalate additions were made, is shown in Figure 4. Clearly the ability to take up zinc is affected by the inorganic ion and this was noted in past work[2].

The organic species present also reduced uptake, presumably due to complex formation. This was not seen when the succinate additions were made.

Results for the mordenite and ZSM-5 scavenging of iron are in Figure 5. The sensitivity of cation removal from solution to both inorganic and organic anions was

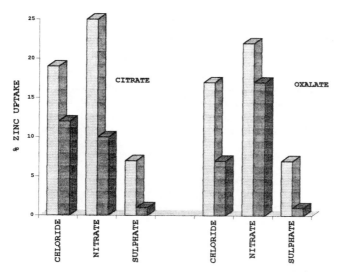

**Figure 4.** Uptake of zinc onto ZSM-5. Right hand column with, and left hand column without, citrate/oxalate additions in the presence of inorganic anions.

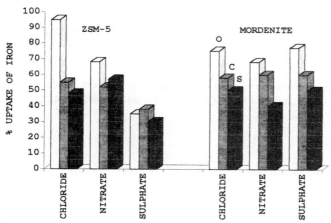

**Figure 5.** Uptake of iron onto zeolites ZSM-5 and mordenite.Effects of oxalate (O) citrate (C) succinate(S) shown left to right for each inorganic anion.

again evident but it must be commented that this was the only ion to which mordenite

displayed differential exchange. The high capacity for iron shown by these low
capacity, high -silica, zeolites certainly does not stem entirely from ion-exchange.
Some metal oxide/hydroxide surface precipitation will contribute and the possibility of
iron radiocolliods adhering to the relatively hydrophobic zeolite surfaces is high.
These observations notwithstanding the cation uptake is clearly influenced by anion
driven speciation. Access to the cation sites in these zeolites is via oxygen windows
intermediate in size to those of A and X (51-56 nm for ZSM-5 and 65-70 nm for
mordenite [6] ). The organic species present also reduced uptake,  presumably due to
complex formation. This was not seen when the succinate additions were made.

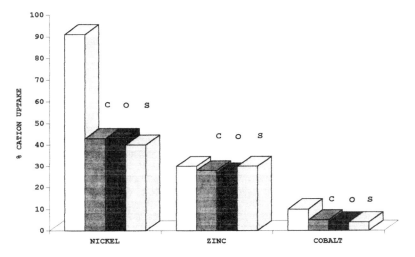

**Figure 6.** Uptake of Ni,Co and Zn on clinoptilolite.The first colunm in each set is
from metal nitrate solutions followed by uptakes in nitrate with citrate (C) , oxalate(O)
and succinate (S) additions.

### 3.4 Clinoptilolite

Clinoptilolite resembles mordenite in Si/Al ratio and pore size, and the results
obtained in this work reflect this. Cation uptakes are low and little influenced by the
presence of organic species . An exception to this comes when nickel nitrate is dosed
with citrate, oxalate or succinate anions. The consequence can be seen in Figure 6
with Zn and Co results shown for contrast. No explanation of this unusual result can
be put forward. The mixed cation content of the natural clinoptilolite seems unlikely
to be of consequence. Replacement of the sodium ions present initially in the zeolite
by divalent nickel could be expected but the potassium and calcium ions would be less

likely to exchange. It is difficult to see that any subtle selective cation exchange accounts for these results.

## 4  SUMMARY

The broad premise that zeolites can be used as model exchangers to "sense" the influence of organic species on the uptake of Co, Ni, Zn. and Fe cations has been confirmed. The influence of the sodium cation inevitably added with the organic anion can be ignored on two grounds. Firstly that the concentrations added (ppb) are orders of magnitude lower than the concentrations of transitional metal cation present ($10^{-2}$ M) and ,secondly, the sodium cation is the least preferred of all the common cations by the zeolite frameworks [8].

The sensitivity of zeolite cation exchange to speciation seen herein mirrors that already reported for changes in inorganic ion nature and concentration. In some of the systems studied here the influence of the very low concentration of organic anion addition is beyond that expected for the creation of complex species able to withdraw the cation for exchange by, say, formation of an uncharged moiety. Perhaps the drops in cation uptake are helped by the adsorption of organic species onto the external zeolite surface, so blocking entry to the cation sites rather than the cation uptake being entirely reduced by steric effects.

This work also demonstrates that scavenging of the radioisotopes studied from aqueous nuclear waste treatment is best carried out by a zeolite, such as zeolite A, which has a high substitution of Al present in the aluminosilicate framework.

## 5 ACKNOWLEDGEMENTS

Thanks are due to the Minister of Science and Technology, Pakistan, for financial support to TS and MZ. Gifts of zeolites from Laporte Inorganics, Widnes, Cheshire, UK, were greatly appreciated.

## 6 REFERENCES

1.    A.Dyer in 'Inorganic Ion Exchangers in Chemical Analysis', Eds. M.Qureshi and K.G.Varshney, CRC Press,Boca Raton,Florida, USA, 1991, p.33.

2.    A.Dyer and T.Shaheen, Sci.Total Environ., 1995, **173/174**, 301.

3.    J.D.Wilkins, DoE Report No. DOE/RW/89/032, UKAE, Harwell, Oxon,UK,1988.

4.      A.P.Toste, L.J.Kirby, and T.R.Pahl, in 'Geochemical Behavior of Disposed
        Radioactive Waste', Eds., G.S.Barney, J.D.Navratil and W.W.Schultz, ACS
        Symp.Ser., 1984, **246**, 251.

5.      A.Dyer and J.K.Abou-Jamous, J.Radioanal.Nucl.Chem.,Articles, 1994, **183**,
        225.

6.      W.M.Meier and D.H.Olson, Zeolites, 1996, **17**, 1.

7.      E.Rasquin, A.Maes and A.Cremers, in 'Proc.6th Int. Zeolite Conf.' Eds.,
        D.Olson and A.Bisio, Butterworths, Guildford, Surrey, UK, 1984, p.641.

8.      J.D.Sherman, AIECh.,Symp.Ser., 1978, **74**, p. 98.

# THE NOVEL ABSORBER EVALUATION CLUB - A REVIEW OF RECENT STUDIES

E. W. Hooper, P. R. Kavanagh and A.F.Wells

AEA Technology
Harwell Laboratory
Harwell, Didcot
Oxon OX11 ORA

1.      INTRODUCTION

The separation processes of ion exchange and sorption have been widely used for many years throughout the nuclear industry to decontaminate aqueous waste streams. The two processes are analogous, both removing ions from solution either by exchange of a counter ion (ion exchange) or by direct physical removal (sorption). Standard formulations of sorbents are used worldwide to treat waste streams, but as new materials are manufactured and discharge authorisations are lowered, it becomes increasingly difficult for operators to adhere to the governing principles of ALARP *(as low as reasonably practicable)* and BATNEEC *(best available technology not entailing excessive cost)*. The objective of the Novel Absorber Evaluation Club is to remain aware of new sorbent developments and produce decontamination data on these products for direct comparison with standard 'benchmark' materials.

The Novel Absorber Evaluation Club (NAEC) was inaugurated in 1988 by AEA Technology on behalf of the UK Nuclear Industry as part of an overall requirement to minimise any environmental impact resulting from operations.

Membership of the NAEC is open to all parties with an interest in the decontamination of liquid radioactive effluents. The work programme of the NAEC is guided by the membership with a remit to provide quantitative experimental information of the ability of new and novel absorbent materials to remove radionuclides from solution.

Manufacturers and suppliers from industry and research are invited to submit materials for testing. Members prioritise the testing which comprises an initial screening of the material in a series of batch contacts with up to five standard NAEC reference waste streams. Further evaluation at suppliers request may include flow through packed beds or use as a finely divided addition in combination with membrane filtration. Test conditions may be varied by change of pH (in the range 3 - 11) and additions of organic contaminants or complexants.

With a membership covering the spectrum of the nuclear industry, the NAEC is in a unique position as an independent test facility with input from acknowledged experts in the field. Present members include British Nuclear Fuels plc, Nuclear Electric plc, Magnox Electric plc, Ministry of Defence (represented by the Atomic Weapons Establishment, Aldermaston and Director General Submarines, Bristol) and UKAEA . A parallel contract also exists with the Her Majesty's Environment Agency for full exchange of information.

Features of the Club programme include:
- Independent test facility for manufacturers of sorbents
- Utilisation and access to a database of results and reports generated since 1988
- Use of full radioactive facilities, experimental techniques and operator expertise
- NAEC methods and reference waste streams internationally recognised by IAEA
- Reference waste streams cover the range of radionuclides expected to be found in virtually any nuclear stream
- Worldwide contacts through Club experience, membership and exchange of information with research centres
- Capability of tailoring an absorber recipe to optimise the decontamination of specified waste streams

This paper will detail the tests carried out since 1991 and summarise the results obtained for the sorption of a range of elements including Mn, Co, Cr, Zn, Ru, Cs, Sr, Cd, Hg, Ag, Fe, Tc, U and alpha emitters (Pu + Am).

## 2.    TEST DETAILS

### 2.1    Reference Waste Streams

Five reference waste streams are generally used in the test programme, known as NAEC (Novel Absorber Evaluation Club) S1, S2, S3, S4 and S5. Other streams that have been used during this period are S6 and also S-Cs, which is used as a screening basis for hexacyanoferrate absorbers. All waste streams are prepared from standard Amersham sources in 0.05M sodium nitrate and are adjusted to the test pH immediately before use. The standard test pH is 9, but other values may be used according to requirements and the specified absorber operational range.

The standard isotopic compositions and methods of analysis are as follows:

NAEC S1 contains $^{137}Cs$, $^{60}Co$, $^{65}Zn$, $^{51}Cr$, $^{59}Fe$, $^{54}Mn$, $^{106}Ru$, $^{203}Hg$, $^{109}Cd$, and $^{110m}Ag$ at a nominal level of 100 Bq/mL and is analysed by gamma spectrometry.

NAEC S2 contains $^{99}Tc$ as the pertechnetate ($TcO_4^-$) at a nominal level of 40 Bq/mL and is analysed by beta scintillation counting.

NAEC S3 contains $^{239}Pu(IV)$ at 2 Bq/mL, $^{241}Am$ at 1 Bq/mL and $^{90}Sr$ at 5 Bq/mL. Analysis is by both alpha and beta scintillation counting.

NAEC S4 is NAEC S1 plus 0.25 g/L ethylene diamine tetra-acetic acid (EDTA) added as the sodium salt, and 0.15 g/L citric acid. Analysis is by gamma spectrometry.

NAEC S5 contains $^{239}Pu(IV)$, $^{241}Am$ and $^{237}Np$ (V) at 1 Bq/mL plus1g/L NaHCO$_3$ (to produce anionic carbonato species in alkaline solution.)   Analysis is by alpha scintillation counting.

NAEC S6a contains 0.1 mg/L  U (as uranyl nitrate) with 1 g/L NaHCO$_3$ prepared in demineralised water and is analysed for total uranium by ICP-MS.

NAEC S6b is the same as S6a but prepared in tap water (to introduce competition from Ca in particular) and is analysed for total uranium by ICP-MS.

NAEC S-Cs contains 100 Bq/mL Cs-137 and may be analysed by either gamma spectrometry or beta counting.

The waste streams used in the test will depend on the type of sorbent being examined.  NAEC S1, S3 and S4 will be generally used for cation absorbers and S2, S4 and S5 for anion absorbers.

## 2.2    Sorbent Conditioning

Prior to use, sorbents are conditioned by washing with water adjusted to the experiment pH value using sodium hydroxide solution or dilute nitric acid. The washings are continued until the pH of the wash remains at its original value for 2-3 hours or preferably overnight.

The absorber is used in wet condition after decanting the wash liquor and removing excess moisture with a cellulose tissue.

## 2.3    Batch Contact Experiments

1 mL portions of the conditioned absorber are measured into 100 mL borosilicate conical flasks using a hypodermic syringe which has had the conical end removed to provide a 'full-bore' syringe. 50 mL of the reference waste stream is added and the flask sealed with a polythene stopper before agitating gently at room temperature. 1.5 mL portions of the liquid are removed after 1, 2, 4, 6 and 24 hours, centrifuged and then 1 mL removed for counting.

When magnetic absorbers are being tested, the conical flask is allowed to stand for 2 minutes on a disk permanent magnet (rare earth) to settle the absorber prior to sampling.

The analytical results are presented as Bq/mL and also as a calculated decontamination factor with reference to the original waste stream activity level:

$$\text{Decontamination Factor (DF)} = \frac{\text{activity per mL in feed}}{\text{activity per mL in sample}}$$

No correction is made for the volume changes resulting from sampling.

## 2.4    Column Experiments

Glass columns of 1 cm bore are packed with 5 mL absorber. 500 mL of water adjusted to the test pH is washed through each column to condition the absorber. The columns are operated at a throughput of 10 bed volumes per hour (approximately 400 mL per day) using reference streams NAEC S1, S2 and S3, as appropriate. The pooled effluent is sampled daily for analysis.

## 3.    ABSORBERS TESTED

The results of tests carried out during 1988-1991 were reported at IEX'92[1] and those during 1992-1995 at IEX'96[2].

The following absorbers have been tested since that time with various NAEC reference streams at appropriate values of pH:

- Activated Carbons from Loughborough University, Department of Chemical Engineering.
- "Fezhel" Hexacyanoferrate based materials from Russia , "Compomet Cantea" Yekaterinburg
- Hexacyanoferrate based matreials from Comenius University, Bratislava, Slovak Republic.
- IXE inorganic absorbers manufactured by the Toagosei Company, Nagoya R&D Institute, Nagoya, Japan.
- Purolite chelating resins, Purolite International Ltd., Hounslow, UK.
- A natural clinoptilolite from AllDeco s.r.o, Jaslovske Bohunice, Slovak Republic.

**Table 1:** *Decontamination Factors after 24 hour contact*

| Absorber | pH | U | $^{109}$Cd | $^{203}$Hg | $^{51}$Cr | $^{106}$Ru | $^{137}$Cs | $^{54}$Mn | $^{110m}$Ag | $^{59}$Fe | $^{65}$Zn | $^{60}$Co | $^{99}$Tc | $^{90}$Sr | α |
|---|---|---|---|---|---|---|---|---|---|---|---|---|---|---|---|
| **Miscellaneous** | | | | | | | | | | | | | | | |
| OPC S1, S3,S6a | 12 | 7000 | >80 | 39 | >550 | 31 | 1.3 | 43 | 11 | | | 8.5 | | 1.4 | >550 |
| UOP Na titanate S1,S3 | 9 | | >560 | 8.8 | >550 | 75 | 470 | 23 | 320 | 320 | 87 | 6.1 | | 38 | 110 |
| Macrosorb CT100 S1,S2 | 9 | | >580 | 2.0 | 4.5 | 110 | 2.7 | >960 | 17 | >1400 | >2000 | 830 | 1.0 | | |
| CT100 S4 | 4 | | 1.1 | 1.3 | 1.6 | 1.4 | 2.5 | 1.0 | 20 | 1.2 | 1.2 | 1.2 | | | |
| DiphonixSiO$_2$ S1,S3 | 9 | | >1200 | 5.4 | 2.4 | 15 | 1.4 | 111 | 2.8 | 28 | 73 | 16 | | 6.2 | 22 |
| Alldeco zeolite S1,S3 | 9 | | >400 | 2.1 | 1.4 | 48 | 50 | 11 | 3.4 | 70 | 120 | 4.9 | | 7.4 | 5.5 |
| Reillex HPQ S2 | 9 | | | | | | | | | | | | 42 | | |
| **Hexacyanoferrates** | | | | | | | | | | | | | | | |
| Fezhel AN1 S1 | 3 | – | 1.6 | 1.3 | 1.4 | 2.4 | 26 | 1.0 | 21 | 2.5 | 1.7 | 2.1 | | | |
| Fezhel AN4 S1 | 3 | – | 2.1 | 1.5 | 1.3 | 3.3 | >52 | 1.0 | >88 | 2.7 | 2.9 | 1.5 | | | |
| Fezhel AN1 S1 | 9 | – | >292 | 1.3 | 2.1 | 15 | 52 | 75 | 3.8 | 9.5 | 78 | 147 | | | |
| Fezhel AN4 S1 | 9 | – | >292 | 1.9 | 2.1 | 15 | 48 | 76 | 3.9 | 9.5 | 73 | 137 | | | |
| FCN-SiO$_2$ 3 S1 | 9 | – | >292 | 1.3 | 2.1 | 6.9 | 54 | 11 | 76 | 3.1 | 129 | 17 | | | |
| PAN1-NiFCC S1,S3 | 9 | | 250 | 4.0 | 1.4 | 9.3 | 120 | 3.1 | 270 | 7.1 | 59 | 5.6 | | 2.5 | 8.3 |
| PAN-NiFCC S4 | 9 | | 1.4 | 4.4 | 1.3 | 2.4 | 15 | 1.8 | 56 | 1.9 | 1.4 | 1.1 | | | |
| PAN-NiFCC S4 | 4 | | 77 | 1.9 | 1.6 | 3.4 | 58 | 7.2 | 220 | 1.3 | 150 | 1.1 | | | |
| NiFSil S1 | 9 | | >110 | 1.3 | 1.2 | 5.4 | 82 | 7.9 | 95 | 4.2 | 9.8 | 7.8 | | | |

---

1 PAN materials are finely-divided absorbers incorporated into polyacrylonitrile beads, which enables them to be used in packed columns.

**Table 1: Decontamination Factors after 24 hour contact**

| Absorber | pH | U | $^{109}$Cd | $^{203}$Hg | $^{51}$Cr | $^{106}$Ru | $^{137}$Cs | $^{54}$Mn | $^{110m}$Ag | $^{59}$Fe | $^{65}$Zn | $^{60}$Co | $^{99}$Tc | $^{90}$Sr | $\alpha$ |
|---|---|---|---|---|---|---|---|---|---|---|---|---|---|---|---|
| Loughborough University Carbons: | | | | | | | | | | | | | | | |
| SCNTiSi S1 | 9 | - | 475 | >875 | 110 | 50 | 150 | 84 | 270 | 67 | 300 | 270 | - | - | - |
| SK3H+ S1 | 9 | | 72 | 21 | 14 | 12 | 3.2 | 6.8 | 8.6 | 12 | 25 | 119 | | | |
| F400H+ S1 | 9 | | 460 | 56 | 26 | 37 | 210 | 29 | 96 | 41 | 250 | 80 | | | |
| KAVNa+ S1,S3 | 9 | | 720 | 110 | 16 | 21 | 2.7 | 230 | 27 | 69 | 820 | 810 | | 13 | 190 |
| F400Na+ S1,S3 | 9 | | 410 | 86 | 30 | 43 | 3.1 | 50 | 58 | 46 | 230 | 140 | | 5.1 | 44 |
| SK3H+ S4 | 9 | | 1.3 | 1.6 | | 1.3 | 3.7 | 1.2 | 115 | 1.2 | 1.2 | 1.1 | | | |
| F400H+ S4 | 9 | | 1.3 | 1.4 | | 1.6 | 1.6 | 1.1 | 150 | 1.4 | 1.1 | 1.2 | | | |
| KAVNa+ S4,S5 | 9 | | 1.3 | 1.4 | | 1.5 | 4.6 | 1.2 | 189 | 1.5 | 1.2 | 1.2 | | | 25 |
| F400Na+ S4,S5 | 9 | | 1.3 | 1.3 | | 1.5 | 1.6 | 1.1 | 330 | 1.6 | 1.0 | 1.2 | | | 54 |
| Purolite Resins | | | | | | | | | | | | | | | |
| A830 S1 | 9 | | >450 | 45 | 2.1 | 3.9 | 1.4 | 20 | 78 | 19 | 72 | 1.4 | | | |
| S950 S1 | 9 | | 86 | 49 | 2.2 | 2.6 | 4.4 | 105 | 20 | 11 | 143 | 67 | | | |
| A830 S4 | 9 | | 1.7 | 6.3 | 1.6 | 1.6 | 1.4 | 1.3 | 4.0 | 1.2 | 1.3 | 1.3 | | | |
| S950 S4 | 9 | | 1.2 | 2.5 | 1.2 | 1.1 | 1.4 | 1.1 | 17 | 1.1 | 1.1 | 1.1 | | | |
| Toagosei Absorbers | | | | | | | | | | | | | | | |
| IXE 100Na S1,S3 | 9 | | >560 | 13 | 1.5 | 64 | 4.7 | 810 | 350 | >120 | >145 | 910 | | 25 | 650 |
| IXE 110Na S1,S3 | 9 | | >40 | >35 | 1.1 | 15 | >94 | >185 | >88 | 140 | 79 | 200 | | 15 | 39 |
| IXE 400 S1,S3,S6b | 9 | 10 | >40 | >35 | 1.2 | >32 | 47 | >185 | >88 | >200 | 140 | 170 | | 72 | 220 |
| IXE 700 S5 ,S2 | 9 | | | | | | | | | | | | 1.1 | | >170 |
| Zirconium Phosphates | | | | | | | | | | | | | | | |
| MEL ZrP S1,S3 | 9 | | >500 | 5.8 | 3.1 | 6.8 | 180 | 600 | 44 | 5.8 | 845 | 780 | | 22 | 18 |
| MEL ZrP S1 | 3 | | >82 | 6.2 | | 6.7 | 160 | >900 | 11 | | | 900 | | | |

Table 1 Decontamination Factors after 24 hour contact

| Absorber | pH | U | $^{109}$Cd | $^{203}$Hg | $^{51}$Cr | $^{106}$Ru | $^{137}$Cs | $^{54}$Mn | $^{110m}$Ag | $^{59}$Fe | $^{65}$Zn | $^{60}$Co | $^{99}$Tc | $^{90}$Sr | α |
|---|---|---|---|---|---|---|---|---|---|---|---|---|---|---|---|
| Termoxid 3A S1 | 3 | | >82 | 1.9 | | 2.3 | 107 | 69 | 5.0 | | | 3.0 | | | |
| PAN-ZrP S1 | 3 | | >82 | 2.7 | | 18 | 250 | 3.1 | 11 | | | 3.0 | | | |
| Magnetic Absorbers[2] | | | | | | | | | | | | | | | |
| Magnetite S1,S3 | 9 | | 7.1 | >36 | >84 | >34 | 1.1 | 5.3 | 9.6 | 3.1 | 9.4 | 1.4 | | 5.5 | >330 |
| Magnetic ZrP S1 | 9 | | >1200 | 180 | >1600 | 190 | 220 | >840 | >1200 | >1300 | 700 | 100 | | | |
| *ZrP S1* | 9 | | *>1200* | *17* | *13* | *11* | *270* | *130* | *80* | *8.1* | *520* | *150* | | | |
| Magnetic NiFC S1 | 9 | | >1200 | 3.7 | 1.5 | 95 | >1335 | 73 | 1100 | 1300 | >1100 | 93 | | | |
| *NiFC S1* | 9 | | *>230* | *4.6* | *1.2* | *42* | *76* | *150* | *64* | *180* | *420* | *75* | | | |
| Magnetic HTiO S1 | 9 | | >1200 | >1000 | 5.1 | 62 | 4.9 | >840 | 1000 | >1300 | >1100 | 250 | | | |
| *HTiO S1* | 9 | | *>1200* | *473* | *1.6* | *64* | *5.2* | *>840* | *860* | *960* | *>1100* | *250* | | | |
| Magnetic Carbon S1 | 9 | | >390 | 19 | 2.2 | 50 | 1.6 | 755 | 370 | >140 | 390 | 490 | | | |
| Magnetic Carbon S4 | 9 | | 1.7 | 1.9 | 2.5 | 4.3 | 19 | 1.6 | >830 | 3.1 | 1.5 | 1.5 | | | |
| Magnetic Carbon S4 | 4 | | 1.2 | 1.2 | 1.5 | 3.2 | >43 | 1.1 | >73 | 1.4 | 1.2 | 1.2 | | | |
| Texas A&M Research Materials | | | | | | | | | | | | | | | |
| RC–5–23A S1 | 9 | | 62 | 29 | 1.3 | 14 | >140 | 460 | 850 | 30 | 540 | 110 | | | |
| RC–5–27A S1 | 9 | | 31 | 10 | 1.3 | 18 | 450 | >885 | 550 | 87 | 550 | 144 | | | |
| KTiSi S1 S1 | 9 | | 180 | 100 | 1.4 | 53 | 77 | 600 | 590 | 160 | 400 | 160 | | | |
| GMG-I-39 S1 | 9 | | >110 | >875 | 1.5 | 40 | 5.3 | 110 | 230 | 84 | 68 | 60 | | | |
| ALPilc S1 | 9 | | 9.0 | 1.6 | 1.9 | 69 | 7.8 | 3.2 | 8.5 | >130 | 20 | 2.9 | | | |

[2] Results in italics are for the "as-received" absorber that was incorporated into a magnetic base to form the composite material.

- Titanium and zirconium phosphate absorbers from Termoxid Co., Zarechny, Sverdlovsk, Russia
- Ordinary Portland Cement and a Chinese alumina cement
- Cellulose sponge manufactured by Dynaphore Inc., Richmond, Virginia, USA
- Brimac from Tate&Lyle, Greenock, Scotland. UK.
- Goethite prepared at the Department of Chemistry, University of Reading
- Hydrous titanium oxide prepared at the Harwell Laboratory of AEA Technology
- Kanemite from the University of Reading, Department of Chemistry
- A sodium titanate manufactured by UOP Corporation, Des Plaines, Illinois, USA.
- Reillex HPQ strong-base polyvinylpyridine anion resin, Reilly Industries, Indianapolis, USA.
- Macrosorb CT100, a hydrotalcite supplied by Crosfields, Warrington, UK.
- Magnetic sorbents supplied by Bio-separations Ltd.,Enfield, UK.
- Silica based Diphonix (Diphosil) supplied by EiChrom Europe, Paris, France.
- Sorbents supplied by Texas A & M University, Department of Chemistry, Texas, USA.

## 4.     RESULTS

Table 1 shows a selection of the results obtained since with the above materials from batch contact trials with reference waste streams. A > sign indicates that the nuclide was removed to below the limit of detection of the analytical method used, which varies according to a number of factors. Shading is used to indicate particularly high DFs (a DF of 100 represents 99% removal).

It is clear from these results that there are selections of materials available that show either high selectivity for particular species or high overall nuclide removal.

## 5. SUMMARY

The results presented in this paper show only a fraction of the data collected by the Novel Absorber Evaluation Club. A regularly updated Database provides Members of the Club with relevant information on sorbents for the decontamination of radioactive waste streams. The material of choice for any particular application would depend on both the species to be removed and a number of factors including the pH of the waste stream and its inactive composition.

The information obtained from the Club's Test Programme is also of value in designing treatment processes for the clean-up of industrial wastes to meet discharge authorisations, recover metal values and to allow water recycle.

## References

1.     "Ion Exchange Advances" Edited by M.J.Slater. Published by Elsevier Applied Science. pp 310-317 (1992)

2.     "Ion Exchange Developments and Applications" Edited by J.A.Grieg. Published by The Royal Society of Chemistry. pp 143-150 (1996)

# Metal Ion Removal and Separation

# A COLUMN SYSTEM FOR SEPARATION AND RECOVERY OF PRECIOUS METALS FROM LEACHING SOLUTIONS WITH THE CHELATING RESIN DUOLITE GT-73

M. Iglesias, E. Anticó, V. Salvadó*.

*Dept. Química, Facultat de Ciències, Universitat de Girona. 17071 Girona. Spain.*

## ABSTRACT

The synthetic polymer resin Duolite GT-73, bearing thiol groups, has been investigated for the separation and recovery of Pd(II) and Au(III) from leaching solutions which result from the treatment of copper/nickel sulphide ores. The capacity of the resin under different experimental conditions has been determined for the metals Pd(II), Au(III), Cu(II) and Ni(II) in batch operations.

Column experiments have been also carried out. Taking into account the results previously obtained, breakthrough curves for Pd(II), Au(III) and several mixtures of the metals were obtained as a function of the pH of the feed solution and the flow rate.

Previous batch experiments tested several stripping solutions for the recovery of the metals loaded in the resin under column operation, i.e. HCl 1.2 M, HCl 3.2 M and thiourea 0.5 M in HCl 1 M. When such solutions were tested in a mixture of Pd, Au and Cu adsorbed in the resin the following results were obtained: Cu(II) recovery with HCl 1.2 M was superior to 80%, compared to a 3% for Pd(II) and less than a 1% for Au(III). On the other hand, thiourea 0.5 M in HCl 1 M was shown to be the best eluting solution for the simultaneous recovery of Pd(II) and Au(III) yielding a value of 60% each metal.

Finally, the column system was assayed for the separation of Ni(II), Cu(II), Pd(II) and Au(III) simulating a leached solution obtained after the treatment of a copper/nickel sulphide ore.

---

* to whom correspondance should be adressed.

## INTRODUCTION

Precious metals (PM) occur at trace levels in the copper/nickel sulphide ores which are treated by different techniques finishing with an electrolytic refining; in this operation precious metals are collected in the anode slimes. The first stage in the treatment of the concentrates is extraction with aqua regia. This aqua regia solution contains basically gold, palladium and base metals [1].

Several methods are available for the concentration and separation of metals. Among them sorption methods are widely used in spite of the low selectivity of conventional ion-exchange resins. Chelating resins having specific chelating groups attached to a polymer have been developed taking advantage of the intrinsic selectivity provided by the functional group [2]. Moreover, in agreement with the hard and soft acids and bases theory, functional groups containing S and N donor atoms interact strongly with soft acids like PM [3]. In this sense the chelating resin Duolite GT-73, a macroporous resin based on a cross-linked polystyrene matrix bearing thiol groups that was developed for the removal of mercury from waste water[4], looks promising for the recovery of PM.

This work was undertaken to study both the adsorption and elution of Pd(II) and Au(III) from hydrochloric acid solutions using the chelating resin Duolite GT-73. A comparison of the Au(III), Pd(II), Cu(II) and Ni(II) adsorption and the behaviour of the metals in mixed solutions has been made in order to design their separation. The influence of different parameters such as pH and flow rate was investigated under dynamic conditions as well as the elution efficiency taking into account the results obtained under batch conditions.

Finally, the separation of metal mixtures containing precious and base metals, in the same concentration ratio as the solutions obtained from extracting operations of copper and nickel sulphide ores was carried out in a chromatographic column by using the best experimental conditions.

## EXPERIMENTAL

### Reagents and solutions

Duolite GT-73 macroporous resin (Supelco, USA) with a crosslinked polystyrene matrix bearing -SH groups .

Aqueous Pd(II), Au(III), Cu(II) and Ni(II) stock solution were prepared from solid $PdCl_2$ and $HAuCl_4$ ( both for synthesis; Merck, Germany), $CuCl_2.2H_2O$ (A.R., Probus, Spain) and $NiCl_2.6H_2O$ (A.R., Probus, Spain) respectively. The pH was adjusted using standardised HCl solution. Working solutions were prepared by dilution and appropriate volumes of HCl standardised solution were added to adjust the $H^+$ concentration to the desired value.

### Apparatus

A Labinco rotary mixer was employed to shake the solid and liquid phases in batch experiments. A Gilson Minipuls 2 peristaltic pump (Gilson Medical Electronics (France), Villiers-le-Bel, France) was employed to propel the solutions through the column. A Gilson FC 203 fraction collector (Gilson Medical Electronics, Middleton, WI, USA) was used to collect the effluent samples at the outlet of the column. A Varian Spectra A-300 atomic absorption spectrophotometer was used for the determination of metals in the aqueous phase. pH measurements were carried out by a pH-meter model micro pH 2000 (Crison, Spain).

### Experimental procedure

#### Batch experiments

Batch sorption procedures were carried out by mixing weighed amounts of dry resin with the aqueous metal solution for 24 h until the equilibrium was reached. The remaining metal in the aqueous phase was determined by atomic absorption spectrometry and the amount of metal loaded on the resin phase was calculated by applying the mass balance.

For the elution of the loaded metals, the resin was filtered, washed with distilled water and brought in contact with 10 ml of stripping solution for 24 h (equilibrium time). The metal concentration in the stripping solution was then determined.

*Column experiments*

For column procedure glass columns of 0.6 mm inner diameter were packed with weighed samples of dry resin. Then metal solution volumes were processed through the column at a given flow rate and effluent samples were collected for analysis. Breakthrough curves were obtained by representing the ratio of metal concentration in the effluent vs. bed volume.

The absorbed metals were then eluted at constant flow rate and the percentage of metal eluted was determined.

Experiments were carried out at a controlled room temperature of 22±1 °C and reproducibility was verified by duplication.

**RESULTS**

Batch experiments

Fig. 1 shows the adsorption isotherms of Pd(II) , Au(III), Cu(II) and Ni(II) single solutions for various initial metal concentrations at pH 2.0 in chloride media. It can be seen that the resin can effectively adsorb Au(III), Cu(II) and Pd(II); only traces of Ni(II) were retained at low concentrations of this metal. From these data, maximum capacity of the resin for each metal has been calculated, being for Au(III) the highest value. The results are presented in Table 1.

**Table 1.** Maximum capacity for each metal

| metal | Capacity(mmols/g resin) |
|---|---|
| Pd(II) | 0.262±0.015 |
| Au(III) | 0.58±0.03 |
| Cu(II) | 0.25±0.03 |
| Ni(II) | 0.05±0.06 |

In order to apply this system to the precious metal concentrates already mentioned, the adsorption of metal mixtures was studied using a feed solution with the same concentration ratio as the aqua regia solution, this means 5:25:20 of Au:Pd:Cu respectively[1]. Ni(II) was not considered due to its low adsorption (see Fig.1).

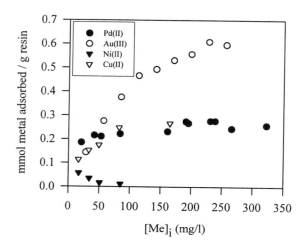

**Figure 1.** Adsorption of Cu(II), Ni(II), Pd(II) and Au(III) from 0.01 M HCl vs. initial metal concentration.

In table 2, the results of the binary mixtures at different pH values show the complete adsorption of gold in all cases, indicating the higher stability of the complex formed with the thiol group in the resin phase. On the other hand, the Cu(II) extracted decreases at more acidic pH values as a consequence of the H⁺ competition toward the -SH group. The results obtained for the mixtures of the three metals in the same conditions are shown in Table 3. The same tendency is observed, the lower the pH the more Cu(II) remains in the aqueous phase, while Au(III) and Pd(II) adsorption are not affected by the acidity of the feed solution.

**Table 2.** Separation of metal couples.(Metal remaining in the aqueous phase).

| | Pd/Au ($[M]_i$=103/19.4) | | Pd/Cu ($[M]_i$=110/100) | | Au/Cu ($[M_i]$=20/100) | |
|---|---|---|---|---|---|---|
| pH | $[Pd]_{aq}$ | $[Au]_{aq}$ | $[Pd]_{aq}$ | $[Cu]_{aq}$ | $[Au]_{aq}$ | $[Cu]_{aq}$ |
| 0.5 | - | - | 4.7±0.4 | 79.1±0.7 | 0.2±0.1 | 87±2 |
| 1 | 36±1 | 0 | 0.2±0.4 | 53.7±0.7 | 0.1±0.1 | 69±2 |
| 2 | 12±1 | 0 | 0.2±0.4 | 11.9±0.7 | 0.1±0.1 | 29±2 |

Concentration in mg/l

**Table 3.** Mixture separation of the three metals. (Metal remaining in the aqueous phase).

| pH | $[Pd]_{aq}$ | $[Au]_{aq}$ | $[Cu]_{aq}$ |
|---|---|---|---|
| 0.5 | 5.95±0.2 | 0.12±0.1 | 80.20±0.2 |
| 1 | 0.67±0.2 | 0.13±0.1 | 51.75±0.2 |
| 2 | 0.72±0.2 | 0.19±0.1 | 11.2±0.2 |

$[Pd]_i$=120 mg/l, $[Au]_i$=20 mg/l, $[Cu]_i$=100 mg/l

The elution of the metal ions from the resin was carried out with acidic solutions as well as with complexing agents like thiocyanate and acidified thiourea. The results are presented in Table 4. As observed the HCl shows the most effective elution for Cu(II). In contrast this stripping agent is not effective for Au(III) and Pd(II) allowing their separation from copper. With respect to palladium and gold the best stripping conditions were obtained using acidified thiourea.

**Table 4.** Elution efficiency

| eluent solution | %Pd(II)eluted | %Au(III)eluted | %Cu(II)eluted |
|---|---|---|---|
| SCN⁻ 0.5M, HCl 0.01M | 26.6% | 21.9% | 57.21% |
| thiourea 0.5M, HCl 0.01M | 48.2% | 39.2% | 47.07% |
| thiourea 0.5M, HCl $10^{-5}$M | 38.9% | | |
| thiourea 0.5M | 30.2% | 9.5% | 24.03% |
| thiourea 0.1M, HCl 1 M | 45.47% | 62.59% | 68.24% |
| thiourea 0.5M, HCl 1M | **60.33%** | 75.52% | 52.52% |
| thiourea 0.8M, HCl 1 M | 55.51% | 78.83% | 49.94% |
| thiourea 0.8M, HCl 3 M | 57.81% | **80.72%** | 49.13% |
| HCl 3.2M | 13.3% | 0.7% | **94.67%** |
| $Na_2S_2O_3$ 0.2M | 9.23% | 16.63% | 67.17% |

Column experiments

To characterise the Duolite GT-73 resin under dynamic conditions the breakthrough capacity of Pd(II), Au(III) and Cu(II) was determined. The breakthrough capacity is defined as the amount of metal ion that can be adsorbed per unit mass of solid before being detected in the outlet of the column; in this case was determined from the volume of metal solution where the sorbent gives a removal efficiency higher than 95% under specified conditions.

Breakthrough curves obtained for Pd(II) solutions with 0.5 g of air-dried resin at a flow rate of 0.5 ml/min (38.59 BV/hour) and pH 2.0 and 1.0. In Fig. 2, the metal concentration in the effluent outlet divided by the initial concentration is plotted against the bed volume (BV), defined as the ratio of the volume of effluent to the volume of resin. From this plot it can be seen that a lower pH reduces the breakthrough capacity of the metal, this behaviour has been explained [5] by the formation of different chlorocomplexes of palladium as consequence of the variation of the pH and $Cl^-$ concentration.

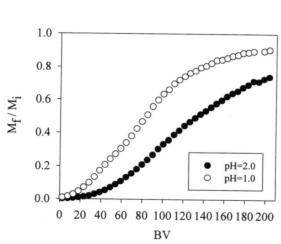

**Figure 2.** Breakthrough curves for palladium adsorption at different pH of the initial metal concentration.
- ● Breakthrough capacity = 0.074±0.004 mmol Pd/g resin
- ○ Breakthrough capacity = 0.033±0.002 mmol Pd/g resin

The breakthrough curves of Au(III) at two different flow rates are shown in Fig. 3. As expected, a decrease in the flow rate causes an increase of the breakthrough capacity. Such an increase can be ascribed to a higher residence time which affects mostly the gold adsorption due to the previously observed slow adsorption kinetics of this metal [5].

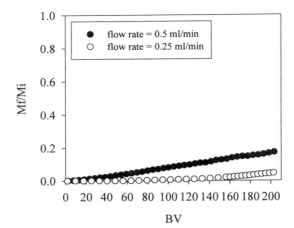

**Figure 3.** Breakthrough curves for gold adsorption at different flow rates, (pH=2.0, 0.5 g resin).

●    Breakthrough capacity = 0.0576 mmol Au/g resin
○    Breakthrough capacity > 0.0860 mmol Au/g resin

Under dynamic conditions, sorption studies of metal mixtures of palladium, gold and copper were carried out with regard to a potential application of the system to the separation of PM metals from leaching solutions containing base metals. In this sense feed solutions with the same concentration ratio as the precious metals concentrates, i.e. 100ppm Cu(II), 120 ppm Pd(II) and 25 ppm Au(III),were used.

Since the most important parameters affecting the loading of metals on Duolite GT-73 were found to be the kinetics of the process and the pH of the feed solution, experiments at several values of pH and flow rate were performed. The breakthrough curves obtained are shown in Fig. 4-7. In these experiments the amount of air-dried resin was kept constant at 0.5 g.

As can observed in the figures, an increase in the breakthrough capacities is obtained when the flow rate is decreased due to an increase of the residence time. On the other hand, the acidification of the feed solution mostly affects the Cu(II) capacity. These results are in agreement with those obtained under batch conditions.

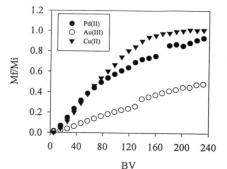

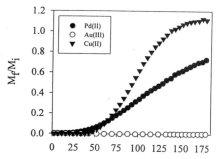

**Figure 4.** Breakthrough curves for a mixture of Pd(II), Au(III) and Cu(II).Flow rate=1ml/min (77.17 BV/hour) and pH=2.
- ● Breakthrough capacity=0.024±0.009 mmols/g resin
- ○ Breakthrough capacity=0.0049±0.0007 mmols/g resin
- ▼ Breakthrough capacity=0.0414±0.0016 mmols/g resin

**Figure 5.** Breakthrough curves for a mixture of Pd(II), Au(III) and Cu(II). Flow rate=0.25 ml/min (19.29BV/hour) and pH=2.0
- ● Breakthrough capacity = 0.084±0.005 mmols/g resin
- ○ Breakthrough capacity > 0.038 mmols/g resin
- ▼ Breakthrough capacity = 0.135±0.002 mmols/g resin

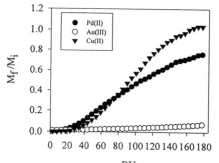

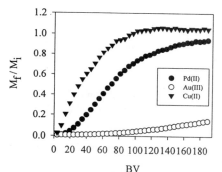

**Figure 6.** Breakthrough curves for a mixture of Pd(II), Au(III) and Cu(II). Flow rate=0.5ml/min (38.59BV/hour) and pH=2.0.
- ● Breakthrough capacity=0.057±0.001 mmols/g resin
- ○ Breakthrough capacity= 0.026±0.001 mmols/g resin
- ▼ Breakthrough capacity=0.089±0.004 mmols/g resin

**Figure 7.** Breakthrough curves for a mixture of Pd(II), Au(III) and Cu(II). Flow rate=0.5ml/min (38.59BV/hour) and pH=1.0.
- ● Breakthrough capacity= 0.034±0.002 mmols/g resin
- ○ Breakthrough capacity=0.020±0.004 mmols/g resin
- ▼ Breakthrough capacity=0.014±0.006 mmols/g resin

Elution efficiency was also investigated under dynamic conditions using the stripping agents which gave the best results in the previous batch experiments. The results presented in Fig. 8 (a and b) show  that the desorption process using HCl 1.2 M provides a selective elution for

Cu(II) with 80% of the total metal retained in the resin after 120 ml of eluting solution percolated through the column (Fig. 8b). For palladium and gold elutions thiourea 0.5 M at HCl 1.0 M provides a high degree of metal concentration in the eluting effluent. In this case the recovery of the metals is 60% (Fig. 8a).

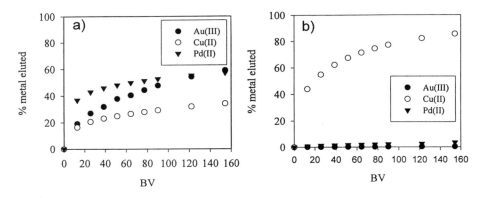

**Figure 8.** Elution efficiency a) with thiourea 0.5 M at HCl 1.0 M and b) with HCl 1.2 M at a flow rate of 0.5 ml/min (38.59 BV/hour).

Finally, a complete scheme for the separation and recovery of a mixture of Pd(II), Cu(II), Au(III) and Ni(II), has been assayed. Two different experiments were carried out under different experimental conditions. In experiment A, 125 ml of feed solution were passed through a 0.6 cm inner diameter column with 1 g of air-dried resin at pH=2.0 and flow rate 0.5ml/min (18.95 BV/hour). After the loading was accomplished, two elution steps were carried out. First with HCl 1.2 M and then with thiourea 0.5 M in HCl 1.0 M. The results of this experiment are shown in Table 5. The data are expressed as percentage of metal eluted calculated with respect to the total metal in the feed solution. As seen Ni(II) is separated in the loading step. Afterwards, the HCl 1.2 M elution provides an almost selective separation of Cu(II) (62.16%). The following elution with acidic thiourea gives a mixture of palladium and gold, being the palladium content the highest one (51.13%).

**Table 5.** Separation of a mixture of four metals. Experiment A.

|         | [M]$_i$ (ppm) | %M$_f$ | %M elution HCl 1.2M | %M elution tiourea 0.5M |
|---------|------|------|------|------|
| Cu(II)  | 99.55  | 5.2  | 62.16 | 8.69  |
| Ni(II)  | 97.7   | 94.6 | 5.18  | 1.28  |
| Pd(II)  | 117.4  | 2.45 | 1.22  | 51.13 |
| Au(III) | 25.40  | 0.04 | 0     | 36.15 |

1g dry resin, pH = 2.0, flow rate = 0.5ml/min, 125ml of feed solution.

In experiment B, two glass columns of 0.6 cm inner diameter were filled with 0.5 g (C1) and 1.5 g (C2) of air-dried resin respectively; 225 ml of feed solution passed sequentially through them at pH=1.0 and flow rate 0.5 ml/min (38.58 BV/hour (C1) and 13.63 BV/hour (C2) respectively). In this case, the first column (C1) was eluted only with thiourea 0.5 M in HCl 1.0 M and the second one (C2) with both HCl 1.2 M and thiourea 0.5 M in HCl 1.0 M.

The results summarised in Table 6 show that the four fractions, i.e., the effluent, the elution solution of C1, the first elution solution of C2 and the acidic thiourea elution of C2, mixtures of metals are obtained. Nevertheless, Ni(II) predominates in the first fraction, Au(III) in the second fraction and Cu(II) and Pd(II) in the third and the fourth fractions respectively. A better separation and recovery of gold is obtained in this way but losses of copper in the effluent are important compared to experiment A. These losses could be overcome by using more resin in the second column or by increasing the pH after the first one.

**Table 6.** Separation of a mixture of four metals. Experiment B.

|         | [M]$_i$ (ppm) | %M$_f$ | %M elution C1 tiourea 0.5M | %M elution C2 HCl 1.2M | %M elution C2 tiourea 0.5M |
|---------|------|-------|------|------|------|
| Ni(II)  | 99.4  | 94.71 | 0.15  | 42    | 0     |
| Cu(II)  | 101.3 | 24.61 | 3.71  | 45.82 | 6.19  |
| Au(III) | 28.1  | 0.2   | 55.0  | 0     | 1.13  |
| Pd(II)  | 124.0 | 3.54  | 17.84 | 1.04  | 35.35 |

0.5 g resin (C1) + 1.5 g resin (C2), flow rate = 0.5ml/min, pH = 1.0, 225ml of feed solution.

## CONCLUSIONS

The chelating resin Duolite GT-73 can readily used to adsorb Pd(II) and Au(III) from chloride solutions with a maximum capacity of 0.262±0.015 mmols/g resin and 0.58±0.03 mmols/g resin respectively at pH=2.0. Cu(II) and Ni(II) as a representative base metals have also been studied; only traces of Ni(II) were retained; in contrast Cu(II) was extracted with a maximum adsorption of 0.25 mmols/ g resin at pH=2.0.

The adsorption of mixtures of the metals at different pH, shows a complete adsorption of gold in all cases, indicating the higher stability of this metal in the resin phase. On the other hand, the Cu(II) extracted decreases at more acidic pH values as a consequence of the $H^+$ competition toward the -SH group.

Under dynamic conditions, the increase in the flow rate causes a decrease of the breakthrough capacity of the three metals studied ascribed to a lower residence time. The acidification of the feed solution mostly affects the Cu(II) capacity, which is minor at lower pHs, in agreement with the results obtained under batch conditions.

HCl 1.2 M provides the best selective elution for copper, arising 80% of the metal retained in the resin. With respect to palladium and gold elution, the best conditions were obtained using thiourea 0.5 M at HCl 1.0 M .

The separation of a mixture of Pd(II), Au(III), Cu(II) and Ni(II) present in a ratio similar to that of the concentrate obtained in the treatment of Copper/Nickel sulphide ores has been assayed under different experimental conditions. In all cases, the separation of Ni(II) was nearly complete (95% of the metal was found in the effluent of the feed solution); Cu(II) can be obtained using 1.2 M HCl as eluting solution with few amount of Pd(II) and Au(III) being also present, whereas Pd(II) and Au(III) recovery and separation is more difficult to achieve and can be improved by using two separated columns and two elution steps.

## ACKNOWLEDGEMENTS

This work was carried out under CICYT (Spanish Commission for Research and Development) Project No. QUI96-1025-CO3-03.

## REFERENCES

1. B. F. Rimmer, Chemistry and Industry, 19 January 1974, 63.

2. M. Ahuja, A.K. Rai, P.N. Mathur, Talanta **43**, 1996, 1953.

3. R. G. Pearson, J. Am. Chem. Soc. **85**, 1963, 3533.

4. G.J. De Jong and C.J.N. Rekers, J. Chromat. **102**, 1974, 443.

5. M. Iglesias, E. Anticó and V. Salvadó. Accepted for publication in Analytica Chimica Acta.

# DETERMINATION OF OPTIMUM CONDITIONS OF A WEAK CARBOXYLIC ION-EXCHANGE RESIN FOR REMOVAL AND RECOVERY OF CHROMIUM(III)

Sevgi Kocaoba, Goksel Akcin

Yildiz Technical University, Faculty of Art and Science, Department of Chemistry, 80270-Sisli, Istanbul- TURKEY

## ABSTRACT

A weak carboxylic cation exchange resin Amberlite IRC 76 was used for removal and recovery of chromium(III). This resin was prepared in two different cationic forms as $Na^+$ and $H^+$. In these two conditions, exchange capacities and the other optimum conditions of the resin were determined in batch system. These optimum conditions were concentration, pH, stirring time, sample amount, moisture content and exchange capacity. The pH range studied was between 1 to 8, the concentration range was between 5 to 100 ppm, stirring time between 5 to 60 minutes and the sample amount of resin was between 50 to 1000 mg. Stirring speed was 2000 rpm during all these experiments. The resin chosen Amberlite IRC 76 had shown better performance in $Na^+$ form. This resin was also used in a column system for removal and recovery of chromium(III) in another study. The results obtained showed that Amberlite IRC 76 weak carboxylic resin was suitable for removal and recovery of chromium(III).

## INTRODUCTION

Amberlite IRC 76 is a high capacity weakly acidic cation exchange resin containing carboxylic functionality within a macroporous crosslinked acrylic matrix. The pK value of Amberlite IRC 76 enables the resin to remove monovalent cations such as sodium as well as multivalent cations like calcium and magnesium. Amberlite IRC 76 combines a high capacity with a smaller volume variation than that of conventional carboxylic resins (1).
Industries which employ trivalent chromium directly in manufacturing processes include glass, ceramics, photography, inorganic pigments (including anodizing compounds), textile dye, animal glue manufacture and tannery industry (2). Although chromium is an essential nutrient for humans, there is no doubt that its compounds, at higher concentrations, are both acutely, and chronically, toxic (3). Conservative technologies for metal control have an increasing interest, able to remove pollutants and reuse valuable byproducts resulting from the wastes and/or side streams from manufacturing processes (4). Removal of metals from wastewaters is achieved by the application of separation processes such as adsorption, cementation, electrochemical processes, ion-exchange, separation using membranes, precipitation and solvent extraction (5). The resins generally are produced in their $H^+$ ionic forms. For applications, they are used either as $H^+$ or the $Na^+$ form. In this study, Amberlite IRC 76 was prepared in two different cationic forms as $Na^+$ and $H^+$ for comparison. The equivalent to this resin (Purolite C106) was used for removal and recovery of chromium(III) from tannery wastes (6).

# MATERIAL AND METHODS

A weak acid carboxylic cation exchange resin (Amberlite IRC 76, Rohm and Haas Company) was used in this study. Table 1 shows the main physicochemical properties of the resin and Table 2 shows the suggested operation conditions. This resin was prepared in two different cationic forms ($Na^+$ and $H^+$). In these two conditions, exchange capacities and the other optimum conditions were determined by the batch system. The optimum conditions were concentration, pH, stirring time, sample amount, moisture content and exchange capacity. The results are given in Table 3-11 and Figures 1-4. Chromium(III) concentrations were determined by Atomic Absorption Spectrophotometry on a Model Spect. AA 20 from Varian. Reagent grade basic chromic sulphate from Merck, Germany was used. HCl, NaOH, NaCl were also used from Merck, Germany. Freshly prepared solutions were used throughout the experiments.

## Conditioning of the resin

After preliminary recycling of the resin, three times with 1 M HCl and NaOH solutions to remove chemicals residues (solvents, functionalizing agents) trapped in the resin matrix during their preparation, the samples were finally converted to the sodium or hydrogen form using 1 M NaCl or HCl.

## Determination of the resin moisture content

Accurately weighed samples of 1 g of ion exchanger in the $H^+$ and $Na^+$ form were dried at 110 °C, cooled in a desiccator and weighed. This was continued until the attainment of constant weight.

## Determination of ion exchange capacity

For the weak acid cation resin Amberlite IRC 76, the ion exchange capacity was determined by reference to both hydrogen and sodium forms by using the column technique. Accordingly, after loading the sample (3 g) into a glass column the resin was eluted with an excess quantity of chromium(III) solution.

## Operation conditions for resin Amberlite IRC 76

## Concentration

Chromium(III) concentrations were selected in the range of between 5 to 100 ppm for two different ionic forms of resin.

## pH

pH range was between 1 to 8. Since precipitation of chromium, in the high values of pH were not studied.

## Stirring time

Stirring time range was between 5 to 60 minutes.

## Resin amount

Resin amounts were taken between 50 to 1000 mg.

Exchange capacities and moisture contents of the resin were determined for two different ionic forms.

**TABLE 1.** Physicochemical properties of the resin investigated

| Resin | Functional group | Matrix | Crosslinking | Total Exchange Capacity |
|-------|------------------|--------|--------------|--------------------------|
| Amberlite IRC 76 | - COOH | acrylic | 20 % DVB | $\approx$ 4 eq/L (H+ form) |

**TABLE 2.** Suggested operation conditions of Amberlite IRC 76 resin.

| Operating temperature | 120 $^\circ$ C (max.) | |
|-----------------------|------------------------|---|
| Service flow rate | 5 to 70 BV/h | |
| Regenerate | HCl | $H_2SO_4$ |
| Concentration | 2 to 5 % | 0.5 to 0.8 % |
| Flow rate | 2 to 8 BV/h | 15 to 25 BV/h |
| Slow rinse | 2 BV/h | |
| Fast rinse | 4 to 8 BV/h | |

## RESULTS AND DISCUSSION

### Concentration
The results are given in Table 3,4 and Figure 1.

**TABLE 3.** Concentration values for $H^+$ form resin

| Initial concent. of $Cr^{3+}$ (mg/L) | Final concent. of $Cr^{3+}$ on the resin ($\mu$g/L) | $Cr^{3+}$ retained on the resin (%) |
|---------------------------------------|------------------------------------------------------|--------------------------------------|
| 5 | 145.5 | 58.20 |
| 10 | 292.9 | 58.58 |
| 20 | 472.0 | 47.20 |
| 30 | 804.0 | 53.60 |
| 50 | 1212.0 | 48.48 |
| 60 | 1376.0 | 45.86 |
| 75 | 1217.0 | 32.45 |
| 100 | 1507.0 | 30.14 |

**TABLE 4.** Concentration values for $Na^+$ form resin

| Initial concent. of $Cr^{3+}$ (mg/L) | Final concent. of $Cr^{3+}$ on the resin (μg/L) | $Cr^{3+}$ retained on the resin (%) |
|---|---|---|
| 5 | 239.4 | 95.78 |
| 10 | 481.4 | 96.28 |
| 20 | 954.0 | 95.40 |
| 30 | 1432.9 | 95.52 |
| 50 | 2376.8 | 95.07 |
| 60 | 2840.8 | 94.69 |
| 75 | 3553.9 | 94.77 |
| 100 | 4646.7 | 92.93 |

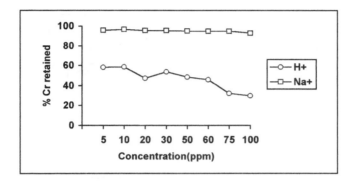

**FIGURE 1.** Effect of initial chromium concentration on the recovery of chromium by $Na^+$ and $H^+$ form resin.

## pH

The results are given in Table 5,6 and Figure 2.

**TABLE 5.** pH values for $H^+$ form resin

| pH | $Cr^{3+}$ retained on the resin(μg/L) | $Cr^{3+}$ retained on the resin (%) |
|---|---|---|
| 1 | 19.5 | 3.09 |
| 2 | 93.4 | 18.69 |
| 3 | 226.2 | 45.24 |
| 4 | 285.5 | 57.11 |
| 5 | 291.7 | 58.35 |
| 6 | 432.9 | 86.59 |
| 7 | 457.5 | 91.51 |
| 8 | 459.3 | 91.86 |

**TABLE 6.** pH values for Na$^+$ form resin

| pH | Cr$^{3+}$ retained on the resin($\mu$g/L) | Cr$^{3+}$ retained on the resin (%) |
|---|---|---|
| 1 | 406.6 | 81.32 |
| 2 | 406.5 | 81.31 |
| 3 | 410.3 | 82.06 |
| 4 | 411.8 | 82.37 |
| 5 | 470.9 | 94.18 |
| 6 | 490.9 | 98.18 |
| 7 | 487.8 | 97.57 |
| 8 | 489.8 | 97.96 |

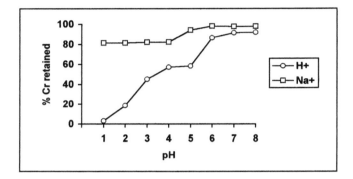

**FIGURE 2.** Effect of pH on the recovery of Chromium by Na$^+$ and H$^+$ form resins

## Stirring time
The results are given in Table 7,8 and Figure 3.

**TABLE 7.** Stirring time values for H$^+$ form resin

| Time(min.) | Cr$^{3+}$ retained on the resin($\mu$g/L) | Cr$^{3+}$ retained on the resin (%) |
|---|---|---|
| 5 | 186.4 | 37.28 |
| 10 | 219.4 | 43.89 |
| 15 | 238.4 | 47.68 |
| 20 | 259.6 | 51.92 |
| 30 | 252.9 | 50.59 |
| 45 | 266.9 | 53.39 |
| 60 | 271.2 | 54.54 |

**TABLE 8.** Stirring time values Na$^+$ form resin

| Time(min.) | Cr$^{3+}$ retained on the resin($\mu$g/L) | Cr$^{3+}$ retained on the resin (%) |
|---|---|---|
| 5 | 452.9 | 90.58 |
| 10 | 459.2 | 91.85 |
| 15 | 471.0 | 94.20 |
| 20 | 478.0 | 95.60 |
| 30 | 466.3 | 93.27 |
| 45 | 472.6 | 94.52 |
| 60 | 481.7 | 96.34 |

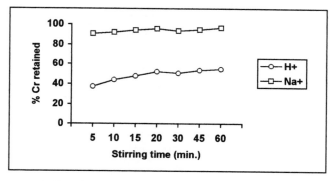

**FIGURE 3.** Effect of stirring time on the recovery of chromium by Na$^+$ and H$^+$ form resins

## Resin Amount
The results are given in Table 9,10 and Figure 4.

**TABLE 9.** Resin amount values for H$^+$ form resin

| Resin amount (mg) | Cr$^{3+}$ retained on the resin($\mu$g/L) | Cr$^{3+}$ retained on the resin (%) |
|---|---|---|
| 50 | 190.3 | 38.07 |
| 100 | 245.8 | 49.17 |
| 150 | 279.8 | 55.97 |
| 200 | 340.2 | 68.04 |
| 250 | 292.4 | 58.48 |
| 500 | 320.6 | 64.12 |
| 750 | 320.6 | 64.12 |
| 1000 | 327.7 | 65.54 |

**TABLE 10.** Resin amount values for Na$^+$ form resin

| Resin amount (mg) | Cr$^{3+}$ retained on the resin($\mu$g/L) | Cr$^{3+}$ retained on the resin (%) |
|---|---|---|
| 50 | 465.1 | 93.03 |
| 100 | 473.2 | 94.64 |
| 150 | 477.7 | 95.55 |
| 200 | 483.2 | 96.65 |
| 250 | 483.9 | 96.79 |
| 500 | 488.9 | 97.78 |
| 750 | 491.5 | 98.31 |
| 1000 | 490.5 | 98.10 |

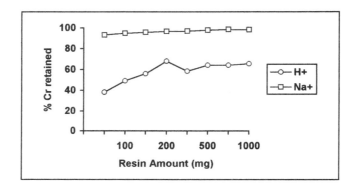

**FIGURE 4.** Effect of resin amount on the recovery of chromium by Na$^+$ and H$^+$ form resins

The results of exchange capacities and moisture contents of the resin are given Table 11.

**TABLE 11.** Exchange capacities and moisture contents of the resin

| Resin(Amberlite IRC 76) | Na$^+$ form | H$^+$ form |
|---|---|---|
| Capacity (eq/L) | 2.96 | 1.69 |
| Moisture content (%) | 66.48 | 54.35 |

In principle, by using a macroporous carboxylic resin (Amberlite IRC 76), it was demonstrated that the resin had capability and efficiency to remove and separate chromium from wastewaters. All experiments showed that, Amberlite IRC 76 had better performance in the Na$^+$ form. Tables 3 and 4 for Na$^+$ form resin, show that chromium percentages retained were increased in the concentration range studied but for the H$^+$ form resin, above 10 ppm concentration, the %'s chromium retained were decreased. Tables 5 and 6 showed that percentages of chromium retained increased with increasing pH values for Na$^+$ form resin. H$^+$ form resin did not show the same performance. Above pH 4, percentages of chromium retained were higher in the two forms but in $Na^+$ form these values were higher than the $H^+$ form. Tables 7 and 8 showed that percentages of chromium retained were not greatly increased with stirring time. 15 minutes was a suitable time for both ionic forms but in $Na^+$ form, the amount of chromium retained was higher. Tables 9 and 10 showed that percentages of chromium retained were low in $H^+$ form. Above 500 mg resin, percentages of chromium retained were stable for $H^+$ form but in $Na^+$ form, initial amounts of resin, percentages of chromium retained were higher a constant.

# ACKNOWLEDGMENT

The authors would like to thank Dr. Domenico Petruzzelli, Water Research Institute- CNR, Bari, Italy for valuable discussion and guidance .

## REFERENCES

1.1989, Rohm and Haas *Separation Technologies*, Product Data Sheet.

2.1992, Patterson, J.W., *Industrial Wastewater Treatment Technology*, 77.

3.1992, Beccari, M., Campanella, L., Majone, M., Petronio, M., Rolle, E., *Oriental J. Chem.*, **8**, 2.

4.1991, Macchi, G., Pagano M., Pettine, M., Santori, M., Tiravanti, G., *Water Research* **25**, 8.

5.1995, Ramos, R.L., Rubio, L.F., Coronado, R.M.G., Barron, J.M., *J. Chem. Tech. Biotechnol.*, **62**, 64.

6.1991, Petruzzelli, D., Santori, M., Passino, R., Tiravanti, G., New Developments in Ion-Exchange Proceedings of the International Conference on Ion-Exchange, ICIE' 91, Tokyo, Japan., 383, Elsevier Science.1991, Abe, M., Kataoka, T., Suzuki, T.

# BATCH AND COLUMN EXTRACTION STUDIES OF GOLDCYANIDE WITH IMPREGNATED AND ION EXCHANGE RESINS

E. Meinhardt[1], R.M. Kautzmann[2], N. Ruiz[1], A.M. Sastre[1], C. H. Sampaio[2] and J.L. Cortina[1].

[1]Chemical Engineering Department, Universitat Politecnica de Catalunya, Diagonal, 647, Barcelona, 08028, Spain. Fax:3434015814,
[2]Department of Metallurgy, Mineral Processing Laboratory (LAPROM), Universidad Federal Rio Grande do Sul, Osvaldo Aranha, 99, s.611, Porto Alegre, Brazil

## 1 INTRODUCTION

The application of hydrometallurgy to the recovery of precious metals from dilute ore sources is increasing in use (1). For example, dump, heap and vat leaching of selected ores are employed commercially in the gold mining industry, where greater use has been reached. In this case cyanidation leaching steps are combined with subsequent steps of gold cyanide adsorption with activated carbon, where a lack of selectivity is found due to the presence of mixtures of other metal cyanides (2).

Although, this processing is performed using activated carbon, in recent years, research efforts have attempted its substitution by Ion Exchange (IX). The main objective with the application of IX is solving existing problems associated in the activated carbon use by improving the selective recovery of goldcyanide; increasing loading and stripping efficiencies; and developing an integrated process of leaching and extraction as resin-in-leach (RIL) and resin-in-pulp (RIP). Commercially available resins have been unable to compete with activated carbon due to poor selectivity, mechanical breakdown of the beads and the requirement for complex elution and regeneration processes. However, resins offer some chemical advantages over activated carbon and have excellent technical potential for application in gold extraction systems (3).

Since typical gold leaching solutions contain gold as $Au(CN)_2^-$ at pH about 10, it was reasoned that the objectives might be accomplished if ion-exchange resins could be developed with a functional group (R), which would operate on the hydrogen ion cycle shown in the following reaction:

$$P\text{--}R_{res} + H^+ \Leftrightarrow P\text{--}RH^+_{res} \qquad [1]$$

where the equilibrium lies far to the right at pH < 10 and far to the left a pH>12.5. Gold cyanide extraction will be accomplished by the following reaction:

$$P\text{--}RH^+_{res} + Au(CN)_2^- \Leftrightarrow P\text{--}RH^+ Au(CN)_2^-{}_{res} \qquad [2]$$

where P represents the polymeric support.

As none of the materials commercially available fit these conditions, efforts to prepare solid materials for gold cyanide extraction have promoted the evaluation of

"Solvent Impregnated Resins" (SIR) by using extractants developed for solvent extraction applications (4-6). SIR are prepared principally by simple immobilisation procedures of complexing reagents onto conventional macroporous polymeric supports and have been presented as a very attractive approach, combining the features of both ion exchange and solvent extraction and with potential advantages in a hydrometallurgical process, due to the high selectivity and easie preparation and due to the wide choice of reagents of desired selectivity. Available single extractants do not meet the above objectives on the pH cycle with the exception of the recently developed reagent at Henkel Co (U.S.A) containing a guanidine ion pairing functionality and known as LIX79 (7-8).

This paper deals with recent investigations in the application of ion exchange and impregnated resins in the extraction, separation and recovery of gold cyanide from leaching solutions. The impregnated resins prepared were obtained by direct impregnation of polymeric macroporous polystyrene divinylbenzene supports with LIX79. This new reagent is accomplishing the chemical conditions described previously. Solid-liquid extraction studies of goldcyanide using batch and column experiments have been performed. The influence of both reagent and polymer functionality structure in their extraction ability and the influence of the aqueous composition have been studied.

2 EXPERIMENTAL

**2.1 Reagents**

LIX79 was used without further purification. The structure of this reagent is:

$$R_1 \underset{R_2}{\diagdown} N - \underset{\overset{\displaystyle N''}{\|}}{C} \underset{R_4}{\diagup} N' \diagup R_3 \qquad R_5$$

having a pKa value of 12 (7), or greater, depending the R groups attached to the various nitrogens. N,N'-bis(2-ethylhexyl)guanidine is being used under the name of LIX79. Samples of polymeric macroporous supports were obtained from Purolite and Rohm and Haas. Resin properties are listed in Table 1

**Table 1** *Polymeric Macroporous Supports and Chemical Properties(9)*

| Resin | Manufacturer | Functionality | Matrix * | Structure | SurfaceArea(m²/g) |
|---|---|---|---|---|---|
| Amberlite XAD2 | PUROLITE | non-functionalized | ST-DVB | macroporous | 300 |
| Macronet MN200 | Rohm and Haas | non-functionalized | ST-DVB | macro/microporous | 800-1000 |
| Macronet MN100 | Rohm and Haas | weak base | ST-DVB | macro/microporous | 800-1000 |

* ST-DVB: styrene divinylbenzene.

Hypersol Macronet™ sorbent resins (MN100 and MN200) are a new family of synthetic adsorbents with an appreciable portion of small micropores which create very high internal surface areas (800-1000 m²/g). These new series of Macronet materials contain both macropores and micropores, the later providing the high surface area and the former providing rapid access to the internal surfaces, as is the case of MN200. Functional groups are added as desired, to make polymers hydrophilic and to provide a degree of selectivity as is the case of MN100 containing weak base groups.

Stock solutions of $Au(CN)_2^-$, $Ag(CN)_2^-$, $Fe(CN)_6^{4-}$ and $Cu(CN)_3^{2-}$ ($1g.dm^{-3}$) were prepared by dissolving the corresponding salts (Johnson Mathey, A.R. grade) in sodium cyanide solution. Real cyanide leach liquors were obtained by leaching of a gold mineral ore from Brazil. The composition of this solution is shown in Table 2.

**Table 2** *Composition of the Metal Cyanide Solutions*

| Metal | Mining Leach Solution Conc. (ppm) | Synthetic Solution Conc.(ppm) |
|---|---|---|
| Gold (Au) | 11.6 | 10.0 |
| Silver (Ag) | 0.7 | 5.0 |
| Coper (Cu) | 1.4 | 30.0 |
| Iron (Fe) | 10.7 | 30.0 |
| Nickel (Ni) | 0.1 | 2.0 |

## 2.2 Impregnation Process

XAD2-LIX79 and MN200-LIX79 resins were prepared according to a modified version of the dry impregnation method described previously (10). The amount of LIX 79 impregnated was evaluated by washing a known amount of resin with ethanol, which completely elutes these extractants for subsequent determination by potentiometric titrations. Impregnates prepared achieved loadings of 0.33 g LIX/g MN200 and 0.20 g LIX/g XAD2.

## 2.3 Experimental Procedure

*2.3.1 Batch Experiments.* Samples between 0.05-0.2 g of resins, were mixed mechanically in special glass stoppered tubes with an aqueous solution (20-250 ml) until equilibrium was achieved. The composition of the aqueous solutions varied depending on the nature of the experiment. After phase separation with a high-speed centrifuge, the equilibrium pH was measured using a Methrom AG 9100 combined electrode connected to a CRISON digital pH-meter. Metal content in both phases was determined by Atomic Absorption Spectrophotometry (Perkin-Elmer 2380 AAS with air-acetylene flame) or Inductively Coupled Plasma Spectrophotometry (ICP) (SpectroFlame, Kleve, Germany) depending on solution composition.

*2.3.2 Column Experiments.* Known amounts of swollen resins were slurry-packed in an Omnifit borosilicate glass column fitted with porous 25 micron polyethylene frits and teflon end pieces. A peristaltic pump at the column entrance delivered solution at constant flow rate of 0.5 to 1 mL/min. Metal ions were determined by following the change in concentration with throughput of the samples collected to follow the extraction histories. After each metal extraction experiment, the flow of the metal solution is stopped and the resin washed successively with water, stripping solution at flow rate of 1 mL/min through the resin bed in the column. Metal ion concentration was determined in all the samples.

## 3 RESULTS AND DISCUSSIONS

The metal extraction studies on the efficiency of the resins used in this work were performed simulating the experimental conditions expected from the leach solutions obtained in the cyanidation step (pH, free cyanide concentration and presence of other metal cyanide interference).

### 3.1 Extraction pH dependence

The extraction pH behaviour of LIX79 impregnated resins (Amberlite XAD2 and MN200) and a weak base ion exchange resin MN100 has been studied. As it is shown in Figure 1, a quantitative extraction in the basic range 9-10, showing a typical S function shape for which moving the pH to values of 12 the reaction is reversed allowing the elution of the goldcyanide complex was observed. Both type of sytems, LIX79 impregnates and the weak base resin (MN100) will fit the process operation when applying in Resin in Pulp and Resin in Leach, in terms of the extraction and stripping steps. On the other hand the kinetics of the process was shown to be faster than that typically found when using solvent impregnated resins and compatible with those of the leaching steps.

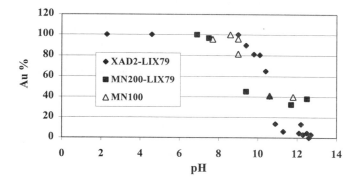

**Figure 1** *Gold extraction percentage (%E) for the different ion exchange and impregnated resins as a function of pH in the aqueous phase. To Xad2-LIX79[Au(CN)$_2$]=10 ppm and MN200-LIX79 and MN100 [Au(CN)$_2$]=20 ppm.*

### 3.2 Goldcyanide loadings

Goldcyanide loadings of MN100 resin were close to 10 times higher than those obtained with LIX79 impregnated resins as it is shown in figures 2a-b. This big difference could be basically explained due to the differences on capacity of the weak base resin MN100 in comparison with the LIX79 loadings achieved with the impregnate.

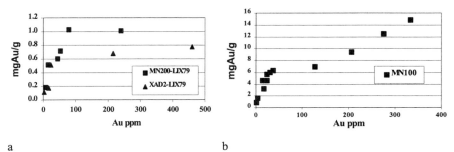

a                                                                          b

**Figure 2** *Metal cyanide loadings as a function of goldcyanide concentration in the aqueous phase for LIX79 impregnated resins and MN100 resin. Phase ratio was 0.2g resin/20ml of solution.*

### 3.3 Cyanide concentration effect

The effect of total cyanide concentration on goldcyanide extraction was evaluated in the concentration range expected from the processing of mineral ores. Cyanide concentration changes through the mineral leaching step from values of 1000 ppm (starting point) to 100-200 ppm (closing point). Figure 3a shows the equilibrium gold distribution isotherms for XAD2-LIX79 impregnates as a function of cyanide concentration in the aqueous phase. As can be seen, the extraction of $Au(CN)_2^-$ is affected by the presence of total cyanide content in concentrations higher than 250 ppm. For cyanide concentrations higher than 250 ppm and lower than 1000 ppm, simulating the final and initial leach solutions, the extraction percentage reach a constant value. This effect shows the cyanide competition of $CN^-$ on $Au(CN)_2^-$ extraction. A similar trend was observed for the weak base support, as is shown in figure 3b.

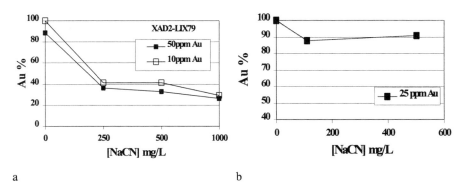

a                                                                          b

**Figure 3** *Gold extraction percentage (%E) for the different ion exchange and impregnated resins as a function of cyanide concentration in the aqueous phase ([CN]) . Phase ratio was 0.2g resin/20mL of aqueous solution.*

### 3.4 Extraction efficiency and selectivity

The extraction efficiency and selectivity of $Au(CN)_2^-$ and other metal cyanides complexes in mining leach solutions under different experimental conditions was evaluated. Figures 4a -b show the metal concentration changes in the leaching solution as a function of time for the MN 100 resin. As a general trend, the resin shows a high efficiency extraction of $Au(CN)_2^-$, with extraction percentages close to 100%, and high selectivity for $Au(CN)_2^-$ extraction when compared with the other metal cyanide complexes (Fe, Ag ,Cu and Ni).

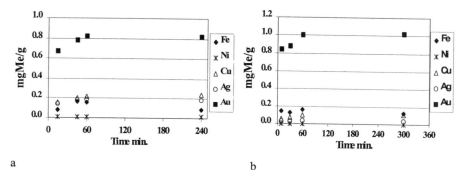

a                                b

**Figure 4** *Loading for the different metal cyanide complexes for MN100 resins in (a) synthetic solution (b) mining leach solution.*

The selectivity factors (SAu/Me) of MN100 resin were calculated and shown are shown in Figure 5.

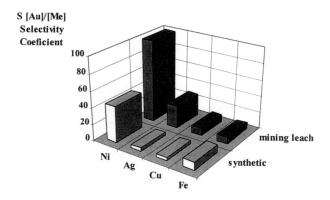

**Figure 5** *Selectivity coeficient of MN 100 to synthetic and mining leach solution(2).*

## 3.5 Stripping studies

The loaded phases obtained in the extraction experiments were used for the study of the back extraction of Au(CN)$_2^-$. The stripping data given in Table 2 point out that mixtures of aqueous solutions of NaOH and NaCN with organic solvents as ethanol and acetone are required to recover Au(CN)$_2^-$ quantitatively from the resin phase. These types of stripping agents are compatible with subsequent operations of electrowinning.

**Table 2** *Stripping of Au(CN)$_2^-$ from loaded LIX 79 impregnated resins and MN100.*

| Eluent | Stripping % | |
|---|---|---|
| | MN100 | MN200-LIX79 |
| NaOH 1M | 7.6 | 37.3 |
| NaOH 1M / ethanol 40% | 64.9 | |
| NaCN 5gL$^{-1}$ | 20 | 61.5 |
| NaCN 5gL$^{-1}$ / ethanol 40% | 100 | |
| NaCN 5gL$^{-1}$/ acetone 40% | 99.5 | 96.8 |
| Thiourea 1gL$^{-1}$/ H$_2$SO$_4$ 0.5M | 2 | 25 |

## 4 CONCLUSIONS

Hypersol Macronet Sorbent Resin MN100 with a weak base functionality shows promising results when applied to the extraction of goldcyanide from cyanide media in terms of pH behaviour, loadings and separation factors. This behaviour shown by MN100 appear to be determined by a combination of factors, among then the following could be pointed out: a) the suitable acid-base properties of the functional group, b) the balanced degree of hydrophobicit of the polymer matrix and c) the high surface area of the resin.

The impregnated resins prepared containing a guanidine functionality (LIX79) also gave promising results in terms of goldcyanide extraction at alkaline pH and good elution properties when used impregnated onto Amberlite XAD2 and MN200 supports, however improvements in goldcyanide loadings should be pursued.

## 5 ACKNOWEDGMENTS

We wish to acknowledge to Dr. M. Virning (Henkel) for their support in LIX 79 sample supply and to Purolite and Rohm and Haas Hispania for resin supply. Finally, this work was supported by CICYT Project **QUI96C02-002** (Ministerio de Educación y Ciencia de España).

## 6 REFERENCES

1. M.B. Moiman, J.D. Miller, J.B. Hiskey, and A.R. Hendricksz in "Comparative of Process Alternatives for Gold Recovery from Cyanide leach Solutions, Au and Ag Heap and Dump Leaching Practice", ed. J.B. Hiskey, AIME 1984, **93**, 108.

2.  C.A. Fleming and G. Gromberge, *J. S. Afr. Inst. Min. Metall.*, 1984, **84**, 369.
3.  M. Akser, R.Y. Wan and J.D. Miller, The Metallurgical Society (AIME), The Reinhardt Schuhmann International Symposium on Innovative Technology and Reactor Design in Extraction Metallurgy, Colorado Springs, Co, 1986, **87,** 92.
4.  B.R. Green, Mintek 50-Proceedings of the International Conference on Mineral Science and Technology, Edited by L.F. Haughton, Johannesburg, South Africa, vol.2, p. 627 (1985).
5   M.B. Moiman, and J.D. Miller, Proc. of ISEC, *AIChE*, 1983, 530-531.
6   J.L. Cortina, E. Meinhardt, O. Roijals, V. Martí, *React. and Funct. Polym.*, 1998, **36**, 149.
7   G.A. Kordosky, J.M. Sierakoski, M.J. Virning, and P.L. Mattison, *Hydrometallurgy*, 30(1992) 291.
8   M.J. Virning, and G.A. Wolfe, Proc. Int. Solvent Extraction Conference, ISEC96, Melbourne, 1996, vol 1, p 311.
9   Purolite and Rohm and Haas, Technical Bulletins.
10  J.L. Cortina, A.M. Sastre, N. Miralles, M. Aguilar, A. Profumo, and M. Pesavento, *React. Polym.*, 1992, **18**, 67.

# List of Delegates

ABE M — Tsuruoka National College of Technology, 104 Sawada, Inooka, Tsuruoka, Yamagata 997, Japan

ADAM A — Dionex UK Ltd, 4 Albany Court, Camberley, Surrey, GU15 2PL, UK

ADLARD E — Delryn, Vicarage Lane, Burton, S. Wirral, L64 5TJ, UK

CARTER M — Danisco Sugar, Danisco Sugar Development Centre, Nakskov, Maribovej 2, Postbox 119, DK 4900 Nakskov, Denmark

CLEARFIELD A — Dept of Chemistry, Texas A+M University, P O Box 300012, College Station, TX 77842-3012, USA

CORTINA J L — Dept of Analytical Chemistry, Universitat Politecnica de Catalunya, E T S E I B, Diagonal 647, E-08028 Barcelona, Spain

DIVAN K — Dionex UK Ltd, 4 Albany Court, Camberley, Surrey, GU15 2PL, UK

DYER A — University of Salford, Salford, M5 4WT, UK

EKMAN K — Smoptech, Virusmakivagen 65, Abo, FIN-20300, Finland

GLENNON J D — Dept of Chemistry, University College, Cork, Ireland

GARA D — Metrohm UK Ltd, Unit 2, Top Angel, Buckingham Industrial Park, Buckingham, MK18 1TH, UK

GREENE J — Laboratory Services, Nort West Water, Langley Green Avenue, Great Sankey, Warrington, Cheshire WA5 3QT

HATHI P — Whatman International Ltd, Springfield Mill, Maidstone, Kent, ME14 2LE, UK

HAMANN H C — Purolite Co, 3620 G St, Philadelphia, PA 19134, USA

HADDAD P — Dept of Chemistry, University of Tasmania, GPO Box 252-75, Hobart 7001, Australia

HEALY L — Analytical Division, Dept of Chemistry, University College Cork, Ireland

HÖLL W H — Forschungszentrum Karlsruhe GmbH, Postfach 3640, D-76021 Karlsruhe, Germany

HATSUSHIKA T — Dept of Analytical Chemistry and Biotechnology, Faculty of Engineering, Yamanashi University, Takeda, Kofu 400, Japan

HUGHES H — The North East Wales Institute, Plas Coch, Mold Road, Wrexham, LL11 2AW, UK

IRVING J — Purolite International Ltd, Cowbridge Road, Pontyclun, CF72 8YL, UK

JONES W — Dept of Chemistry, University of Cambridge, Lensfield Road, Cambridge, CB2 1EW, UK

| | |
|---|---|
| JYO A | Dept of Applied Chemistry and Biochemistry, Faculty of Engineering, Kumamoto University, Kumamoto 860, Japan |
| JUNCA M I | Universitat de Girona, Facultad de Ciencias, Campus de Montilivi, E-17071 Girona, Spain |
| JEGLE U | Hewlett Packard Gmbh, Analytical Division, Hewlett-Packard Strasse 8, 76337 Waldbronn, Germany |
| JONES K | Affinity Chromatography Ltd, Freeport, Ballasalla, Isle of Man, IM9 2AP, UK |
| JONES P | Dept of Environmental Sciences, University of Plymouth, Drake Circus, Plymouth, Devon, PL4 8AA |
| KAVANAGH P | AEA Technology, Dorset, UK |
| KUMADA N | Yamanashi University, Miyamae-7, Kofu 400-8511, Japan |
| KINOMURA N | Yamanashi University, Miyamae-7, Kofu 400-8511, Japan |
| KANEYOSHI M | Dept of Chemistry, University of Cambridge, Lensfield Road, Cambridge, CB2 1EW, UK |
| KODAMA H | NIRIM, Namiki 1-1, Tsukuba City, Ibaraki 305, Japan |
| KASHIWAGI T | Mitsubishi Chemical Corporation, Kurosaki Plant, Kurosaki, Yahatanishi-ku, Kitakyusyu 806-0004, Japan |
| KARKI A | Finex OY, Huumantia 5, FIN-48230 Kotka, Finland |
| KOCAOBA S | Yildiz Technical University, Faculty of Art and Science, Department of Chemistry, 80270 Sisli-Istanbul, Turkey |
| LEVISON P | Whatman International Ltd, Springfield Mill, Maidstone, Kent, ME14 2LE, UK |
| MAKITA Y | Shikoku National Industrial Research Institute, 2217-14 Hayashi-cho, Takamatsu 761-0395, Japan |
| MIJANGOS F | UPV-EHU, Chemical Engineering Dept, University of the Basque Country, Apdo 644, Bilbao 48080, Spain |
| MYERS P | X-tech, 19 Woodlea Close, Bromborough, Wirral, L62 6DL, UK |
| MOLLER T | Lab. of Radiochemistry, Dept of Chemistry, University of Helsinki, P O Box 55, 00014 University of Helsinki, Finland |
| MATEJKA Z | Institute of Chemical Technology, Technicka 5, 166 28 Prague 6, Czech Republic |
| MURAVIEV D | Dept of Analytical Chemistry, Autonomous University of Barcelona, E-08193 Bellaterra, Barcelona, Spain |
| MURPHY G | AEA Technology, Dorset, UK |
| McLAUGHLIN A | The Biocomposites Centre, University of Wales, Bangor, Gwynedd LL57 2UW |

NEWTON J                    University of Salford, Salford, M5 4WT, UK

NONAKA T                    Kumamoto University, Kurokami 2-39-1, Kumamoto-shi,
                            Kumamoto 860-8555, Japan

NESTERENKO P                Lomonov State University, Moscow, Russia

NADEN D                     Purolite International Ltd, Kershaw House, Great west Road, Hounslow,
                            Middlesex, TW5 0BU, UK

O'CONNELL M                 Dept of Chemistry, University College, Cork, Ireland

O'MAHONY T                  Dept of Chemistry, University College, Cork, Ireland

PAULL B                     University College Dublin, Ireland

PILLINGER M                 University of Salford, Salford, M5 4WT, UK

PLANT S                     Purolite International Ltd, Cowbridge Road, Pontyclun, CF72 8YL, UK

PETRUZELLI D                Instituto di Ricerca sulle Acque, 5 via de Blasio, 70123 Bari, Italy

REGNIER F E                 Purdue University, 1393 Brown Building, Dept of Chemistry, West
                            Lafayette, IN 47907, USA

SYLVESTER P J               Dept of Chemistry, Texas A+M University, P O Box 300012, College
                            Station, TX 77842-3012, USA

STAFFORD R G                Laboratory Services, Nort West Water, Langley Green Avenue, Great
                            Sankey, Warrington, Cheshire WA5 3QT

SUTTON R                    University of Cincinatti, USA

TITTLE K                    40 Tarnbeck Drive, Mawdsley, Lancs, L40 2RU, UK

TISCHENKO G                 Institute of Macromolecular Chemistry, Academy ofSciences of the Czech
                            Republic, Heyrovsky Sq 2, 162 06 Prague 6, Czech Republic

WILLIAMS P A                The North East Wales Institute, Plas Coch, Mold Road, Wrexham,
                            LL11 2AW, UK

WHITE K                     Merck Ltd, Merck House, Poole, Dorset, BH15 1TD, UK

WALL P                      Merck Ltd, Merck House, Poole, Dorset, BH15 1TD, UK

# Subject Index

7